W9-CCP-996

The Repair and Maintenance of Small Gasoline Engines

GEORGE R. DRAKE

RESTON PUBLISHING COMPANY, INC.
Reston, Virginia
A Prentice-Hall Company

Library of Congress Cataloging in Publication Data

Drake, George R
The repair and maintenance of small gasoline engines.

Includes index.
1. Gas and oil engines—Maintenance and repair.
I. Title.
TJ785.D68 621.43 75-29228
ISBN 0-87909-724-8

Reston Publishing Company, Inc.
A Prentice-Hall Company
Reston, Virginia 22090

10 9 8 7 6 5 4 3 2 1

Printed in the United States of America.

Contents

PREFACE ix

ACKNOWLEDGMENTS xi

1 SMALL GASOLINE ENGINES 1

1-1. What is Powered by a Small Gasoline Engine?, 4
1-2. The Future of Small Gasoline Engines, 12
1-3. Basic Operation of Small Reciprocating Gasoline Engines, 12
1-4. Safety, 15

2 THE BASIC SMALL GASOLINE ENGINE 17

2-1. Operation of Two-Stroke Cycle Engines, 18
2-2. Operation of Four-Stroke Cycle Engines, 20
2-3. Cylinder Block, 24
2-4. Cylinder Head, 25
2-5. Crankcase, 26
2-6. Crankcase Breather, 26
2-7. Reed, Rotary, and Poppet Valves, 28
2-8. Piston, 29
2-9. Piston Rings, 34
2-10. Piston Pin (Wrist Pin), 35
2-11. Valves, 35
2-12. Connecting Rod, 37

2-13. Crank and Crankshaft, 38
2-14. Flywheel, 40
2-15. Camshaft, 41
2-16. Bearings, Bushings, and Journals, 42
2-17. Muffler, 43
2-18. Cooling Systems, 44
2-19. Summary of Engine Operation, 47
2-20. Engine Maintenance, 49
2-21. Cylinder Head Maintenance, 50
2-22. Connecting Rod Maintenance, 51
2-23. Piston, Piston Ring, and Piston Pin Maintenance, 53
2-24. Cylinder Maintenance, 59
2-25. Camshaft Maintenance, 62
2-26. Valve Maintenance, 64
2-27. Crankshaft Maintenance, 71
2-28. Main Bearing Maintenance, 72
2-29. Oil Seal Maintenance, 73
2-30. Reed Valve Maintenance, 73
2-31. Cooling System Maintenance, 74

3 OPERATION AND PERIODIC MAINTENANCE PROCEDURES 75
3-1. Engine Starting Procedures, 76
3-2. Engine Operating Procedures, 78
3-3. Engine Shutoff Procedures, 79
3-4. Winter or Long Duration Storage, 80
3-5. Saltwater Operation, 81
3-6. Returning Engine to Service after Storage, 81
3-7. Periodic Maintenance and Minor Engine Tune-Ups, 82
3-8. Cleaning the Engine, 84
3-9. Cleaning Carburetor Air Cleaners, 86
3-10. Cleaning Fuel Filters (Strainers), 90
3-11. Cleaning Fuel Tank Vent Caps, 93
3-12. Cleaning Crankcase Breathers, 94
3-13. Checking Oil Level and Changing Oil in the Four-Cycle Engine, 94
3-14. Lubrication of Two-Cycle Engines, 97
3-15. Lubrication of Nonengine Parts, 99
3-16. Adjusting Belt Tensions, 99
3-17. Checking, Cleaning, and Regapping Spark Plugs, 99
3-18. Simple Compression Check, 103
3-19. Checking the Battery, 104

4 BASIC ELECTRICITY AND ELECTRICAL COMPONENTS 109

4-1. Atoms, 109
4-2. Electricity, 111
4-3. Electrical Units, 112
4-4. Magnetism and Induced Voltage, 113
4-5. Electrical Components, 114
4-6. Continuity, 127
4-7. Test Equipment, 127
4-8. Summary, 127

5 IGNITION SYSTEMS 128

5-1. Magneto Ignition Systems, 129
5-2. Solid-State Ignition Systems, 132
5-3. Battery-Ignition Systems, 136
5-4. Timing Advance Mechanisms, 137
5-5. Multicylinder Engines, 139
5-6. Ignition System Maintenance, 139
5-7. Spark Plugs, 141
5-8. Flywheel Removal/Replacement, 145
5-9. Breaker Points 148
5-10. Condenser, 151
5-11. Ignition Coil, 152
5-12. Flywheel Magnets, 154
5-13. Setting Ignition Timing, 155
5-14. Final Check of Ignition System, 160

6 STARTING SYSTEMS 161

6-1. Manual Starters, 161
6-2. Starter Motor, 165
6-3. Generators, 169
6-4. Alternator, 171
6-5. Magneto-Generator, 172
6-6. Motor-Generator, 173
6-7. Regulators, 174
6-8. Electrical System—A Summary, 176
6-9. Starter Maintenance, 177
6-10. Manual Starter Maintenance, 177
6-11. Starter-Motor Maintenance, 180
6-12. Generator Maintenance, 184
6-13. Regulator Maintenance, 186

7 FUEL SYSTEMS 189
7-1. Carburetors, 190
7-2. Governors, 200
7-3. Fuel Tanks, 204
7-4. Fuel Filters, 207
7-5. Air Cleaners (Filters), 208
7-6. Summary—Fuel Systems, 208
7-7. Fuel System Adjustments, 208
7-8. Carburetor Adjustments, 208
7-9. Governor Adjustments, 213
7-10. Fuel System Maintenance, 213
7-11. Carburetor Maintenance, 213
7-12. Carburetor Float Valve Maintenance, 221
7-13. Diaphragm Fuel Pump Maintenance, 222
7-14. Governor Maintenance, 224
7-15. Fuel Tank Maintenance, 224
7-16. Gasoline, 225
7-17. Preignition and Detonation, 227

8 LUBRICATING SYSTEMS 229
8-1. Purpose of Lubricating Oils, 229
8-2. Oils, 231
8-3. Flow of Lubricating Oils, 233
8-4. Two-Cycle Engine Lubrication, 234
8-5. Four-Cycle Lubrication, 235
8-6. Maintenance, 237

9 TROUBLESHOOTING, TUNING, AND OVERHAULING 238
9-1. The Customer and the Engine, 239
9-2. Four Basic Engine Tests, 240
9-3. Test No. 1—Ignition System, 240
9-4. Test No. 2—Spark Plug Condition, 241
9-5. Test No. 3—Fuel Supply, 242
9-6. Test No. 4—Compression, 242
9-7. Troubleshooting, 244
9-8. Engine Tune-Up and Overhaul, 254

10 MATH AND MEASUREMENTS 256

10-1. Work, 257
10-2. Potential Energy, 257
10-3. Power, 258
10-4. Torque, 258
10-5. Horsepower, 259
10-6. Volumetric Efficiency, 262
10-7. Summary, 262
10-8. Torque Value Conversions, 263
10-9. Metric to English and English to Metric Conversions, 263

11 ROTATING ENGINES AND BEYOND 269

11-1. Operation of the Rotary Engine, 269
11-2. Rotary Engine Advantages, 272
11-3. Rotary Engine Problems, 273
11-4. A Brief History of Rotary Engines, 273
11-5. Beyond the Rotary Engine, 274

APPENDIXES 275

A. Hand Tools, 275
B. Metric to English and English to Metric Conversion Factors, 286
C. Inch-Millimeter Equivalents of Decimal and Common Fractions, 288
D. Decimal Equivalents of Millimeters (0.01 to 100 mm), 290
E. English System of Weights and Measures, 291
F. Metric System of Weights and Measures, 293
G. Battery Test Procedure, 295
H. Illustrated Parts of Typical Small Gasoline Engines, 297
I. Clearances, Torque Data, Specifications and Lubrication Charts of Sample Engines, 315

GLOSSARY 325
INDEX 333

Preface

The Repair and Maintenance of Small Gasoline Engines is written for the person who wants to study small gasoline engines so that he can repair his own engines, work in a repair shop, or open his own small engine repair shop. The book is also written for contractors such as lawn servicers or tree trimmers who need to provide periodic maintenance on small gasoline engine-powered equipment to assure its daily use over a long period of time.

The Repair and Maintenance of Small Gasoline Engines contains the fundamentals of operation and maintenance of many types and models of small gasoline engines used to drive machinery and vehicles, including: lawn mowers and edgers; ranchette machinery such as tractors, mulchers, tillers, and sweepers; chain saws; cutters; drills; electric generators; pumps; welding machines; industrial sweepers and scrubbers; carts; tri-carts; small all-terrain vehicles; snowmobiles; outboard motors for boats; minibikes; and lightweight motorcycles. Conscientious study of the fundamentals of operation and maintenance presented in this book will provide you with a thorough background for repairing any small gasoline engine.

The Repair and Maintenance of Small Gasoline Engines is divided into eleven chapters that logically teach the future small gasoline engine technician, the contractor, and the do-it-yourself home owner how small reciprocating gasoline engines work, how to perform periodic maintenance to the engines to ensure their continued usefulness over the rated design period, and how to recognize troubles, make tune-ups, and repair malfunctions. Chapter 11 presents the rotary engine and a look at future engine designs beyond the rotary. Many chapters contain handy Quick Reference Charts that abbreviate the detailed procedures of the text.

Acknowledgments

The author would like to acknowledge and thank the following companies and persons who generously donated many of the photographs and line art in this book.

	Figure No.
AMF Harley-Davidson, Milwaukee, WI 53201, Mr. Delbert L. Ohlschmidt, Technical Services Department.	1-2, 6-2
Atwater Strong Division, Division of Gougler Industries, Inc., Atwater, OH 44201, Mr. J. March Lane, President.	1-19
Champion Spark Plug Co., Toledo, OH 43661, Mr. Dennis H. Bender, Public Relations Department.	3-19, 4-8, 4-9, 5-12 to 5-17
Clinton Engines Corp., Maquoketa, IA 52060.	1-5, 2-8, 2-18, 2-28, 2-41 to 2-48, 3-2, 3-4, 3-5, 3-6, 3-11, 5-22, 6-5, 6-6, 7-13, 7-15, 7-16, 7-17, 7-19, 7-27, 8-1, 8-2, 8-4, 8-6, 8-7, 9-1, 9-4, A-1, Tables I-1, I-2
Delco-Remy, Division of General Motors, Anderson, IN 46011, Mr. K. E. Adams, Advertising and Sales Promotion.	3-21, Table 3-3, 4-12, 4-13, 4-14, 6-15, 6-16, 6-19, Appendix G

	Figure No.
Evinrude Motors, Milwaukee, WI 53201, Mr. Graeme Paxton, Assistant Public Relations Manager.	1-7, 2-1, 2-32, 3-12, 4-2, 4-3, 4-6, 5-1, 5-2, 7-27, 7-28, H-6, H-9, H-10, Tables I-6 to I-9
Gravely, Clemmons, NC 27012, Mr. Jack H. Guthrie, Assistant Advertising Manager.	1-1, 1-4, 1-8, 1-10
Homelite, a division of Textron Inc., Charlotte, NC, Mr. Rod Ferguson, Supervisor, Technical Publications Dept.	1-13, 1-14, 1-17, H-4
Jari, Mankato, MN 56001, Mr. Gerald Ormsky, Sales Manager, Jari Division.	1-15
Johnson Outboards, Waukegan, IL 60085, Mr. Ron A. Pedderson, Public Relations Manager.	1-1, 1-11, 2-11, 2-16, 2-17, 2-26, 3-7, 3-13, 3-14, 3-20, 4-7, 5-30, 7-21, 7-22, 7-23, 7-27, 7-31
Kohler Co., Kohler, WI 53044, Mr. Edward B. Anderson, Advertising and Promotional Services.	2-31, 2-34, 2-38, 2-40, Table 3-2, 3-16, 5-3, 5-11, 5-18, 5-27, 6-3, 6-4, 6-17, 6-18, Table 6-2, 7-14, 7-25, 7-27, A-1, H-5
Lindig Manufacturing Corp., St. Paul, MN 55113, Mr. John Lindig, Vice President.	1-12
McCulloch Chain Saws, Los Angeles, CA 90009, Mr. Dave Kirby, Public Relations Manager.	2-2, 2-4, 2-9, 2-10, 5-21, 5-24, 5-25, 7-1 to 7-7, 7-9, 7-10, 7-11, 7-32, 7-33, 9-2, 9-3, 9-6, 9-7, 9-8
Merry Manufacturing Company, Edmonds, WA 98020.	1-6
O & R Engines, Inc., Los Angeles, CA 90023.	5-19
OMC–Lincoln, a Division of Outboard Marine Corporation, Lincoln, NE 68501, Mr. Ed Large, Manager, Marketing Communications.	1-3, 1-9, 2-3, 2-21, 2-22, 3-1, 3-8, 3-9, 3-15, 5-4 to 5-9, 5-20, 6-12, 7-8, 7-12, 7-20, 7-24, 7-27, 7-29, 8-3, H-1, H-2, H-8, Tables I-3, I-4, I-5

	Figure No.
ONAN, Division of Onan Corporation, Minneapolis, MN 55432, Mr. Virgil C. Gilbertson and Mr. Henry Coursolle.	1-16, 1-18, 2-7, 2-12, 2-19, 2-20, 2-25, 2-27, 2-29, 2-30, 2-33, 2-35, 2-36, 2-37, 2-39, 2-49, 3-3, 3-10, 3-22, 5-29, 5-31, 6-7, 6-9, 6-10, 6-14, 6-20, 6-21, 6-22, 7-18, 7-26, 7-27, 7-30, A-1, H-3
Sears, Roebuck and Co., Chicago, IL 60684, Mr. Larry W. Wheeler, PR Manager, Hardware Category.	Frontispiece
Teledyne Wisconsin Motor, Milwaukee, WI 53246, Mr. J. W. Perschbacher, Vice President–Sales and Marketing.	2-6, 6-1, H-7

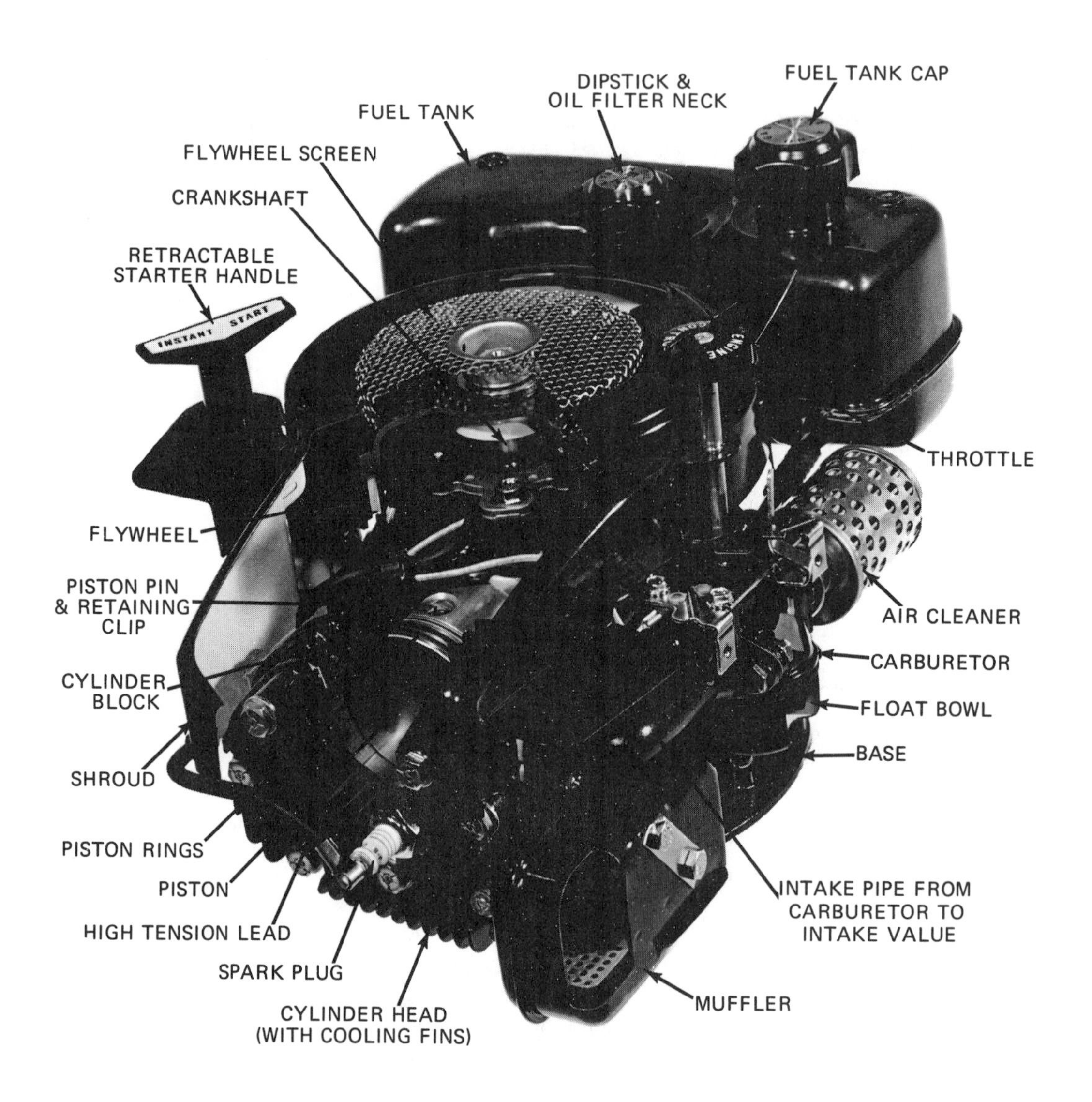

This cutaway of a small gasoline engine shows you some of the major parts and controls.

1

Small Gasoline Engines

An engine is a machine for converting any of various forms of energy into mechanical force and motion. Energy is usually supplied in the form of chemical energy, such as fuel, or electrical energy such as a battery or alternating current used in houses. The mechanical force and the motion is most commonly delivered in the form of rotary motion of a shaft.

Small gasoline engines are used to aid man to work quickly, efficiently, and with less fatigue than if done by hand—to help him to cut grass, till soil, remove snow, cut trees, transport materials, pump liquids, and to sweep floors. Small gasoline engines are used to enable man to enjoy himself in his leisure time by providing power for recreational vehicles such as pleasure boats, racing boats, racing carts, and all-terrain vehicles for transporting him on the golf course or to the hunting lodge. Small gasoline engines are also used to provide power for low cost transportation on light motorcycles and minibikes (Fig. 1-1).

Engines are classified by the:

1. substance used for the driving force—as steam, compressed air and *gasoline.*
2. type of motion of their principal parts—as *reciprocating* and *rotary.*
3. the place where the exchange from chemical to heat energy takes place —as *internal-combustion* (gasoline engines) and external-combustion (steam engines).

FIG. 1-1. Millions of small gasoline engines are used to power recreational vehicles, tools, and machines; many of the owners do not care for them properly, resulting in employment for many small gasoline engine technician/mechanics.

4. method by which the engine is cooled—*air*-cooled, such as lawn mowers and chain saws, and *water*-cooled, such as outboard engines and automotive engines.

5. position of the cylinders of the engines—as V, in-line and radial.
6. number of *strokes* of the piston for a complete cycle—as *two-stroke* and *four-stroke.*
7. type of *cycle*—as *Otto* (used in ordinary gasoline engines), *Diesel,* and *Wankel* (rotary).
8. direction of crankshaft—*vertical,* as a rotary lawn mower; *horizontal,* as a tractor with a drive pulley, or a generator power plant; and *multi-position,* as a chain saw.
9. the use for which the engine is intended—as outboard engine and lawn mower.

The study of small gasoline engines in this text covers the classifications of engines that include a driving force substance of *gasoline,* a motion that is *reciprocating* or *rotary,* an *internal-combustion engine,* cooling methods of both *air* and *water, various positions* of the cylinders of the engines, *two-* and *four*-stroke engines, the *Otto* and the *Wankel* engine, crankshaft directions in the *vertical, horizontal,* and *multi-position* locations, and for *uses* in all types of small power applications discussed in Section 1-1. Engines are sometimes called motors, but the term motor is usually restricted to engines which transform electrical energy into mechanical energy.

This chapter provides an introduction to the subject of small gasoline engines—theory, operation, and repair. The chapter covers subjects including: what is powered by a small gasoline engine, the future of small gasoline engines, basic operation of small reciprocating gasoline engines, and safety.

The remainder of this book details the theory, operation, and maintenance of the reciprocating engine and its supporting systems: Chapter 2, The Basic Small Gasoline Engine; Chapter 3, Operation and Periodic Maintenance Procedures; Chapter 4, Basic Electricity and Electrical Components; Chapter 5, Ignition Systems; Chapter 6, Starting Systems; Chapter 7, Fuel Systems; Chapter 8, Lubricating Systems; Chapter 9, Troubleshooting, Tune-up, and Overhauling; and Chapter 10, Math and Measurements. Chapter 11 provides the reader a basic understanding of the relatively new rotary engine; several future engine designs are also discussed.

Nine appendixes provide essential supplementary information. Appendix A lists the necessary hand tools required for repair and maintenance of small gasoline engines; Appendixes B through F provide metric conversions necessary for the technician if he needs to convert English measurements to metric equivalents, or metric measurements to the English equivalents. Appendix G provides battery tests and Appendix H provides illustrated parts breakdowns of typical small gasoline engines. Appendix I provides sample charts of service part clearances, bolt/nut/screw torque values, engine specifications and lubrication charts of a few typical small gasoline engines; the charts provide a sample of the type of data provided by manufacturers. A glossary of terms is included at the end of the book. Finally, a comprehensive cross-referenced index enables you to easily locate a subject within the text.

1-1. WHAT IS POWERED BY A SMALL GASOLINE ENGINE?

The following small gasoline engine-powered machines are used to aid man to work quickly and efficiently, to enable him to further enjoy himself in his leisure time, or to provide power for low-cost transportation: lawn mowers and edgers; ranchette machinery such as tractors, mulchers, tillers, and sweepers; chain saws; cutters; drills; (power for) electric generators; pumps; welding machines; industrial sweepers and scrubbers; carts; tri-carts; other small all-terrain vehicles; snowmobiles; outboard engines for boats; minibikes; and lightweight motorcycles. New uses of small gasoline engines are constantly being found by engineers, designers, technicians, homeowners, hobbyists, workmen, and leisure time pleasure seekers. Some of the applications of small gasoline engines as power units for machinery, recreational vehicles, and transportation are shown in Figures 1-2 through 1-19.

Fig. 1-2. The motorcycle is an economical mode of transportation as well as a recreational vehicle. This trail bike is powered by a two-stroke cycle single cylinder engine. It has a bore of 1.89 inches, stroke of 1.97 inches, piston displacement of 5.52 cubic inches and a compression ratio of 9.2 to 1.

Fig. 1-3. Most of us are familiar with small gasoline engines used to power lawn mowers. Both engines shown are powered by two-stroke cycle engines. The mower in (B) includes an electric starter and is self-propelled. The crankshafts are vertical.

Fig. 1-4. Small gasoline engines are used to power machines to ease man's workload. This riding tractor with a snowdozer is powered by a 16.5 horsepower engine.

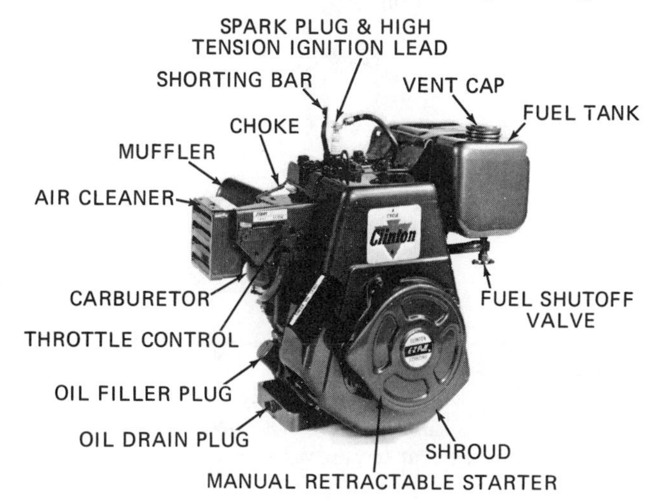

Fig. 1-5. This four-cycle 3.5 horsepower horizontal crankshaft engine can be used for many purposes.

FIG. 1-6. This tiller is powered by a four-stroke cycle engine. The crankshaft is horizontal and drives a roller chain and sprocket transmission.

FIG. 1-7. Small gasoline engines provide many hours of leisure time pleasure for their owners. This water-cooled two cylinder two-cycle engine is 4 horsepower and runs at 4000 to 5000 rev/min. Other leisure time vehicles powered by small gasoline engines include snowmobiles, carts, tri-carts, all terrain vehicles, and mini-bikes.

FIG. 1-8. An 8 horsepower engine powers this lawn tractor. It has a 12-volt electric starter, four speeds forward and reverse, and can travel up to approximately 5.5 miles per hour.

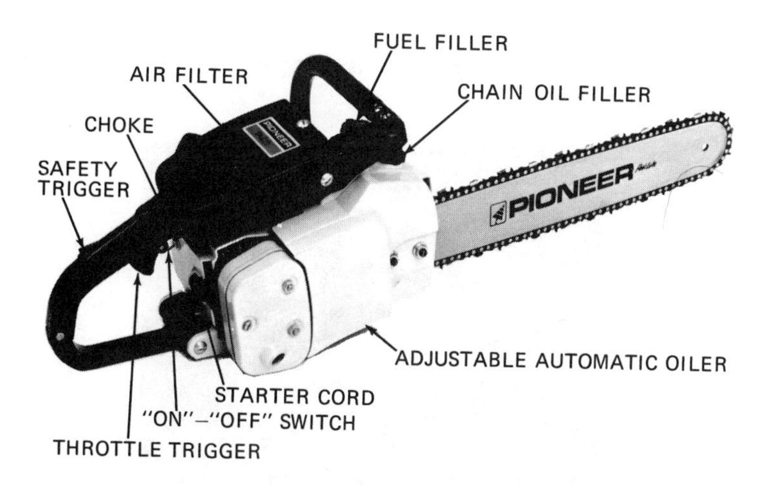

FIG. 1-9. A chain saw with a small internal combustion engine weighs only about 15 pounds. At maximum horsepower, the engine runs at 8,500 rev/min and the chain speed at 3,700 feet/minute. The engine must be able to operate at any angle.

FIG. 1-10. This versatile tractor has attachments used to plow, cultivate, or mow. It is powered by a 10 to 12 horsepower engine.

FIG. 1-11. The sales of small gasoline powered engines are being affected somewhat by advances in the electric motor power field. With the exception of rechargeable units that must be recharged frequently, the electric powered machinery is not portable, and hence, cannot be taken to even the slightest remote location.

FIG. 1-12. Compact chippers replace outdoor burning of branches, stalks, and yard trimmings with power chipping. A double blade rotating at 3600 rev/min reduces branches to mulch. The rotor is powered by a 7 or 10 horsepower small gasoline engine. It is estimated that about 12 million new small gasoline engines are produced per year.

FIG. 1-13. Power generators driven by gasoline engines provide electrical power at remote sites. Oil is mixed with gasoline in this and all two-cycle engines.

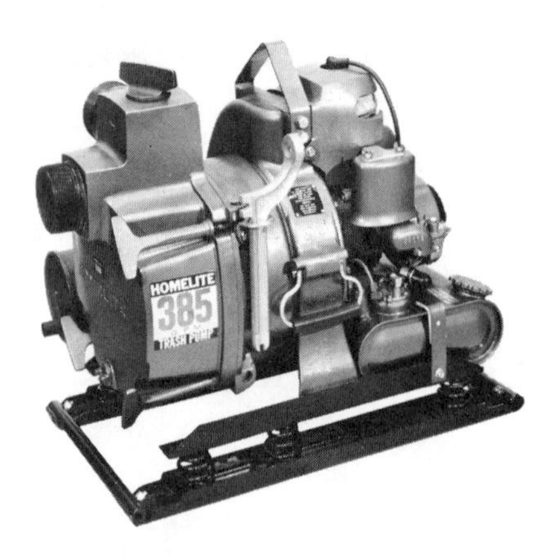

FIG. 1-14. Trash pumps pump water mixed with solids (mud, muck, sand, and light gravel) at capacities of 11,500 to 36,500 gallons per hour.

FIG. 1-15. Sickle mowers cut grass, weeds, and brush. The engine operates on the Otto cycle and has a horizontal crankshaft.

FIG. 1-16. The engine that powers this tractor is 16.5 horsepower, two cylinder, four-stroke cycle, L-head, and air-cooled. The crankcase capacity is 3.5 quarts.

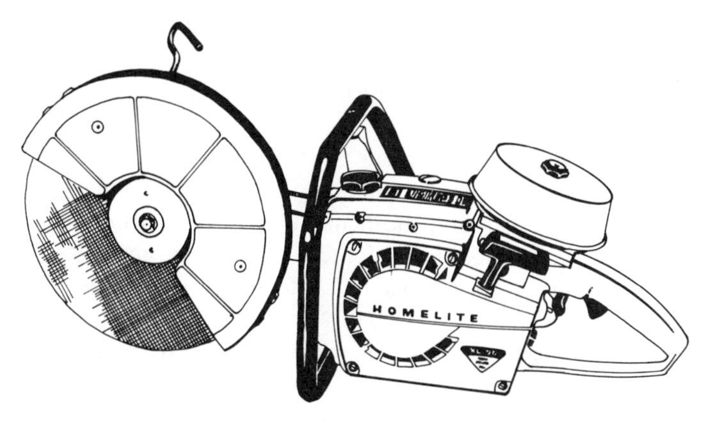

FIG. 1-17. A gasoline engine-powered multipurpose saw is designed for demolition, scrap salvage, forcible entry, and rescue work. Like the chain saw, it must operate at any angle.

FIG. 1-18. The small gasoline engine can be used to provide power for welding machines at remote locations.

FIG. 1-19. A 3 to 7 horsepower engine drives various models of this sweeper/blower. The fan moves 1000 to 2000 cubic feet of air per minute with an air blast of 110 to 150 miles per hour. The sweeper blower removes leaves, litter, and water.

1-2. THE FUTURE OF SMALL GASOLINE ENGINES

There are tens of millions of small gasoline engines in use in the world today and this number is increasing by millions of additional engines built each year to simplify man's work around the home, farm, and industry, and to provide power for recreational pleasure. Each of these millions of small gasoline engines is designed to operate for approximately 1,000 hours; if an engine is used 4 hours per week for 32 weeks out of the year, this would equal 128 hours of operation per year. Thus the life expectancy of an engine is approximately 8 years, but many people don't provide periodic maintenance to the small gas engine. Many engines discontinue operation because of a lack of maintenance or abuse and are junked each year. The engines remaining in service require periodic preventive maintenance, corrective maintenance, and repair. This means work for the technician trained in the servicing and repairing of small gasoline engines.

The sales of small gasoline-powered engines are being affected by advances in the electric motor power field. Electric motor-powered machines include lawn mowers, chain saws, small outboard engines, and rechargeable battery-operated carts, and industrial equipment. With the exception of the rechargeable units that must be recharged frequently, the electric powered machinery is not portable, and hence, cannot be taken to even the slightest remote location.

With the energy shortages of electric power, natural gas, oil, and gasoline that we face in the United States today and will in the years to come, it is evident that more and more people are turning to the smaller gasoline engine powered vehicles for conservation of energy and for economic reasons. This means increased sales of small gasoline engines with subsequent demand for service and repair work by the skilled technician employed in the small gasoline engine field.

1-3. BASIC OPERATION OF SMALL RECIPROCATING ENGINES

Figure 1-20 illustrates the small reciprocating engine and its supporting systems: starting system, ignition system, fuel system, and lubricating system. The small gasoline engine causes a piston to move in a reciprocating motion (up and down) by means of combustion or fire, heat and pressure. It then converts this reciprocating motion to rotary motion of a crankshaft. The engine proper consists of six major parts (Fig. 1-21): a cylinder, cylinder head, piston, connecting rod, crank and crankshaft, and crankcase. These components work together by using a mixture of air and fuel. A spark is used to cause combustion to provide and convert reciprocating motion of a piston to rotary motion at the crankshaft to drive a unit such as a blade on a rotary lawn mower, a pulley on a generator, or gears that drive a machine part such as the propeller on an outboard engine.

The cylinder is stationary and hollow. A piston fits snugly into the cylinder and moves up and down within the cylinder. The cylinder head bolts onto the top of the cylinder and forms the top of the combustion chamber. A connecting rod connects the piston and the crank of the crankshaft in a manner such that the

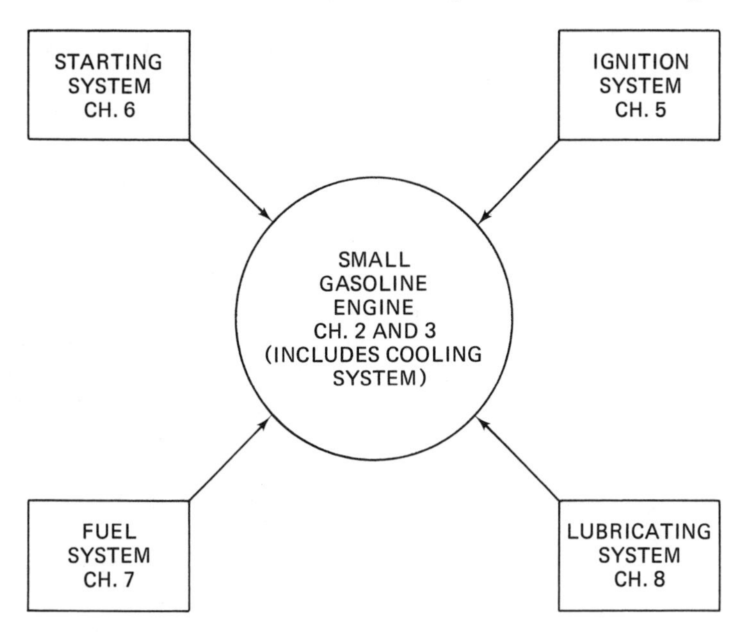

FIG. 1-20. The basic small gasoline engine has five supporting systems.

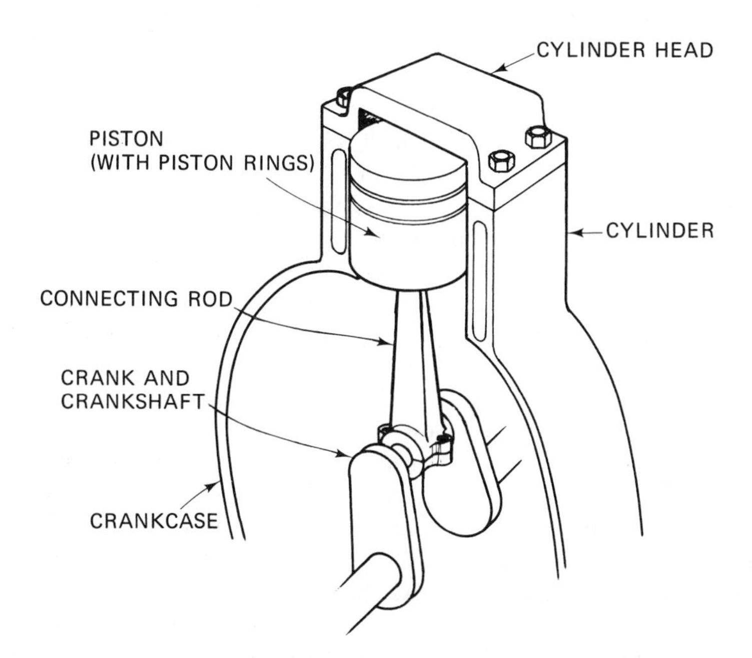

FIG. 1-21. The small gasoline engine has six basic major engine parts.

connecting rod moves up and down with the piston. The connecting rod drives the crank around and around, converting the reciprocal motion of the piston and connecting rod to a rotary motion of the crankshaft. The crankcase houses most of the moving parts of the engine; it is the body of the engine.

In every reciprocating engine, there are four actions that take place: the intake of an air-fuel mixture into the combustion chamber; compression of this air-fuel mixture within the combustion chamber; the power action, in which the air-fuel mixture is ignited and the hot gases expand causing the piston to be pushed down; and finally, the fourth action that is the exhausting of the burned air-fuel mixture from the combustion chamber. These actions of *intake, compression, power,* and *exhaust* comprise the two types of engine cycles that are studied in detail in this text. The two types of cycles are termed the *two-stroke cycle* and the *four-stroke cycle.* A cycle of operation is a complete circle of events; for example, spring, summer, winter, and fall comprise a complete cycle or circle of events. Likewise, a cycle within the small gasoline engine is complete when all actions have taken place one time. This cycle consists of four actions required of every reciprocating engine; the actions of intake, compression, power, and exhaust. In the two-stroke cycle of operation, there are *two strokes* that complete the cycle; compression and power. The actions of exhaust and intake occur between the power and compression strokes. The four-stroke cycle engine uses *four strokes* to complete the four reciprocating engine actions. Thus, the first stroke is used for intake, the second for compression, the third for power, and the fourth stroke of exhaust completes the cycle. The two-stroke cycle and the four-stroke cycle will become more apparent when the complete theory of operation is discussed in Chapter 2.

As shown in Figure 1-20, there are four supporting systems for the small gasoline engine plus a fifth system which is an integral part of the engine. The starting system provides the initial means of starting the engine cycle of operation. The ignition system provides the spark to ignite the air-fuel mixture in the engine at the correct time in the operating cycle. The fuel system provides the correct proportion of air-fuel mixture to the engine and the lubricating system provides the necessary lubrication to reduce friction and hence to extend the life of the moving parts within the engine. The fifth system, an integral part of the small gasoline engine, is the cooling system. The cooling system is either air or water, and its function is to carry away the heat generated by the combustion of the air-fuel mixture within the combustion chamber.

The basic difference between the two- and the four-cycle engine is the method in which the air-fuel mixture *enters* the cylinder combustion chamber and the method in which the burned mixture *exhausts* from the cylinder. As previously mentioned, the number of strokes to complete a cycle is also different—there are two strokes to complete the *two-stroke cycle engine* operation and four strokes to complete the *four-stroke cycle engine* operation. The methods of lubrication to reduce friction within the engine and extend the life of the moving parts are also different within the two- and four-cycle engines.

The two-stroke cycle engine is lightweight, low in cost, powerful, and its operation is simple. There are only three moving parts—piston, connecting rod, and crankshaft. These keep maintenance costs at a minimum and efficiency at a maximum. The two-cycle engine does not require that an oil level be maintained; it is always lubricated by the oil in the oil-gasoline fuel mixture. The engine always remains lubricated regardless of the angle at which it is operated. (For example, the chain saw is a two-cycle engine. It is operated at many angles, but it is continually lubricated because the oil is mixed with the gasoline.) Two-cycle engines power many lawn mowers, edgers, tillers, chain saws, motorcycles, and similar machines.

The four-cycle engine operates more smoothly and quietly than does the two-cycle engine. The small four-cycle engine is the type of engine used in automobiles and trucks. It is not necessary to mix oil with the gasoline, but it is necessary to place oil in the crankcase and to change the oil periodically. The engine must be operated on a fairly level surface to prevent a lack of lubrication of the internal parts. For equivalent horsepower rating, the four-cycle engine is larger and heavier than the two-cycle engine. The four-cycle engine requires a larger cubic piston displacement so a greater amount of fuel can be exploded at each ignition period to drive the engine through a cycle. The four-cycle engine runs cooler because of a longer period between consecutive firing periods.

There are several ways to identify a two-cycle engine from a four-cycle engine. Some of these ways are not readily apparent at this juncture in your course, but will become more apparent as you study this text. The simplest method to tell the difference between the two-cycle and the four-cycle engine is to look at the nameplate or the operating and maintenance instructions. If the nameplate or the operating instructions tell how to mix a proportion of oil into gasoline, then it is a two-cycle engine. Likewise, when the crankcase capacity is mentioned, it means the oil is not mixed with the gasoline, but it is put into the crankcase, and is therefore a four-cycle engine. If you notice an oil filler plug or an oil reservoir, then you will know it is a four-cycle engine. If the instructions require that the oil be drained and refilled periodically, then obviously it is a four-cycle engine. A not-so-obvious means of determining whether an engine is two-cycle or four-cycle is the location of the muffler. If the muffler is installed toward the center of the cylinder at the location of the exhaust *port,* then it is a two-cycle engine. However, if the muffler is installed at the head of the cylinder at the location of the exhaust *valve,* then it is a four-cycle engine.

1-4. SAFETY

It is imperative to emphasize the necessity of safety practices in the small engine repair shop or in your home or garage if you are repairing your own small gasoline engine. Considerations of fire from gasoline and oily rags are the most important. Likewise, it must be emphasized that deadly carbon monoxide exhaust gases are given off by all gasoline engines. You must also be careful in using high-pressure air to clean engines.

Gasoline should never be kept in glass bottles. If a bottle should fall onto the floor and break, it could cause an immediate flash fire if the gasoline is ignited by any spark source. Always keep gasoline stored in small cans and insure that vapors do not leak from the cap. Gasoline vapors can build up within a closed area, and can be ignited, for example, by turning on an electrical switch which could cause a spark to ignite the gasoline vapors. Be sure to check your local city, county, and state regulations regarding the keeping of containerized gasoline. Oily rags should be kept in a closed metal container. Excessively oily rags should be discarded.

Do not run small gasoline engines in a closed environment at any time. Gasoline is made of hydrocarbons (hydrogen and carbon) and when it is mixed with air and burned, three by-products are produced: carbon dioxide, carbon monoxide and water. The molecules of carbon monoxide cannot be seen and cannot be smelled, but they are deadly gases. Therefore always operate gasoline engines in a well ventilated area.

Do not make any adjustments or repairs and do not clean any parts of an engine until the spark plug lead has been disconnected or/and the start switch is placed to off. This will prevent inadvertent engine starting.

Do not fill the fuel tank while the engine is hot unless the fuel tank is located several feet from the engine and there is *no* chance that *any* gasoline could spill onto the engine. Spilling gasoline on a hot engine could cause an explosion and serious injury. For the same reasons, do not fill a fuel tank when the engine is running; ensure that the engine is off before adding fuel.

Some of the procedures within this text recommend the use of high-pressurized air as a method of blowing dirt from engine parts. Be sure to wear goggles or a face shield to prevent dirt particles from getting into your eyes.

2

The Basic Small Gasoline Engine

You have learned the basic theory of operation of the two-stroke cycle and the four-stroke cycle engines. Both the two- and four-cycle engines are internal-combustion engines that have four actions: power, exhaust, intake, and compression. The two-cycle engine, more correctly identified as the two-stroke cycle engine, has what are considered as two strokes—power and compression strokes—with the actions of exhaust and intake in between the power and compression strokes. The four-cycle engine, more correctly identified as the four-stroke cycle engine, has four distinct strokes: power, exhaust, intake, and compression.

The operation of both the two-cycle and the four-cycle engine is basically the same. An air-fuel mixture is brought into the combustion chamber in the intake action. The air-fuel mixture is then compressed and ignition from a spark between the electrodes of the spark plug causes combustion of the air-fuel mixture. The power stroke is initiated driving the piston down and then the exhaust action takes place with the hot burned gases exiting from the exhaust duct. During the power stroke, the piston is driven down coupling the power through the connecting rod and crank to turn the crankshaft. This action converts reciprocating or up and down motion into a rotary action of the crankshaft to which a power takeoff such as a pulley, gear, or mower blade is attached.

This chapter explains the theory of operation and maintenance of the engine portion of the small gasoline engine. The major items discussed include: operation of the two-stroke cycle engine, operation of the four-stroke cycle engine, a description and theory of operation of the major parts within the engine, a summary of engine operation, and maintenance of the major parts of the engine. Diligent study coupled with hands-on-experience in the practical exercises at the end of this chapter will result in a thorough knowledge of the basics of all of the major parts and operation of the engine. Further study in subsequent chapters will give you the total theory of operation and maintenance of not only the engine, but of all of the supporting systems.

2-1. OPERATION OF TWO-STROKE CYCLE ENGINES

The two-stroke cycle engine has two stroke actions, namely, power and compression with the actions of exhaust and intake occurring between the power and compression strokes. During the power stroke (Fig. 2-1A), the air-fuel mixture in the combustion chamber explodes, driving the piston down. As the piston passes the exhaust port (Fig. 2-1B), the hot burned gases begin escaping from the combustion chamber. As the piston continues down, the intake port is opened allowing a new air-fuel mixture from the crankcase to enter the combustion chamber. The piston continues downward driving, by means of the connecting rod, the crank and crankshaft to produce rotary motion. When the piston passes its bottom position, it begins on the upstroke (Fig. 2-1C) compressing the air-fuel mixture in the combustion chamber for the next power stroke. When the piston reaches top center, a spark jumps between the electrodes of the spark plug and the air-fuel mixture again explodes starting the next cycle of operation (Fig. 2-1A).

The theory described can now be expanded to include the introduction of the air-fuel mixture into the crankcase through the reed valve located at the bottom of the crankcase. As the piston travels upward during the compression stroke, there is a slight decrease in pressure in the crankcase that causes a slight vacuum in the crankcase. Atmospheric pressure through the carburetion system causes the air-fuel mixture to flow through the reed valve into the crankcase (the reed valve is open because of the difference in pressure). At the top of the piston stroke, when ignition takes place, the piston is driven downward causing an increase in pressure in the crankcase that closes the reed valve. As the piston continues downward, exhaust gases are exited from the exhaust port; the piston continues downward and opens the intake port. The pressure of the piston going down causes the air-fuel mixture to be pushed through the intake port into the combustion chamber. This new flow of air-fuel mixture aids in forcing the hot exhaust gases out of the exhaust port. After the piston passes bottom center, it then starts on another upstroke closing off the intake port and then the exhaust port. It compresses the new charge of air-fuel mixture in the combustion chamber causing the mixture to be more explosive. This cycle of events of power, compression, and the actions of exhaust and intake taking place between the power and compression strokes, continues as long as the engine continues to run. Rotary valves or poppet valves (Section 2-7) are sometimes used in place of the reed valves.

Some engines use a transfer port (Fig. 2-2) in place of a reed, rotary, or

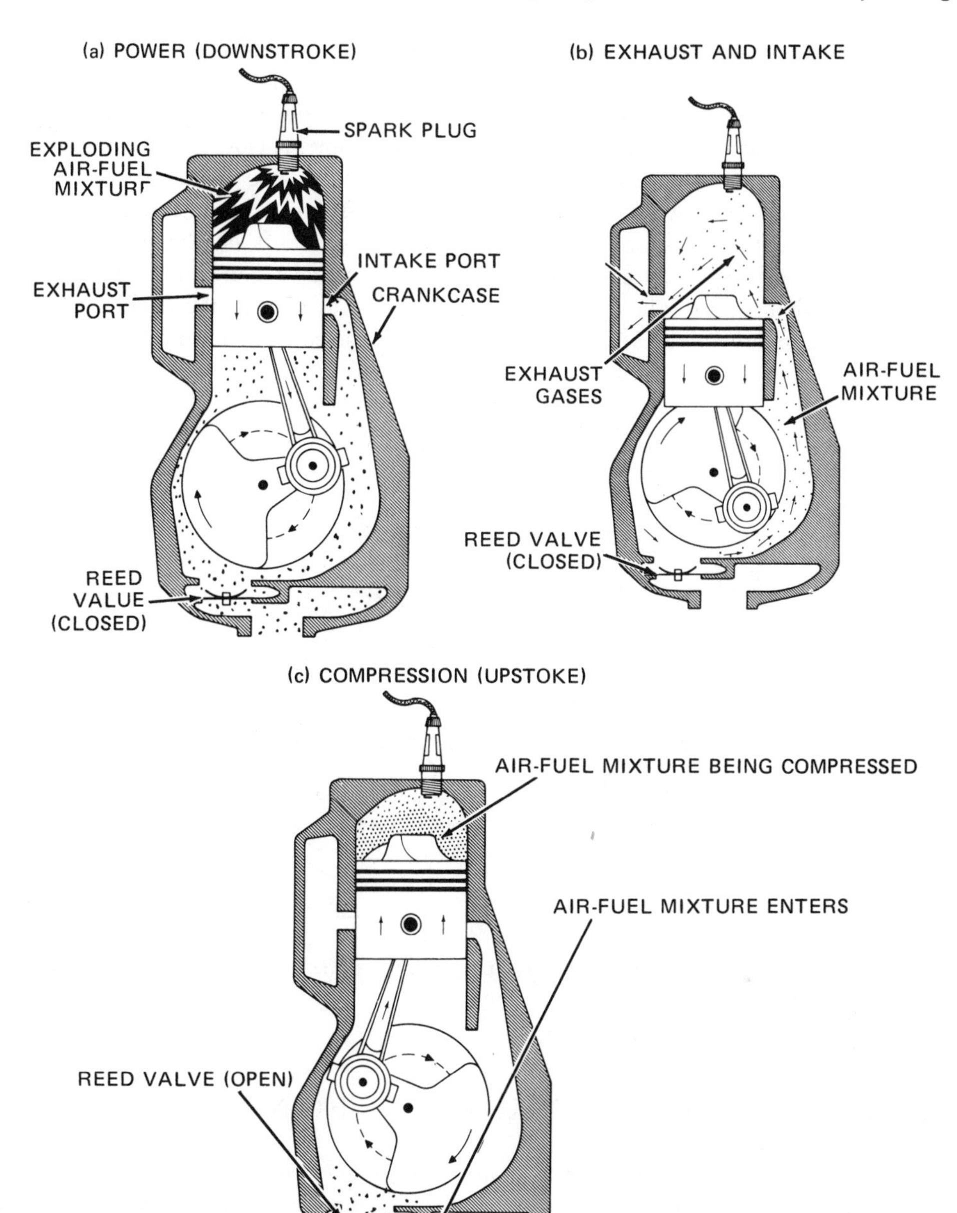

FIG. 2-1. The two-stroke cycle engine has two stroke actions, namely power and compression, with the actions of exhaust and intake occurring between the power and compression strokes.

poppet valve. The transfer port is opened and closed by movement of the piston in the same manner as the intake and exhaust ports. As the piston moves up on the compression stroke, there is a slight vacuum created in the crankcase. Atmospheric pressure causes the air-fuel mixture from the carburetion system to flow into the crankcase. When the piston reaches the top position, ignition takes place and the power stroke is initiated. The piston comes down opening first the exhaust port and then the intake port, and at the same time, it closes off the transfer port. Hot gases are exhausted and the pressure built up in the crankcase caused by the piston coming down forces the new charge of air-fuel mixture through the intake port into the combustion chamber. As the piston starts on the upstroke, it continues to block the transfer port, closes off the intake port, and finally closes off the exhaust port. The cycle continues to repeat as long as the engine runs. Reed valves, rotary valves, and poppet valves are discussed in more detail in Section 2-7.

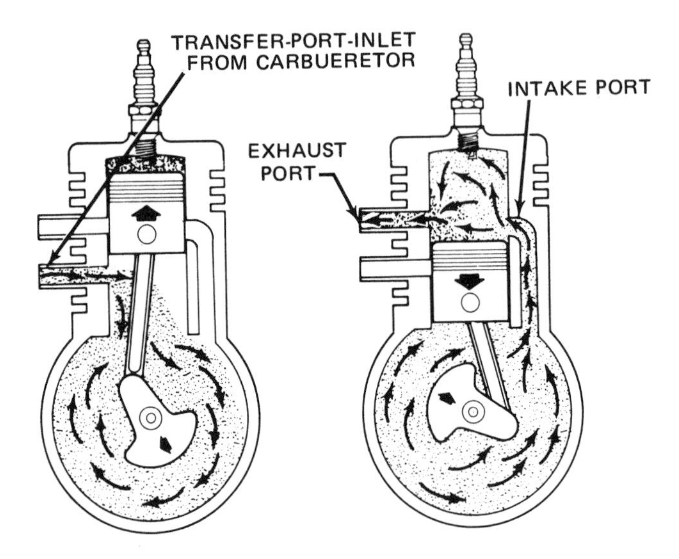

FIG. 2-2. Some two-cycle engines use a transfer port rather than a reed, rotary, or poppet valve to admit the air-fuel mixture into the crankcase.

2-2. OPERATION OF FOUR-STROKE CYCLE ENGINES

Four-cycle engines are used in many small gasoline engines and in most automobiles. The four-cycle engine is basically the same as the two-cycle engine except for the method of introduction of the air-fuel mixture into the combustion chamber and the exhaust of the burned mixture from the chamber. In the two-cycle engine, the intake and the exhaust are through ports in the cylinder walls that are opened and closed as the piston moves past the ports. In the four-cycle engine, intake gases and exhaust gases flow through ports in the top of the cylinder block at the cylinder head. These ports are covered by metal covers known as *valves*. The valves move up and down in guides by means of a gear and cam arrangement to open and close the ports at the proper time.

In the two-cycle engine, two strokes of the piston are required for a complete cycle of operation; this results in one revolution of the crankshaft. In the four-cycle engine, four strokes of the piston are required for a complete cycle of operation. These strokes are the four actions of power, exhaust, intake, and compression; this results in two revolutions of the crankshaft for one complete cycle of operation.

The same major parts of similar design are used in both the two- and four-cycle engines. These parts include the cylinder, cylinder head, crankcase, piston, connecting rod, crank and crankshaft, flywheel, bearings and the muffler. Additional moving parts that are used in the four-cycle engine include timing gears, camshaft, intake and exhaust valves, an oil slinger or oil pump, and springs. A comparison of the moving parts of a two- and four-cycle engine are shown in Figure 2-3.

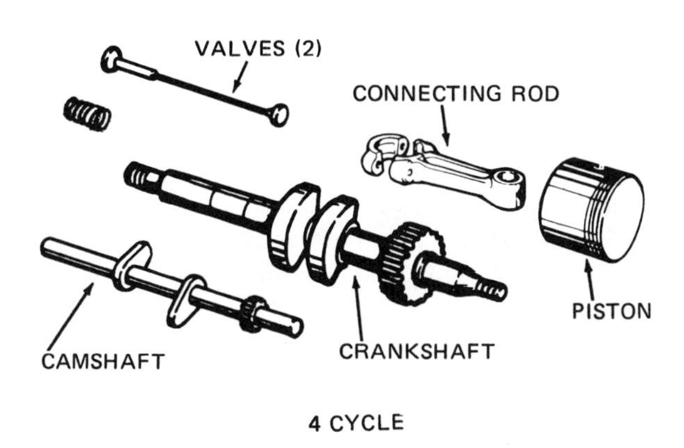

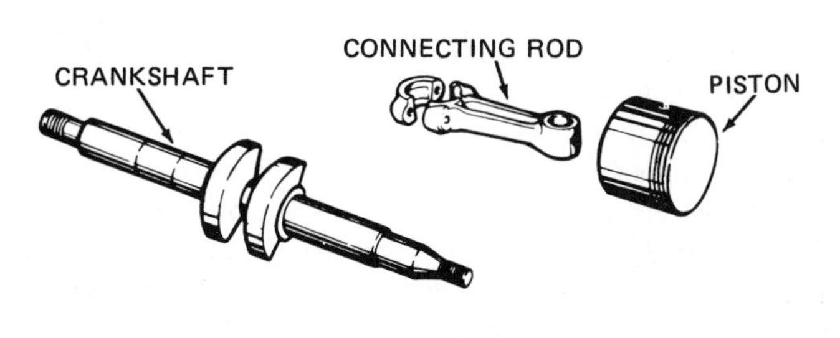

FIG. 2-3. Additional moving parts are used in the four-cycle engine. These additional parts include a camshaft, timing gears, intake and exhaust valves, an oil slinger or pump, and springs.

Four-cycle operation (Fig. 2-4) consists of four strokes. Starting with the power stroke, the ignition system sends a high tension voltage to the spark plug such that a spark jumps between the spark plug electrodes causing a firing, or exploding, of the compressed air-fuel mixture in the combustion chamber. The explosion drives the piston downward. Toward the end of the power stroke, the exhaust valve is opened allowing the exhaust gases to begin to exit. The piston continues down to the bottom of the stroke and then starts upward in the exhaust stroke. As the piston rises in the cylinder, the exhaust gases are pushed out the exhaust valve. Near the top of the exhaust stroke, the intake stroke begins, the intake valve is opened, and a new air-fuel mixture is drawn into the combustion chamber as the piston starts its downward stroke. On the intake stroke, a partial vacuum is created as the piston goes down. Atmospheric pressure through the carburetion system causes the new mixture to flow into the combustion chamber. When the piston reaches a little beyond the bottom of its stroke and begins upward on its compression stroke, the intake valve is closed. Thus the air-fuel mixture in the combustion chamber is compressed by the piston moving upward. At the top of the compression stroke, a spark from the ignition system jumps between the spark plug electrodes and the spark causes ignition or combustion of the air-fuel mixture as another cycle of operation begins. Thus there have been four strokes of the piston to complete one cycle of operation. The four strokes have caused the crankshaft to rotate two revolutions. During the compression and power strokes, the intake and exhaust valves are closed.

The intake and exhaust valves are opened and closed as a result of crankshaft rotation that drives a camshaft. The camshaft and cams that operate the valves are discussed in Section 2-15.

An example of valve timing is shown in Figure 2-5. The left axis of the graph from bottom to top indicates a cycle of operation of a four-cycle engine: power stroke, exhaust stroke, intake stroke, compression stroke, and the beginning of the next power stroke. The horizontal axis shows the degrees of rotation of the crankshaft. Two revolutions of the crankshaft (720°) which occur during one cycle of engine operation are displayed. When the piston is located at 0°, the piston is at its maximum top position. This position is known as *top dead center* (TDC) and occurs at 0°, 360°, and 720°. When the piston is at its bottom most point, it is known as the *bottom dead center* (BDC) position. This occurs at 180°, and 540°. In order for the burned exhaust gases to exit, and for a new air-fuel mixture to flow into the combustion chamber, the valves must open and close at overlapping times in the power and compression strokes. This overlap is shown by the bar graph of Figure 2-5. The power stroke begins at 0°. At some point before the piston hits BDC, the exhaust valve is opened. Most power has been expended, and thus there is not a loss of power. This opening of the exhaust valve before BDC insures that the port is wide open at BDC for the beginning of the exhaust stroke. As the piston moves upward toward the top of the exhaust stroke, the intake valve opens so that it is fully open by the top of the exhaust stroke. The intake stroke begins and a new charge of air-fuel mixture flows into the combustion chamber pushing out the rest of the exhaust gases. The intake valve remains open as the piston passes BDC and the piston starts upward in the com-

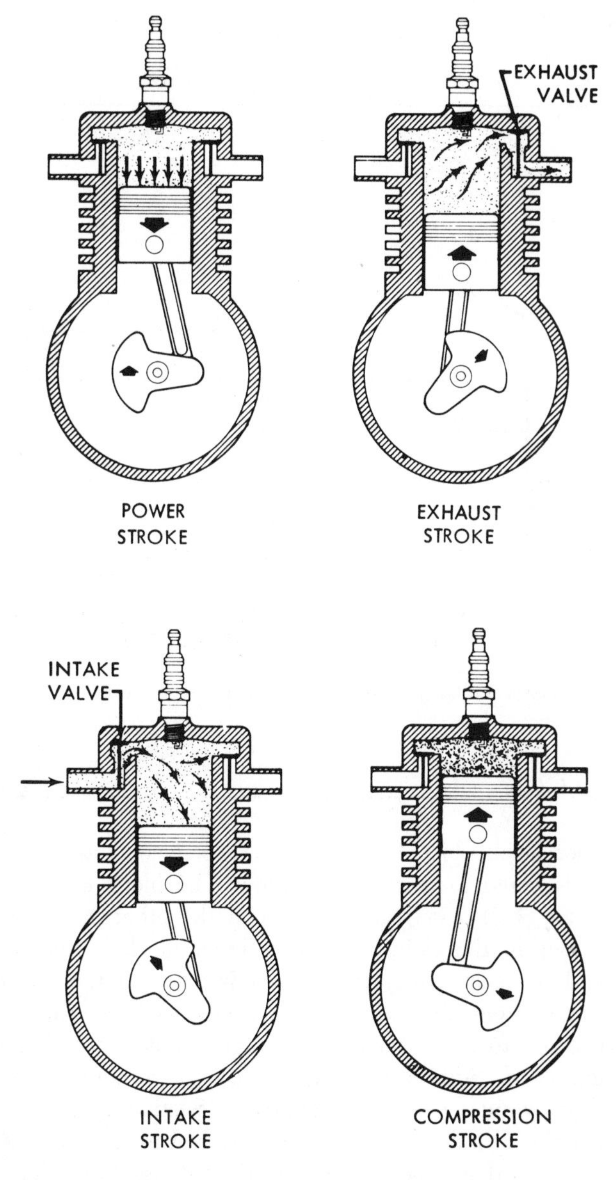

FIG. 2-4. The four-stroke cycle engine has four separate strokes: power, exhaust, intake, and compression.

pression stroke. This additional time beyond BDC allows for a full charge of new air-fuel mixture to enter the combustion chamber. The compression stroke then continues with both valves closed until TDC when a spark from the ignition system causes the air-fuel mixture to explode starting another power stroke.

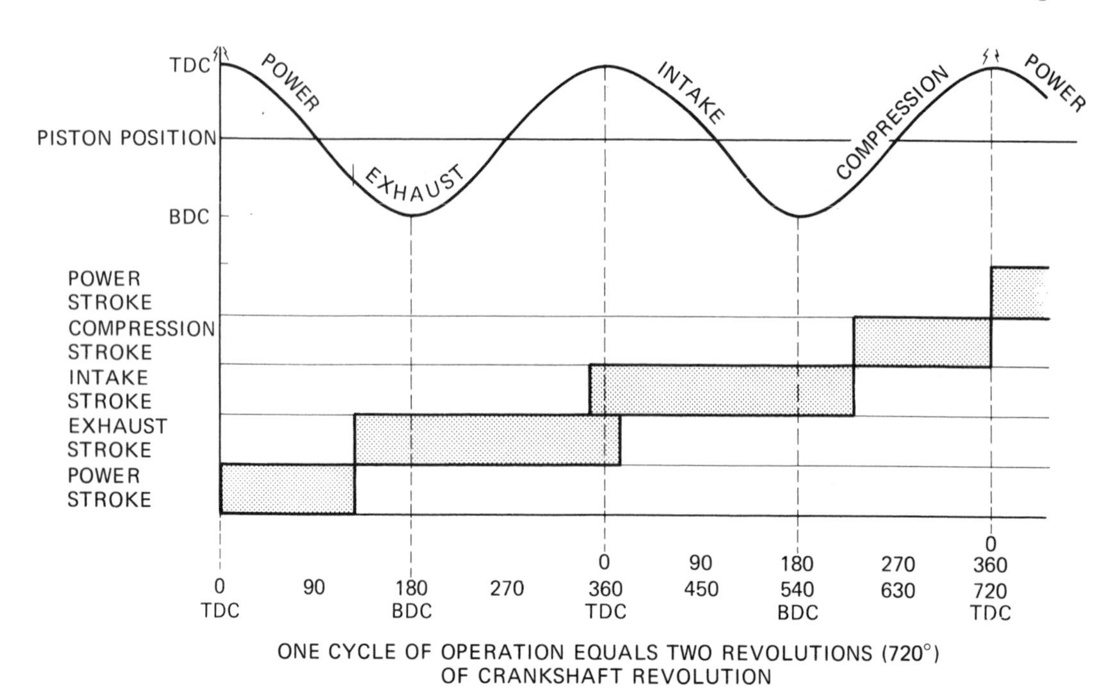

FIG. 2-5. Valve timing is shown graphically along with piston position.

2-3. CYLINDER BLOCK

The cylinder block (Fig. 2-6) is the main body of both the two- and four-cycle engines; it is bored to receive the piston. The block is closed at one end by a cylinder head (bolted on) and is open at the other end. The piston moves up and down within the cylinder. The cylinder block is finely machined, of high quality, and made of aluminum or cast iron (even though the engine may be made basically from aluminum, the *cylinder* itself is usually steel or cast iron and is pressed into place. Some aluminum cylinders are chrome-plated to reduce friction and wear). Aluminum is light and is used more frequently than cast iron or steel. The cylinder block and crankcase (Section 2-5) may be cast as an integral unit or the cylinder block may be attached to the crankcase with bolts. The crankcase is closed on the bottom by a crankcase base plate or oil pan. Some crankcases have both an end plate and a base plate. Gaskets of fiber or cork are placed between mating surfaces of crankcase parts. The cylinder *bore* is the internal *diameter* of an engine cylinder. The cylinder head (Section 2-4) refers to the cylinder cover of an engine.

Cylinders can be arranged either vertically or horizontally. Engines with two cylinders can be arranged such that the cylinders are horizontally opposed or are vertically arranged one behind the other. Multi-cylinder engines have the cylinders placed behind each other in a row or bank. Two banks of cylinders are arranged in a V configuration.

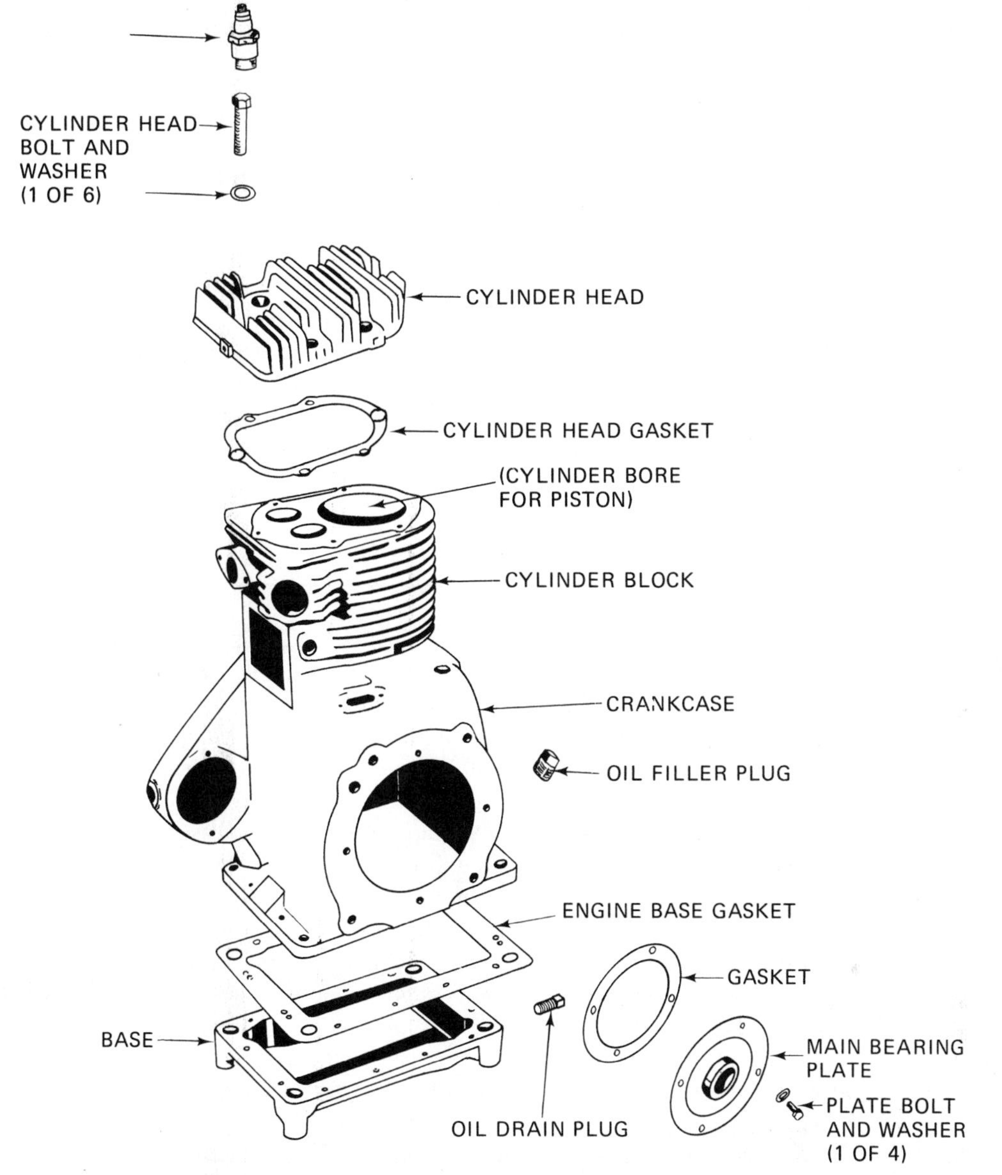

FIG. 2-6. The cylinder bore in the cylinder block is finely machined for the piston. The cylinder head and the cylinder area above the piston form the combustion chamber. The crankcase is below the cylinder block.

2-4. CYLINDER HEAD

The cylinder head (Fig. 2-6) forms the top of the combustion chamber of two- and four-cycle engines. The cylinder head is bolted to the cylinder and the cylinder head bolts are tightened in a specific sequence to a specific torqued value. A gasket of fiber or cork is placed between the cylinder head and the cylinder block so that an air tight seal is made. The spark plug used for igniting the air-fuel mixture in the combustion chamber is located in the cylinder head. The spark plug is also torqued to a specific value.

2-5. CRANKCASE

The crankcase (Fig. 2-6) is the housing for most of the moving parts of two- and four-cycle engines. It houses the crankshaft and the bearings on which the crankshaft turns and therefore maintains alignment of the engine. The internal parts in the crankcase include: a connecting rod, a crankshaft, the bottom of the cylinder, and the bottom of the piston. Four-cycle engine crankcases also include: a camshaft, cams, and cam gears; lubricating mechanism; the oil reservoir; and a crankcase breather.

2-6. CRANKCASE BREATHER

Four-cycle engines require a crankcase breather (Figs. 2-7 and 2-8) to rid the crankcase of pressures caused by blow-by gases escaping past worn piston rings and cylinder walls from the combustion chamber, by the downward force of the piston, and by expansion of the air in the crankcase from increasing heat. The build-up of pressure would eventually cause the oil seals to leak; to alleviate the pressure build-up, a breather is installed in the crankcase. The breather is a spring valve (reed valve) that lets pressure escape from the crankcase to the atmosphere (sometimes the gases are vented through the air cleaner). When the piston begins an upstroke, a slight vacuum is created in the crankcase. The atmospheric pressure pushes against the spring valve and closes it off. In addition to the valve, the crankcase breather includes a filter which prevents dust from coming into the crankcase as the piston goes up. A slight amount of air is drawn through the valve into the crankcase until the valve closes.

Blowby gases containing carbon from the combustion chamber travel past the piston and piston rings into the crankcase where they can cause corrosion of the engine parts, thereby shortening engine life. If raw gasoline passes by into the crankcase, the gasoline dilutes the lubricating oil. Therefore the breather is needed to let these gases out of the crankcase. If any oil is seen leaking through the crankcase breather or oil seals, it is an indication that the breather is clogged and should be cleaned. Excessive pressure in the crankcase that cannot be vented will cause oil to leak from the seals.

Other types of crankcase breathers include a ball check and a floating disk. They are opened by crankcase pressure and closed by atmospheric pressure. The filter elements used in crankcase breathers are metal mesh, fiber, or a polyurethane material. The breather is located so that splashing oil does not hit it; the breather is often incorporated into the valve access cover (Fig. 2-27).

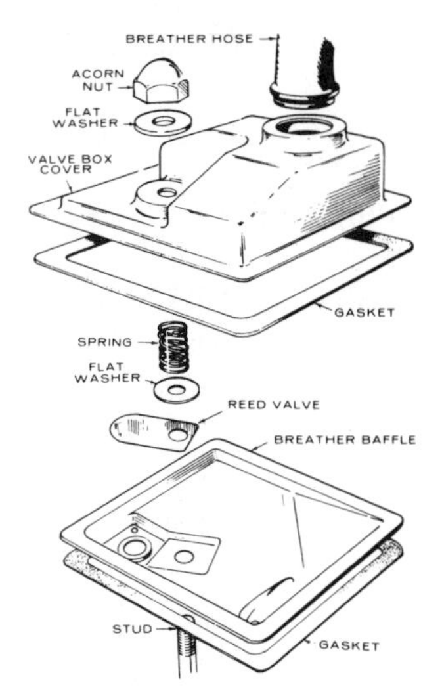

FIG. 2-7. Crankcase breathers rid the crankcase of pressures caused by blow-by gases.

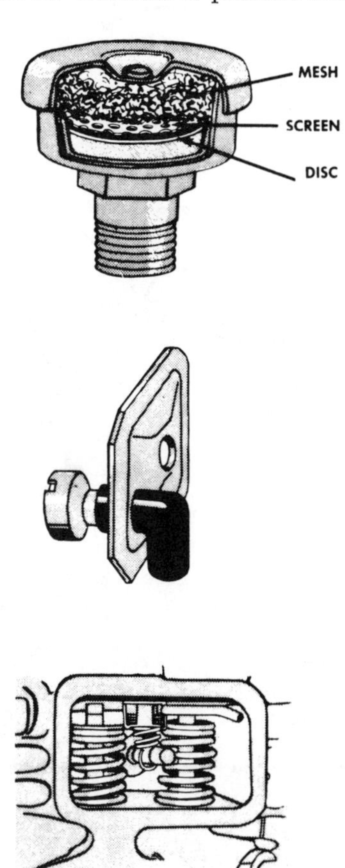

FIG. 2-8. Crankcase breathers take on several different designs. Some breathers are incorporated into the valve access covers.

2-7. REED, ROTARY, AND POPPET VALVES

Reed valves, which are also called leaf valves, are made of a thin flexible steel alloy. The reed valves are located between the carburetor and crankcase interface of two-cycle engines. There may be one or several reed valves in one assembly working together (Fig. 2-9). The reed valve opens to allow the air-fuel mixture from the carburetion system to enter the crankcase when the pressure in the crankcase is lower than atmospheric pressure. The reed valves close when the pressure in the crankcase, due to the downstroke of the piston, is greater than the atmospheric pressure. The closed valve prohibits the entry of the air-fuel mixture into the crankcase.

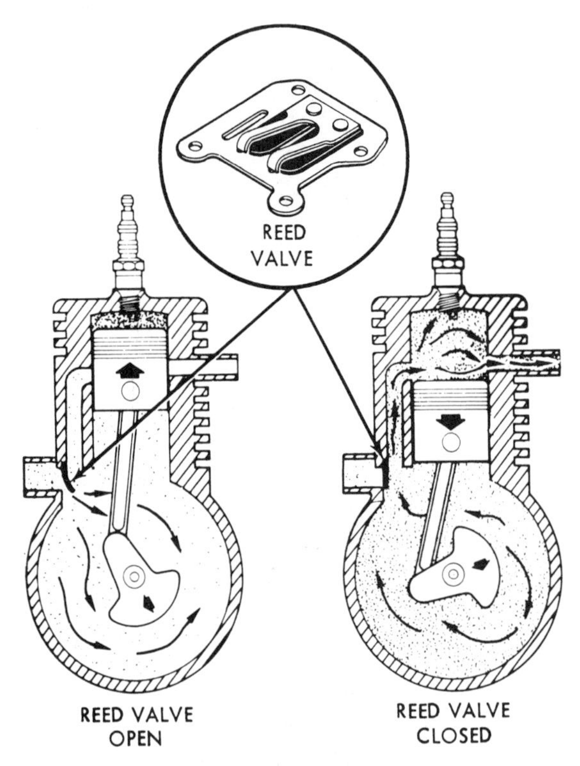

Fig. 2-9. The reed valve opens to allow the air-fuel mixture from the carburetor to enter the crankcase of a two-cycle engine.

Sometimes a rotary valve (Fig. 2-10) is used in place of the reed valve. The rotary valve is built into the crankshaft. Upon rotation of the crankshaft, the rotary valve port lines up with the crankcase port as the piston approaches TDC. This allows the air-fuel mixture from the carburetor to flow into the crankcase the same as previously described. Poppet valves are also sometimes used instead of reed valves. The slight pressure differential created between the outside atmospheric pressure and the slightly lower pressure in the crankcase on the piston upstroke causes the poppet valve to open and admit the flow of air-fuel mixture into the crankcase.

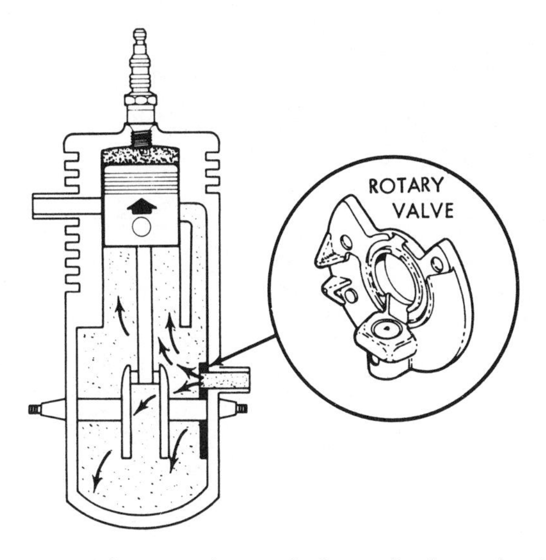

FIG. 2-10. Sometimes a rotary valve is used instead of a reed valve to let the air-fuel mixture into the crankcase.

2-8. PISTON

The piston is a plunger like device (Figs. 2-11 and 2-12) which moves with a reciprocating motion within the cylinder of two- and four-cycle engines. It is made of cast iron, steel, or aluminum but is usually of aluminum because aluminum is lightweight and results in less inertia in the reciprocating parts. The piston has several sections including the head or crown, grooves, lands, and a skirt or base. The head or crown may be flat, convex, concave, or another design shape. The differently shaped heads are used to direct the flow of new gases and exhaust gases in the combustion chamber. The grooves in the piston are for holding the piston rings (Section 2-9). The lands are the areas on the outside of the piston between the grooves; the skirt, or base, of the piston is the lower section of the piston to which the connecting rod is attached by means of the piston pin (Section 2-10).

The piston delivers the driving force through the connecting rod to the crankshaft. It also serves as a valve in the two-cycle engine to open and close the intake and exhaust ports (also the transfer port, if used). The piston must fit loosely within the cylinder so that it can move up and down and can expand from the

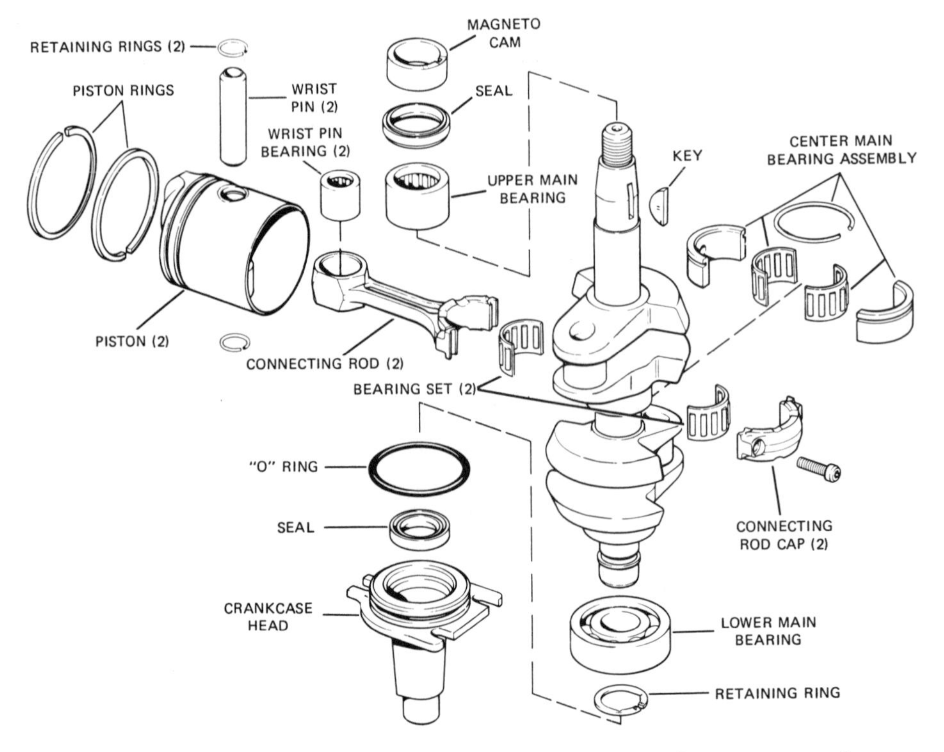

FIG. 2-11. The pistons, connecting rods, crankshaft, and associated components of a two-cycle, two cylinder engine are shown.

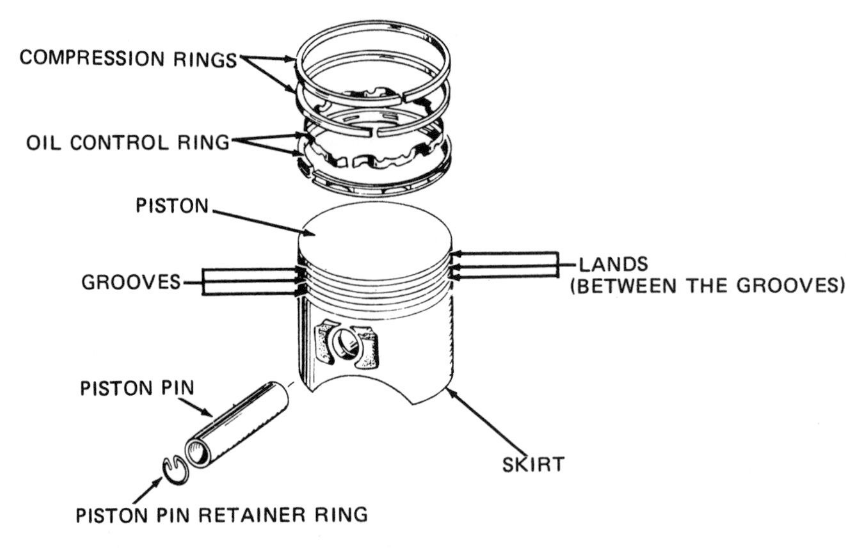

FIG. 2-12. A four-cycle piston has two compression rings and one or two oil control rings.

heat of combustion. The clearance must be approximately 0.003 to 0.004 inch smaller than the cylinder. However the piston cannot be too loose or the pressure from the gases of combustion would slip by the piston causing inefficient operation; there would be less power through the connecting rod to the crankshaft.

The piston head is subjected to extremely high temperatures and mechanical stresses. The piston expands slightly from this heat; the head expands more than the skirt because the heat is more intense on the head. Therefore, the head of the piston is sometimes tapered or the lands of the piston are of smaller diameter to permit expansion and still fit loosely within the cylinder.

A stroke of the piston is movement from the top dead center (TDC) to the bottom dead center (BDC) or from BDC to TDC. The length of the stroke is measured and is expressed in inches as one of the specifications of a gasoline engine. Figure 2-13 shows the stroke distance from TDC to BDC. The bore of the engine, also known as the cylinder bore, is the internal diameter of an engine cylinder. The bore is also measured in inches and is listed on an engine specification sheet.

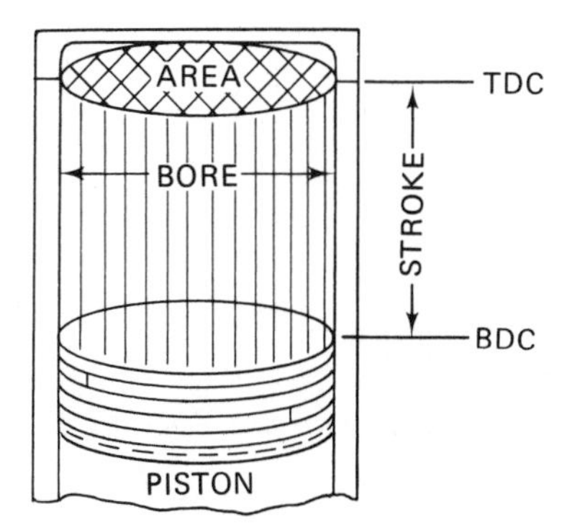

FIG. 2-13. The stroke of the piston is the distance that the piston moves between TDC and BDC. The bore is the diameter of the cylinder.

Piston displacement is another engine specification; it indicates the relative size of the engine. Horsepower is usually in direct proportion to the displacement size. The piston displacement is the volume displaced by the piston in its up and down movement. The larger the piston bore (cylinder inside diameter) and the longer the piston stroke, the greater is the piston displacement.

$$D = \frac{(B)^2}{4} \cdot C \cdot S \cdot \pi \qquad (2\text{-}1)$$

where: D = displacement in cubic inches
B = bore in inches
C = number of cylinders
S = stroke length in inches
$\pi = 3.14$

Example: What is the displacement of a one cylinder engine having a bore of 2 inches and a stroke of 2 inches?

$$D = \frac{(B)^2}{4} \cdot C \cdot S \cdot \pi$$

$$D = \frac{(2)^2}{4} \times 1 \times 2 \times 3.14$$

$$D = 6.28 \text{ cubic inches}$$

The compression ratio is the ratio of the space remaining in the cylinder (Fig. 2-14) at TDC of the piston stroke to the space in the cylinder when the piston is at BDC. For example, a seven to one compression ratio means that there is seven times as much volume in the cylinder at BDC as the area at TDC.

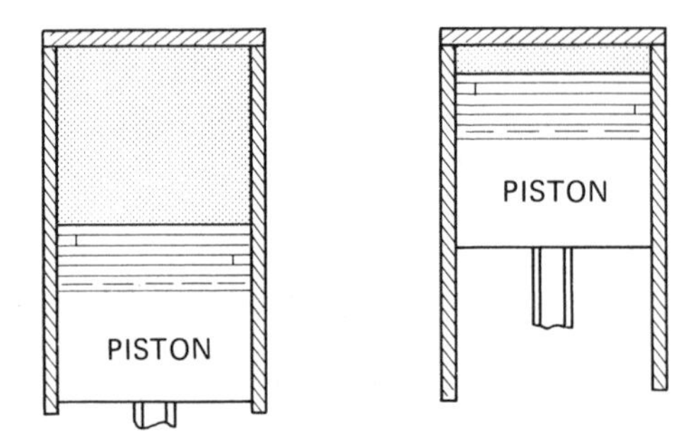

FIG. 2-14. The ratio of the volume above the piston at BDC to that above the piston at TDC is the compression ratio.

The compression ratio, which is another standard listed in engine specifications, does not tell the horsepower of an engine, but it does indicate engine efficiency. The greater the compression ratio, the greater the engine efficiency. However high compression ratios cause increased stresses and loads on the engine parts. Higher compression ratios require premium fuels (Section 7-16). Small gasoline engine compression ratios do not usually exceed ten to one (expressed 10:1).

Piston head designs, particularly of two-cycle engines, vary considerably to direct the flow of intake and exhaust gases to and from the combustion chamber. The intake and exhaust ports are holes that are drilled in the cylinder walls to allow a new charge of air-fuel mixture to enter and to allow exhaust gas to escape as the piston opens and closes the ports as it moves up and down within the cylinder. The intake of the gases and the exiting of the exhaust gases through these ports and across the top of the piston are referred to as the *scavenging* of the gases. Scavenging means removing, or cleaning, burned gases from the cylinder after a working stroke. In two-cycle engines, scavenging is usually done by either the cross flow method, or the loop scavenging method. The piston design determines which of the scavenging methods are used within the engine

(Figs. 2-15, 2-16, and 2-17). The loop scavenging method is identified by the large bores in the piston skirt. The cross flow scavenging piston does not have any bores in the skirt, except for the piston pin.

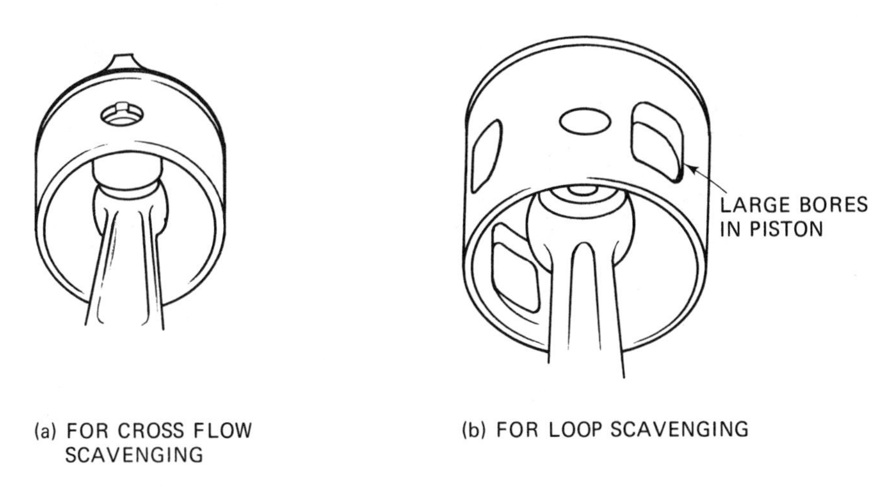

FIG. 2-15. Piston design determines the type of scavenging.

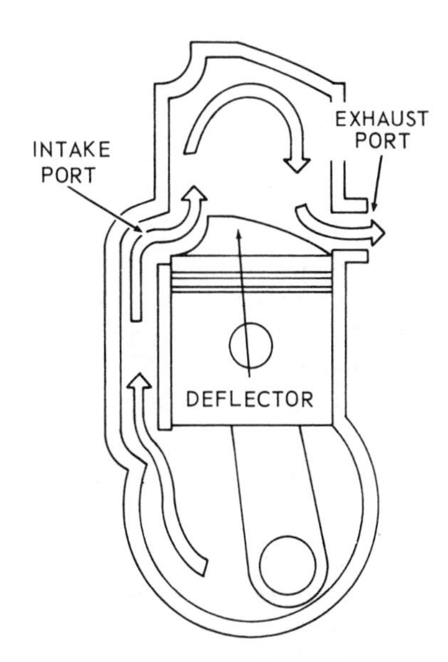

FIG. 2-16. Cross flow scavenging pushes the burned gases out of the exhaust port ahead of the new air-fuel charge.

For cross flow scavenging, the head of the piston is dome-shaped to direct the new charge of air-fuel mixture, as shown in Figure 2-16, to help push the burned gases out of the exhaust port ahead of the new air-fuel charge. The dome shape

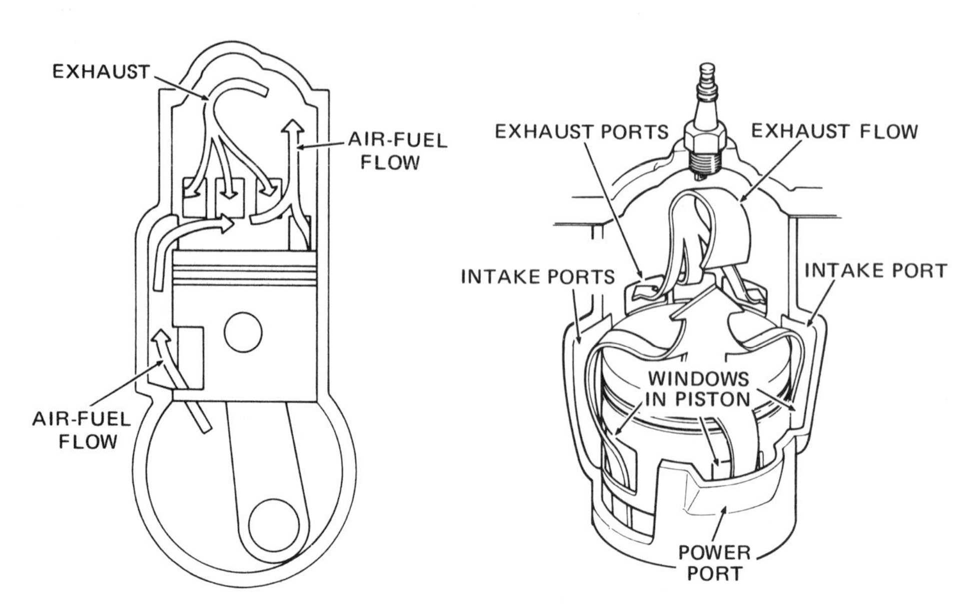

FIG. 2-17. The loop scavenging method through three ports forces the incoming new charges to impinge on one another and flow upward toward the top of the cylinder and then back down meeting and driving the exhaust gases out.

also prevents the new air-fuel mixture from exiting from the exhaust port. In loop scavenging (Fig. 2-17), the piston dome has only a slightly curved contour. There are three intake ports or bores through the cylinder skirt that are slanted upward. This intake method through three ports forces the incoming new charge of air-fuel mixture to impinge on one another and flow upward toward the top of the cylinder and then back down meeting and driving the exhaust gases out. This results in a clean and powerful air-fuel mixture that is ready to be compressed and ignited as the piston reaches BDC. Loop scavenging results in a more complete removal of exhaust gases, and therefore produces somewhat more horsepower. Scavenging is only necessary on two-cycle engines, as the four-cycle engine has separate piston strokes for both intake and exhaust.

2-9. PISTON RINGS

Piston rings (Figs 2-11 and 2-12) are used to give a good seal, or fit, between the piston and the cylinder wall of all small gasoline engines. Piston rings are used to retain maximum power within the cylinder *above* the piston head. To retain this power, the cylinder must be perfectly round and the piston rings must be seated correctly in their grooves. When the rings are installed, they fit tightly against the cylinder wall and against the machined grooves in the piston. The rings are made of cast iron, steel, or other metal; they are frequently chrome-plated. There are many designs of piston rings and some designs, especially the oil rings (oil rings are used only on four-cycle engines), may consist of more than one piece. The rings are split so that they can be installed over the piston lands and into the grooves of the piston and so that the rings can expand and contract with heat and wear.

There are two types of piston rings: compression rings that are used to prevent compression leakage and oil rings to control the amount of oil that passes up the cylinder wall and into the combustion chamber. Generally the grooves for the oil rings have holes in them that permit oil from behind the rings to escape back to the crankcase. The pressure in the combustion chamber forces the rings against the cylinder wall and against the piston groove, thus making a seal. Badly worn rings, or rings damaged by overheating will cause loss of compression and the rings must be replaced.

In two-cycle engines, there are two rings, both used to seal pressure. The pressure that the first ring misses is blocked by the second ring. Both rings are known as compression rings and hold the pressure in the combustion chamber during compression and combustion. In four-cycle engines, the two rings toward the head (top) of the piston are compression rings and are used as described, but as a secondary function they also provide some oil control. A third ring, used in the four-cycle engine, is called an oil control ring. This oil control ring scrapes oil from the cylinder wall to prevent the oil from entering into the combustion chamber where it would burn and cause trouble in the form of carbon deposits. The oil ring is necessary in the four-cycle engine because of the method of lubrication of the four-cycle engine (Chapter 8). Sometimes there is even a fourth ring necessary for additional oil control.

2-10. PISTON PIN (WRIST PIN)

The piston pin, also known as the wrist pin, connects the piston to the connecting rod. The pin (Figs. 2-11 and 2-12) is either solid or hollow and is made of high tensile strength steel. The piston pin may be bolted or clamped to the connecting rod, bolted or pinned to the piston, or held in the piston by retaining rings. The piston pin moves in bronze bushings or needle bearings that are installed in the piston or the connecting rod or both. Where aluminum connecting rods and pistons are used, no bearing or bushing is required. The aluminum serves as the bearing material.

2-11. VALVES

A valve (Figs. 2-18 and 2-19) is a device for regulating the flow of intake and exhaust gases to and from the combustion chamber of a four-cycle engine cylinder. A valve is made of special steel that can withstand the corrosive action of very high temperature exhaust gases. Some valves are of special alloy and are hollow and filled with metallic sodium that melts and helps to conduct heat away from the valve stem into the cylinder block. The valves have to seal well enough and have to be able to withstand pressures of 500 pounds per square inch, and under full load, temperatures between 1200° and 4500°F.

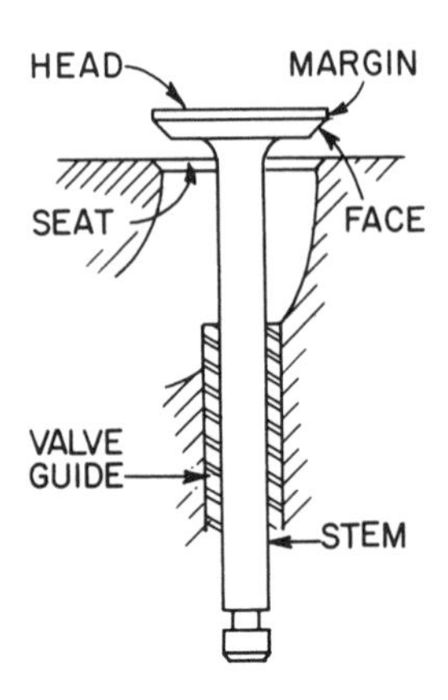

FIG. 2-18. A valve regulates the flow of intake and exhaust gases to and from the combustion chamber of a four-cycle engine.

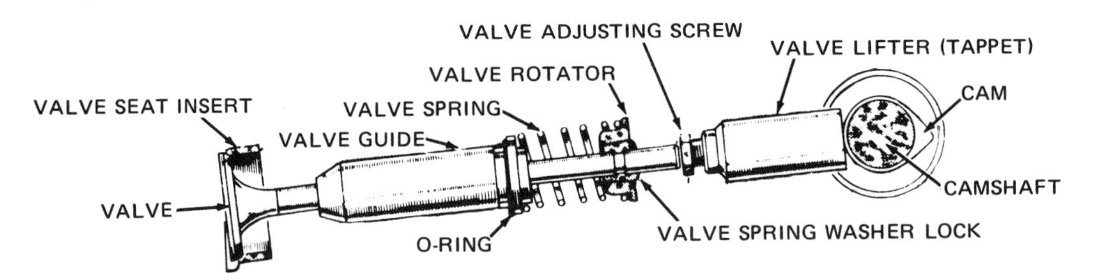

FIG. 2-19. The valve is opened and closed by means of a cam that rotates on the camshaft.

The intake valve is cooled somewhat by the incoming air-fuel mixture, but because of its location, the exhaust valve operates at red heat and is difficult to cool. The cylinder head, the cylinder, and the top of the piston are exposed to the same heat as the exhaust valve, but these parts are cooled by air from the flywheel, fan, and oil from the crankcase.

If the valves do not operate at the correct time, maximum power cannot be developed. If the valves do not seal completely, there is loss of compression resulting in a decrease of engine efficiency.

The intake and the exhaust valve operate separately during one portion of four actions of a complete cycle of operation. The valve is opened and closed by means of a cam that rotates on a camshaft driven by the crankshaft. When the cam lobe (the high point on the cam) passes against the valve lifter (Fig. 2-19), the lifter presses against the valve stem, overcoming the opposing force of the valve spring, lifting the valve face off of the valve seat. This allows the gas to pass into or from the combustion chamber depending upon whether the valve is the intake or the exhaust valve. The valve stem is held in place within the valve guide. The valve spring puts tension on the valve to keep it closed, or seated, at all times. A spring retainer and lock hold the valve and valve spring in place. The valve lifter rides against the cam and rises or falls as the cam lobe rotates against the valve lifter. The valve lifter is also known as the tappet valve.

Each valve (intake and exhaust) operates once per cycle, but the crankshaft rotates two times per cycle. Therefore, a gear arrangement of a two-to-one reduction is required for proper timing. The camshaft, cams, and gears are discussed in Section 2-15.

The valve is held in place in the valve guide by a C-shaped washer, called a keeper. Other types of keepers are a pin that is inserted through a hole in the valve stem or a two-piece collar that fits a groove.

Valves for the four-cycle engine are physically located in four different types of configurations, known as the L-, I-, T-, or F-head. In the L-head configuration, the cylinder in the combustion chamber is in the shape of an inverted L. The valves in the cylinder block are on one side of the cylinder. This L-shaped arrangement is the most common in small gasoline engines. Almost all automotive engines and some small four-cycle engines have valve configurations in the I-head configuration. This is also known as the overhead valve configuration because the valves are located overhead in the cylinder head. This I-head configuration requires *push rods* and *rocker arms* to raise and lower the valves. Both valves (intake and exhaust) open in a downward direction over the cylinder.

Two camshafts are required for the T-head valve configuration; one camshaft is located on each side of the engine. The intake valve is on one side and the exhaust valve is on the other. Both valves open in an upward direction. In the F-head configuration, both valves are located on the same side of the cylinder; one valve opens up and one down.

2-12. CONNECTING ROD

The connecting rod is connected between the piston and the crank on all small gasoline engines. It is used in conjunction with the crank to change the reciprocating motion of the piston to a rotary motion of the crankshaft. The force of combustion causes the piston to push the connecting rod down causing the crank on the crankshaft to rotate. The *small end* (Fig. 2-20) on the connecting rod moves up and down with the piston; the *big end* swings in a circle along with the crank.

The connecting rod is made of forged steel or aluminum, usually aluminum for the smaller engines and steel for the larger engines. The connecting rod is shaped like an I-beam. The lower end, called the big end, is fit with a rod bearing

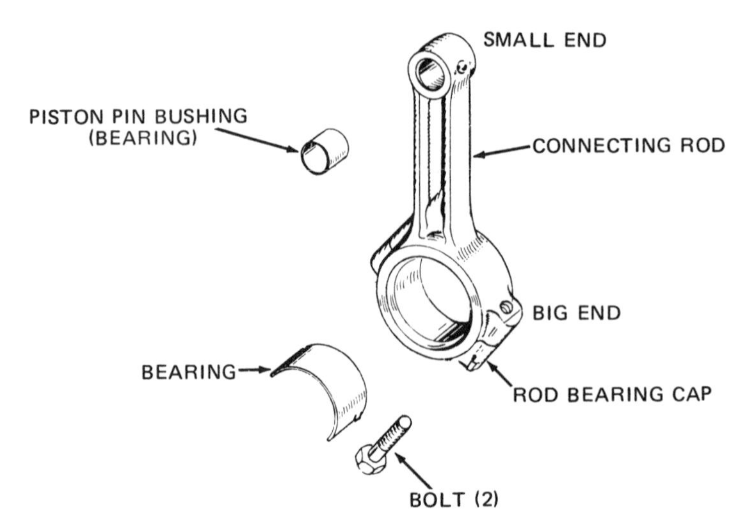

FIG. 2-20. The connecting rod connects between the piston and the crank. It changes the reciprocating motion of the piston to a rotating motion of the crankshaft.

cap that is bolted to the connecting rod. Many connecting rods have replaceable bearings in the small and big ends. The small end of the connecting rod is attached to the piston with the piston pin (Section 2-10). The large end is attached to the crankpin of the crankshaft. The rod bearing cap holds the connecting rod to the crank.

2-13. CRANK AND CRANKSHAFT

The crank is used in conjunction with the connecting rod to change the reciprocating motion of the piston to a rotating motion of the crankshaft. The crankshaft (Fig. 2-21) turns a pulley, blade, etc. to aid man in performing work more easily, and with less fatigue. The crank is an offset section on the crankshaft; when connected to the connecting rod, it cranks the crankshaft around and around. The crankshaft sits in bearings (Section 2-16) so that it rotates with little friction. Counterweights on the crankshaft balance the weight of the crankpin and connecting rod reducing engine vibration; they also reduce the tendency of the crankshaft to go out of round when it is rotating.

The crankshaft is forged and machined steel. One end of the crankshaft is machined to a taper for the flywheel; the other end has the power takeoff. The counterweights may be cast into the crankshaft or bolted onto it. Gears to drive the camshaft may be machined as part of the crankshaft or may be locked on by a nut. The main bearing journals are supported in the crankshaft main bearings that may be attached to the crankshaft or to the crankcase.

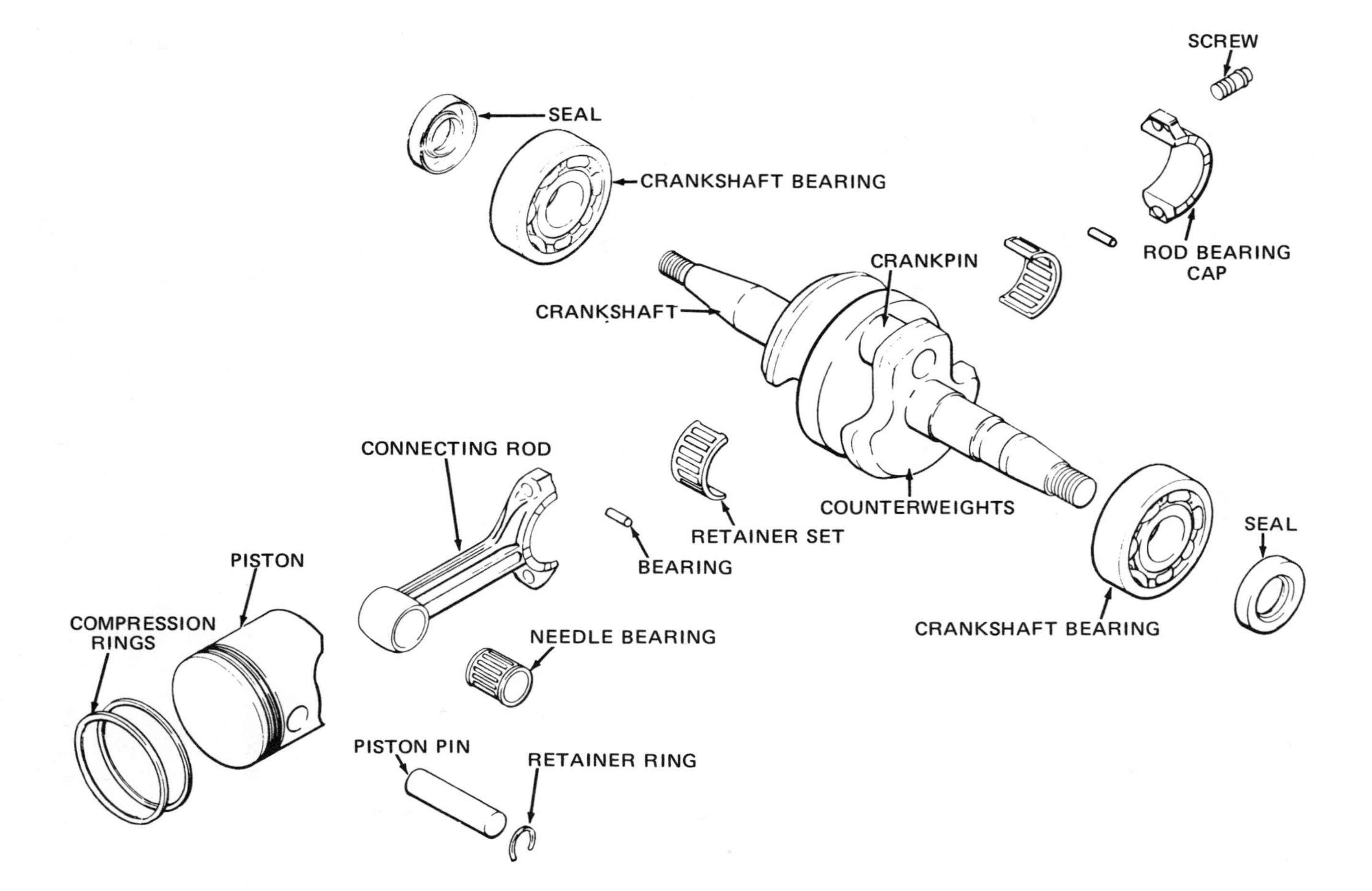

Fig. 2-21. The crankshaft sits in crankshaft bearings. This is a two-cycle single cylinder engine. Four-cycle engine crankshafts have a gear on the crankshaft to drive a camshaft.

2-14. FLYWHEEL

The power stroke occurs one out of two strokes in the operation of the two-stroke cycle engine and one out of four strokes in the operation of the four-stroke cycle engine. Therefore there is a need for a method of continuing uniform crankshaft rotation during all engine actions. The flywheel (Fig. 2-22) continues the crankshaft rotation; once the flywheel is set in motion by the power stroke, its mass causes the rotation of the crankshaft to continue. The momentum of the weight of the flywheel causes the crankshaft, piston, and the other moving parts to continue through the engine cycle to the next power stroke. The engine also has a tendency to speed up during the power stroke portion of the cycle, but the power has to overcome the weight of the flywheel and, hence, the speed of the engine does not appreciably increase. The above are examples of the property of inertia—the property that once something is set in motion, it tends to remain in motion.

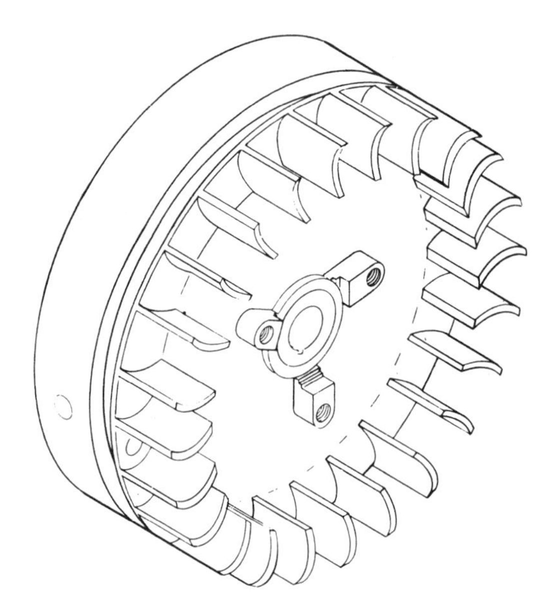

FIG. 2-22. The flywheel is heavy; once set into motion by the power stroke, it causes the crankshaft to continue to rotate until the next power stroke.

The flywheel is mounted on the tapered end of the crankshaft and rotates with it. A key keeps the flywheel and crankshaft aligned during rotation and a nut holds the flywheel on. The flywheel may carry magnets in it for use in a magneto ignition system as described in Chapter 5. The flywheel may also be part of an air cooling system (Section 2-18). Air vanes, or blades, are cast into the flywheel to scoop air as the flywheel rotates. The air is directed to carry away engine heat.

The flywheel may be covered by a shroud that directs the flow of air from the flywheel to the cylinder block and cylinder head.

2-15. CAMSHAFT

The camshaft (Fig. 2-23) with its associated parts controls the operation of the intake and exhaust valves of a four-cycle engine. The valve lifters (tappets) ride on the teardrop shaped cams. The lifters push against the valve stems and operate against the compression of the valve springs to unseat or open the valves. The camshaft can also have an eccentric on it to operate an oil pump. The camshaft is driven by the crankshaft, but must operate at one-half the speed of the crankshaft. As you will recall, the crankshaft of the four-cycle engine rotates two times for a complete cycle of operation, but each valve opens only once during each complete cycle. Therefore a two-to-one gearing arrangement is required so that the camshaft operates at exactly one-half the speed of the crankshaft. The synchronized time is critical and there are timing marks located on the gears that are set during assembly.

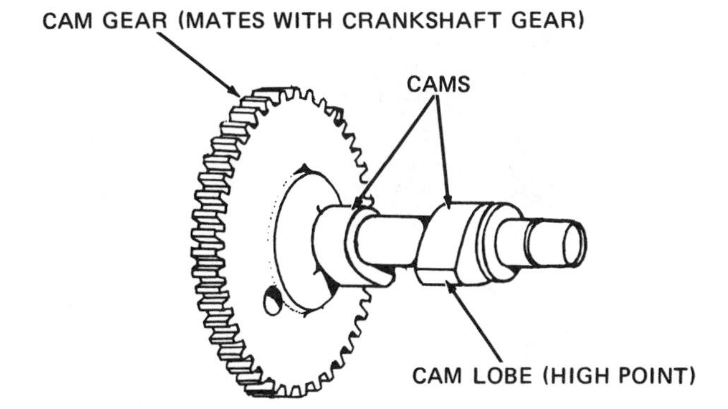

FIG. 2-23. The camshaft with its associated cams operates the intake and exhaust valves of a four-cycle engine. The camshaft is driven by a gear on the crankshaft.

Cranking of an engine is not always easy because of the compression within the engine. At least one manufacturer has devised a method of attaching flyweights onto the camshaft drive gear to try to overcome this compression that causes difficulty in starting the engine. When the engine is not running, the flyweights are close to the driving mechanism, and through other parts prevent the exhaust valve from closing completely. When the engine has started, centrifugal force causes the flyweights to be thrown outward, resulting in complete closure of the exhaust valve. The slight opening of the exhaust valve when the engine is not running reduces the effort required to start the engine because the compression is decreased somewhat.

2-16. BEARINGS, BUSHINGS AND JOURNALS

A *journal* is a rotating machine part (as part of the crankshaft) supported in a bearing. Journals are cylinders that are accurately machined to resist deformation and to prevent excessive pressure in the bearings; the journal is the part of a rotating shaft, axle, roller, or spindle that turns in a bearing. A *bearing* is a support for a revolving shaft or axle. It must be rigid and self-aligning and be designed to take up most of the wear and be replaceable. A *bushing* is a special type of bearing called a *sleeve bearing*. Bearings, bushings, and journals are used to reduce friction between moving parts. A lubricating oil is also used to reduce friction (Chapter 8).

There are three types of bearings: *sleeve bearing, ball bearing,* and *roller bearing*. Sleeve bearings are also referred to as *sliding bearings*. Ball and roller bearings are referred to as *rolling bearings*. Sleeve bearings are less expensive than rolling bearings and are satisfactory for small gasoline engines; however, rolling bearings are sometimes used. Several types of bearings are shown in Figure 2-24.

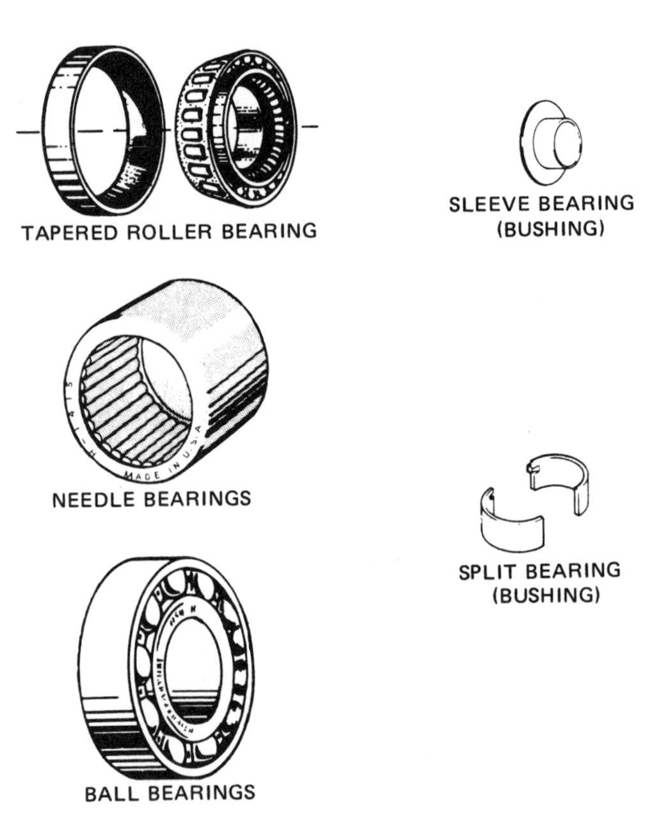

FIG. 2-24. Bearings and bushings are used in machines to reduce friction between moving parts.

Sleeve bearings, also called bushings, are solid or split. A sleeve bearing fits like a sleeve around a rotating journal or shaft. Split bearings are used where the bearing has to be split to enable assembly. An example of this is on the big rod end of the connecting rod where it attaches to the crank. A sleeve bearing is layered like plywood. The outside layer is steel or bronze. Then there is a center layer of bearing alloy followed by a thin layer of a soft bearing alloy such as copper, lead, or tin, on the inside. The thin layer is designed to conform to slight irregularities of the rotating shaft. If wear occurs, it is the bearing that wears out, and the bearing is inexpensive to replace as compared to replacing a shaft or other part.

In a ball bearing, round balls provide a *spot contact* with moving surfaces. Therefore there is less friction than in sleeve bearings. The balls are placed in a grooved spacer between what are known as the *inner* and *outer races*. Some ball bearings are permanently sealed with lubricant and hence require no lubrication. Roller bearings provide *line contact* rather than spot contact as provided by ball bearings. The roller bearings are also placed between inner and outer races.

Bearings are slightly larger than the mating journals to allow for heat expansion of the journal and to provide space between the moving surfaces for lubricating oil. Several points in the small gasoline engine have been mentioned where bearings, bushings, and journals are used: at the connection of the connecting rod and piston, at the connecting rod and the crank, and to support the crankshaft in the crankcase. (Most small engines just use the aluminum of the connecting rod and the steel of the crankshaft as the bearing surfaces.)

The simplest bearing is a machined opening in an aluminum crankcase or end plate. The crankshaft journal rotates in it. In cast iron crankcases, the main bearings are bushings that are pressed into the crankcase openings. An oil seal is installed outside of each main bearing to prevent loss of lubricating oil from the crankcase. The seals also make the crankcase air tight.

2-17. MUFFLER

A muffler (Fig. 2-25) is a mechanical device consisting of a hollow cylinder attached to the exhaust port of two- and four-cycle engines. The noise of the exhaust gases passing through the muffler is partially deadened (muffled). Mufflers come in a variety of sizes and shapes. The muffler is placed against a gasket and is held in place on the engine by cap screws, a threaded fitting, or similar manner. The muffler is located toward the center of the cylinder on two-cycle engines near the cylinder exhaust port; on four-cycle engines, the muffler is located toward the top of the cylinder head at the exhaust valve.

The muffler on the exhaust system of a two-cycle engine does more than muffle the sound and direct exhaust gases away from the engine. The muffler helps scavenge or clean all exhaust gases from the combustion chamber and it helps to get a new air-fuel charge into the combustion chamber more rapidly for cleaner and more complete combustion. Mufflers that are designed for two-cycle

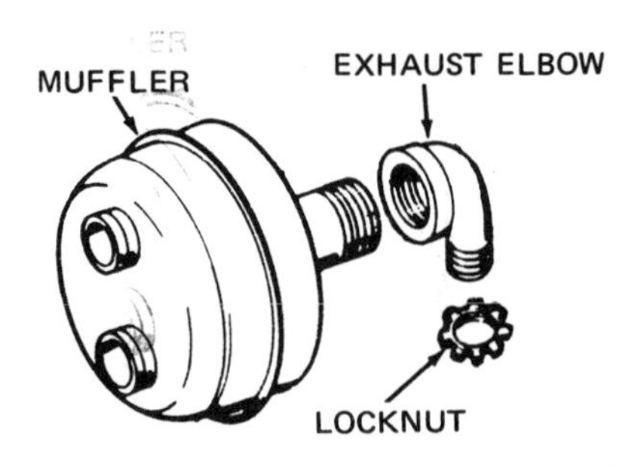

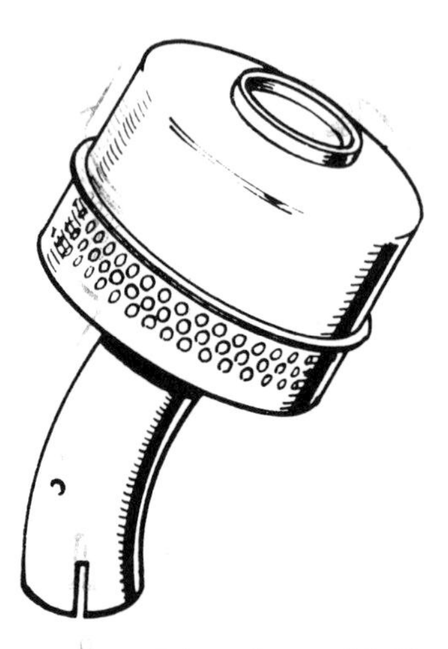

FIG. 2-25. The muffler deadens the noise of the exhaust. Mufflers come in a wide variety of designs, sizes, and shapes.

engines are made with a megaphone-like device. The muffler first amplifies sound to speed up the scavenging of gases and then silences the sound by means of baffles.

A dirty muffler can cause loss of power because it slows the discharge of exhaust gases from the combustion chamber. When the gases are forced to remain in the chamber, they occupy room which should be taken up by a fresh charge. Without a full fresh charge, full development of power is impossible.

2-18. COOLING SYSTEMS

As you can imagine, combustion temperatures are extremely high—to 6,000° Fahrenheit. This heat and the heat caused by the friction of moving parts must be dissipated. If the heat is not dissipated, the oil film will burn and fail resulting in no lubrication and eventual engine failure. The piston, piston rings, and the cylinder wall wear very rapidly without lubrication. Friction caused by dry metals rubbing over each other increases heat and causes rapid wear. Finally *preignition,* that is, ignition before the correct firing time would occur because of the high heat. Preignition can cause holes to be knocked in the piston, and other damage can also result.

There are two methods of cooling the small engine—air and water. There are several advantages to an air-cooled engine; there is no need for a complicated cooling system, the engine is lighter, the engine occupies less space, and the engine that is air-cooled is easy to repair. The air-cooled engine temperature varies greatly with changes in air temperature, load, and the speed of operation. A water-cooled engine is quite constant in maintaining operating temperature.

In the air-cooling system, part of the heat is exited by the exhaust. The cylinder head and block have air-cooling fins machined into them to dissipate the heat as the air circulates past them. The air-cooling fins increase the area of metal thereby permitting rapid dissipation of heat by radiation. Many air-cooled engines also have shrouds and an air fan which forces air over and around the fins. A shroud is a shaped metal device used to direct air flow to the fins. The air fan consists of the air vanes (blades) of the flywheel (Section 2-14).

In a water-cooled engine, water is circulated in jackets, or pockets, that surround the cylinder walls and cylinder head. The water carries the heat away from the engine to a radiator. The radiator dissipates the heat by the air that circulates past the radiator. A water pump, driven by a belt, keeps the water in circulation. The water pump consists of an impeller on a shaft; the impeller throws the water by centrifugal force. In addition to the water pump and the radiator, two other parts in a water-cooling system are a thermostat and a radiator-pressure cap.

A radiator consists of a top and a bottom tank with a series of tubes running between the tanks. Fins surround the tube to aid in dissipating the heat. Air flows around the fins and cools the water. In operation, water is drawn from the bottom tank of the radiator through the water pump, through the engine, and to the top tank of the radiator. The water then flows through the tubes to the bottom tank and is air cooled along the way. In outboard motors, no radiator is required. Water is drawn in through an inlet in the lower portion of the motor casing. It is then passed through a water pump, through the cylinder block and cylinder head jackets and is then exited from the engine. A thermostat is used to control flow and hence operating temperature.

It is initially desirable for an engine to warm up to operating temperature as rapidly as possible; thus, for a time, water circulation is prevented by use of a thermostat. The thermostat is installed between the engine and the top radiator tank. When the engine is cold, the thermostat is closed and prevents the flow of water. When the engine and water temperature rise to the operating temperature (as determined by the thermostat), the thermostat opens and water flows through the system to cool the engine. The thermostat regulates the amount of flow by change in water temperature and, hence, tends to keep the engine at the proper operating temperature.

A radiator pressure cap is used to raise the pressure in a water-cooling system so that the boiling point of the liquid is raised. With pressure, the temperature can go to 250°F without boiling. The radiator cap contains two spring loaded valves. The first is a blowoff valve that opens when the pressure gets too high and allows excessive pressure to escape. The second spring loaded valve is the vacuum valve that operates when the engine cools off. This valve allows air into the system to rid the system of a partial vacuum formed when the steam condenses as the engine cools.

Sometimes a temperature indicator is placed on the engine to indicate that an engine has reached an abnormally high temperature and something is wrong and must be corrected. The temperature indicator is connected by wires to a transducer that operates on heat and sends an electrical signal to a lamp. Damage can occur to the engine if the engine overheats.

Outboard engines are water-cooled (Fig. 2-26). A thermostat opens and closes to allow fresh water to flow throughout the cooling system to maintain the engine temperature within a range of temperatures. By maintaining the engine within the range, the motor life and efficiency are increased.

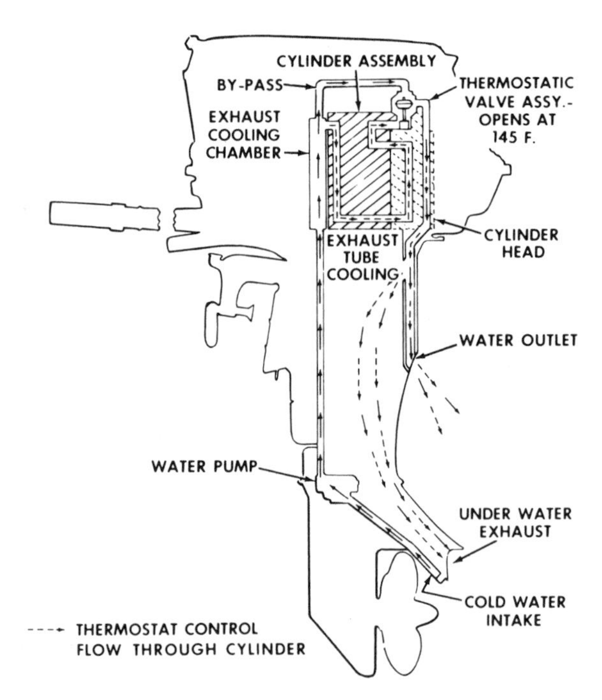

Fig. 2-26. A thermostat controls the flow of fresh water through a water-cooled outboard engine. This keeps the engine's operating temperature within a specified range.

The outboard engine thermostat housing is a part of the cylinder head. When a cold engine is started, the thermostat is closed and prevents the water pump from circulating water in the cooling system. Limited circulation is permitted by a bleed hole in the thermostat valve which also permits discharge of air from the cooling system.

When the power head and the cooling system temperature reach a certain temperature (approximately 145°F), the thermostat valve opens allowing heated water to pass through the water discharge and fresh water to be drawn through the intake. The thermostat opens and closes permitting fresh water to circulate as necessary to keep the engine temperature within the desired range. (Refer to Fig. H-6 in Appendix H).

CYLINDER BLOCK GROUP

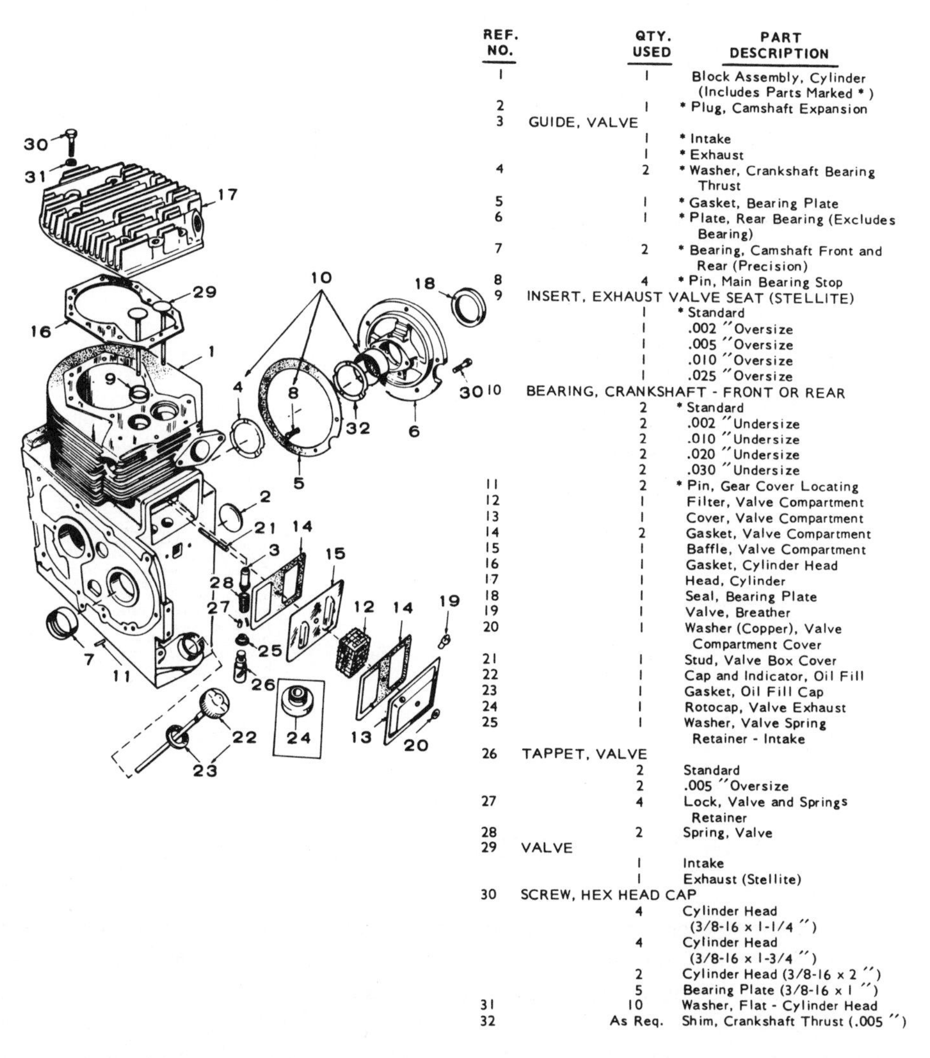

REF. NO.		QTY. USED	PART DESCRIPTION
1		1	Block Assembly, Cylinder (Includes Parts Marked *)
2		1	* Plug, Camshaft Expansion
3	GUIDE, VALVE		
		1	* Intake
		1	* Exhaust
4		2	* Washer, Crankshaft Bearing Thrust
5		1	* Gasket, Bearing Plate
6		1	* Plate, Rear Bearing (Excludes Bearing)
7		2	* Bearing, Camshaft Front and Rear (Precision)
8		4	* Pin, Main Bearing Stop
9	INSERT, EXHAUST VALVE SEAT (STELLITE)		
		1	* Standard
		1	.002 ″Oversize
		1	.005 ″Oversize
		1	.010 ″Oversize
		1	.025 ″Oversize
10	BEARING, CRANKSHAFT - FRONT OR REAR		
		2	* Standard
		2	.002 ″Undersize
		2	.010 ″Undersize
		2	.020 ″Undersize
		2	.030 ″Undersize
11		2	* Pin, Gear Cover Locating
12		1	Filter, Valve Compartment
13		1	Cover, Valve Compartment
14		2	Gasket, Valve Compartment
15		1	Baffle, Valve Compartment
16		1	Gasket, Cylinder Head
17		1	Head, Cylinder
18		1	Seal, Bearing Plate
19		1	Valve, Breather
20		1	Washer (Copper), Valve Compartment Cover
21		1	Stud, Valve Box Cover
22		1	Cap and Indicator, Oil Fill
23		1	Gasket, Oil Fill Cap
24		1	Rotocap, Valve Exhaust
25		1	Washer, Valve Spring Retainer - Intake
26	TAPPET, VALVE		
		2	Standard
		2	.005 ″Oversize
27		4	Lock, Valve and Springs Retainer
28		2	Spring, Valve
29	VALVE		
		1	Intake
		1	Exhaust (Stellite)
30	SCREW, HEX HEAD CAP		
		4	Cylinder Head (3/8-16 x 1-1/4 ″)
		4	Cylinder Head (3/8-16 x 1-3/4 ″)
		2	Cylinder Head (3/8-16 x 2 ″)
		5	Bearing Plate (3/8-16 x 1 ″)
31		10	Washer, Flat - Cylinder Head
32		As Req.	Shim, Crankshaft Thrust (.005 ″)

FIG. 2-27. How many parts of the cylinder, crankshaft, flywheel, camshaft, and piston groups can you identify without looking at the key?

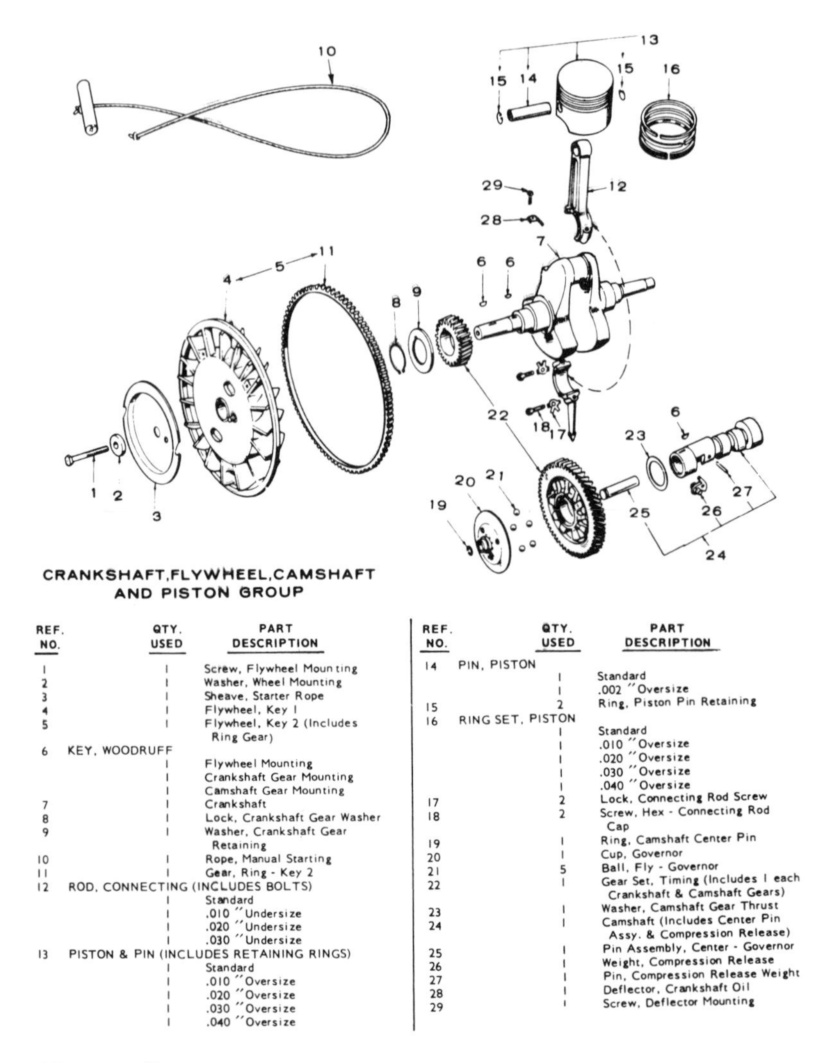

CRANKSHAFT, FLYWHEEL, CAMSHAFT AND PISTON GROUP

REF. NO.		QTY. USED	PART DESCRIPTION
1		1	Screw, Flywheel Mounting
2		1	Washer, Wheel Mounting
3		1	Sheave, Starter Rope
4		1	Flywheel, Key 1
5		1	Flywheel, Key 2 (Includes Ring Gear)
6	KEY, WOODRUFF		
		1	Flywheel Mounting
		1	Crankshaft Gear Mounting
		1	Camshaft Gear Mounting
7		1	Crankshaft
8		1	Lock, Crankshaft Gear Washer
9		1	Washer, Crankshaft Gear Retaining
10		1	Rope, Manual Starting
11		1	Gear, Ring - Key 2
12	ROD, CONNECTING (INCLUDES BOLTS)		
		1	Standard
		1	.010 " Undersize
		1	.020 " Undersize
		1	.030 " Undersize
13	PISTON & PIN (INCLUDES RETAINING RINGS)		
		1	Standard
		1	.010 " Oversize
		1	.020 " Oversize
		1	.030 " Oversize
		1	.040 " Oversize
14	PIN, PISTON		
		1	Standard
		1	.002 " Oversize
15		2	Ring, Piston Pin Retaining
16	RING SET, PISTON		
		1	Standard
		1	.010 " Oversize
		1	.020 " Oversize
		1	.030 " Oversize
		1	.040 " Oversize
17		2	Lock, Connecting Rod Screw
18		2	Screw, Hex - Connecting Rod Cap
19		1	Ring, Camshaft Center Pin
20		1	Cup, Governor
21		5	Ball, Fly - Governor
22		1	Gear Set, Timing (Includes 1 each Crankshaft & Camshaft Gears)
23		1	Washer, Camshaft Gear Thrust
24		1	Camshaft (Includes Center Pin Assy. & Compression Release)
25		1	Pin Assembly, Center - Governor
26		1	Weight, Compression Release
27		1	Pin, Compression Release Weight
28		1	Deflector, Crankshaft Oil
29		1	Screw, Deflector Mounting

FIG. 2-27. (Continued).

2-19. SUMMARY OF ENGINE OPERATION

Two-stroke cycle engine and four-stroke cycle engine operations have been described. All of the major parts that make up two- and four-stroke engines have also been described. If you have any difficulty in recalling the complete operation of the two-stroke cycle engine or the four-stroke cycle engine, you should reread and study sections 2-1 and 2-2 in detail. Figure 2-27 illustrates the cylinder, crankshaft, flywheel, camshaft, and piston groups of a single cylinder, four-cycle engine of 30 cubic inch displacement and 12 horsepower. How many parts can you visually identify? What are their functions?

2-20. ENGINE MAINTENANCE

When the technician's troubleshooting analysis has indicated a problem within the engine proper, it becomes necessary to disassemble the engine. As each part is removed, it should be cleaned in a solvent and then inspected for wear, pitting, cracks, scoring, and wear. Some parts are measured with a micrometer and compared to the manufacturer's specifications to determine worn and/or out of round conditions.

Inspection and measurement can indicate the replacement of several major parts of the engine including the piston, piston rings, connecting rod, bearings, valves, camshaft, and the need to rehone the cylinder walls. Rehoning and replacement of these parts may cost more than the price of a new engine—and your customer must be so advised. Some manufacturers provide a solution by selling a *short block engine*. This engine contains the parts mentioned and is less expensive than a complete overhaul or a new engine. The short block engine does not include the *add on* parts such as a magneto, carburetor, or ignition system.

When disassembling an engine, remove all gaskets and discard them. Remove traces of sealer and gasket material with lacquer thinner—always replace the gaskets with new ones. Reassemble the engine according to the manufacturer's directions and torque all fasteners as specified.

After an engine is reassembled, fill the fuel tank with clean fuel and the crankcase of four-cycle engines with new oil. Start the engine and run it at a moderate speed and without load for an engine break-in period of one hour. Check for oil leaks and strange noises; if noticed, stop the engine and correct the situation immediately. Do not apply a *full* load to the engine until after 5 hours of operating time. (Refer to Chapter 3 for starting, operating, shutoff, fuel mixture, and four-cycle engine lubrication.)

Repair and replacement of one engine part often necessitates disassembly of several engine (and sometimes removal of nonengine parts) parts. For example, repair or replacement of the piston, piston rings, piston (wrist) pin, or connecting rod requires the removal of all of these for repair; the cylinder head and crankcase base or cover must also be removed. Be sure to drain the oil from four-cycle engines before beginning disassembly. As each part is removed, clean it in a solvent and then inspect it for cracks, roughness, and breaks.

The following sections describe the procedures of removal, inspection, measurement for in or out of tolerance and out of round conditions, and reassembly. The sections are sequenced in approximate disassembly order. All of the sections are as detailed as possible considering that they must apply to many models and types of engines. Always use the manufacturer's procedures as the specific guide.

2-21. CYLINDER HEAD MAINTENANCE

The cylinder head bolts to the top of the cylinder and forms part of the combustion chamber. The outside is cleaned during periodic maintenance cleaning procedures (Section 3-8); the inside is cleaned of carbon deposits whenever the head is removed. The cylinder head is removed whenever there is need to inspect or repair the piston, piston rings, valves, piston pin, or connecting rod.

Proceed as follows to remove, clean, inspect, and reinstall the cylinder head:

1. Use a wrench to remove the cylinder head bolts. Some bolts may be longer than others; tag them so that the bolts are returned to the proper locations.
2. Try to remove the head by hand. If you cannot, pry with a screwdriver moving slowly with a prying action around and around the head.
3. Peel off the gasket material from the head and cylinder block. Remove traces of gasket and sealer with lacquer thinner. Do not scratch the metal surfaces with a tool.
4. Remove the carbon off the head using a wire brush in a portable drill or a drill press (Fig. 2-28). Clean the head in a solvent and inspect the head for cracks.
5. When ready to reinstall the cylinder head, insert a new gasket between the head and the cylinder block; do not use shellac or another sealer.
6. Install the cylinder head bolts in the correct holes and tighten them hand tight. Apply graphite grease to the threads of bolts inserted into aluminum cylinders. Then tighten each bolt with a torque wrench in

Fig. 2-28. A small wire brush is used in a drill press or a portable drill to remove carbon from the cylinder head. Do not scratch the surface where the head meets the gasket and cylinder block. Traces of the old gasket are removed with lacquer thinner and a clean cloth.

the sequence recommended by the manufacturer (Fig. 2-29 illustrates the sequence recommended by one manufacturer). Tighten each bolt a little; then tighten in sequence to the torque value recommended by the manufacturer.

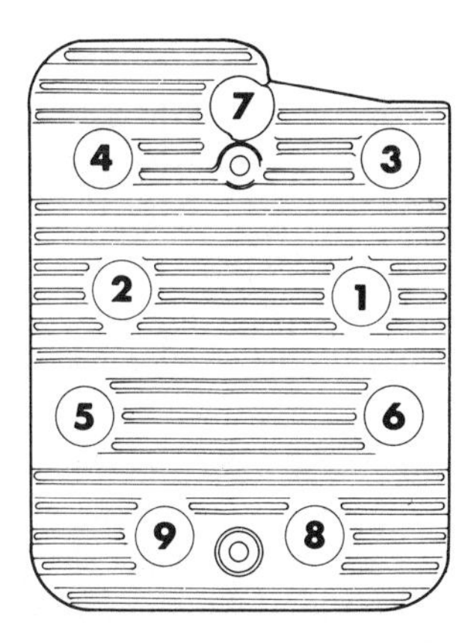

FIG. 2-29. The cylinder head bolts are torqued in a specific sequence to a specific torque value to prevent distorting the head. Refer to the manufacturer's instructions.

QUICK REFERENCE CHART 2-1

CYLINDER HEAD MAINTENANCE

1. Remove cylinder head bolts noting their length and location.
2. Remove head.
3. Remove gasket.
4. Remove carbon from head and traces of gasket material from head and cylinder block. Clean in solvent.
5. Insert new gasket.
6. Reinstall head and bolts. Torque cylinder head bolts in correct sequence to value specified.

2-22. CONNECTING ROD MAINTENANCE

The connecting rod can wear at the bearing where the little end attaches to the piston pin or where the big end attaches to the crankpin of the crankshaft. Some engine connecting rods have bearings at both ends, at one end only, or no bearings at all—the aluminum of the rod is used as a bushing. Some bearings are replaceable; on others, the complete rod has to be replaced. If other parts seize in the engine, the connecting rod may be bent or broken. Inspect, remove, check tolerances, and reinstall the connecting rod as follows:

1. Remove the cylinder head (Section 2-21) and the crankcase base plate and end plate.
2. Inspect the connecting rod for breaks, bends, and cracks. If any of these conditions exist, the connecting rod must be replaced.
3. Remove the connecting rod from the crankpin by removing the bolts and the rod bearing cap. Note the position of the cap; it must be reinstalled in the same configuration.
4. Inspect the rod big end and discard it if it is scored or discolored. If the crankpin is scored, out of tolerance, or out of round, it must be replaced (Section 2-27).
5. Check that the rod to crankpin distance is within the recommended tolerance. This is done by one of three methods: use of shim stock, use of Plastigage, or use of an inside hole gauge and micrometer. In using shims, insert a 0.002 inch shim between the bearing and the crankpin. Install the rod bearing cap and bolts and torque to the value specified by the manufacturer. If the rod does not tighten against the crankpin, then the clearance is greater than 0.002 inch; add another shim and continue until the tolerance is established. If Plastigage is used instead of shims, insert it in the same manner as the first 0.002 inch shim. Torque the rod bearing cap bolts to the required torque; then remove the bolts and cap and measure the thickness of the Plastigage to determine the tolerance. When using the hole gauge and micrometer, remove the rod from the crankpin. Reinstall the cap by aligning it properly and torquing it to the specified requirement. Take measurements from both sides of the rod bearing and 90 degrees to these points to determine out of roundness or a taper. If any of the three methods described for measuring the tolerance between the connecting rod bearing and the crankpin indicate an out of tolerance or out of round condition, replace the part(s).
6. In similar manner to Step 5, check the tolerance between the connecting rod bearing and the piston pin (Section 2-23 describes removal of the piston and piston pin). Oversize piston pins are sometimes available.
7. If roller bearings are replaced, make sure that all are out of the crankcase. Pry out old bearing inserts. Place in new bearing inserts, making certain that the locking tabs are positioned properly. Seat the inserts.
8. To install needle bearings, lay the strip of bearings along your forefinger. Remove the backing from the bearings. Lay the bearings around the crankpin—the grease will hold the bearings in place.
9. Oil the bearing surfaces.
10. Mate the connecting rod with the crankpin. Replace the rod bearing cap aligning the locating marks between the rod and the cap. Position the oil slinger or dipper.

11. Install and torque the bolts to the specified value. A slight amount of drag should be felt as the crankshaft is turned. There should also be a slight amount of end play.

QUICK REFERENCE CHART 2-2

CONNECTING ROD MAINTENANCE

1. Remove cylinder head and crankcase base plate and end plate.
2. Inspect connecting rod.
3. Remove connecting rod from crankpin.
4. Inspect the rod big end, bearing, and crankpin. Check tolerance.
5. Remove rod little end from piston pin.
6. Inspect the little end, bearing, and piston pin. Check tolerance.
7. Replace bearings and other parts, as required.
8. Oil the bearing surfaces and reassemble.
9. Torque the rod bearing cap bolts to the specified torque.

2-23. PISTON, PISTON RING, AND PISTON PIN MAINTENANCE

The piston is inspected for cracks, holes, scoring, excessive wear, damaged ring grooves and lands, and out of tolerance conditions; if any of these conditions are found, the piston is replaced. If the cylinder (Section 2-24) is to be refinished, the piston is replaced with an oversize piston. Proceed as follows to perform piston inspection, cleaning, and tolerance checks:

1. Remove the cylinder head (Section 2-21), crankcase base and/or end plate, and the connecting rod big end from the crankpin. Push the piston through the cylinder.
2. Using a piston ring expander, remove the piston rings (Fig. 2-30); note the top and bottom of each ring and which ring came from each groove.
3. The next step is to remove the piston pin from the piston. First remove the retainer ring(s) by prying it out with an old screwdriver (Fig. 2-31) or by pulling it out with needlenose pliers. The piston pin is then driven out being careful not to distort the pin or the piston (Fig. 2-32). Some manufacturers recommend heating both sides of the piston with a propane torch to expand the piston metal. Then push the piston out.

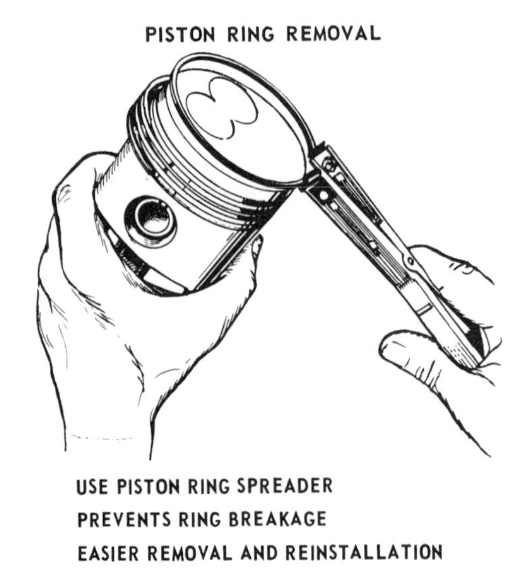

Fig. 2-30. A piston pin expander removes piston rings without breakage of the rings. Inexpensive expanders are available from a number of manufacturers.

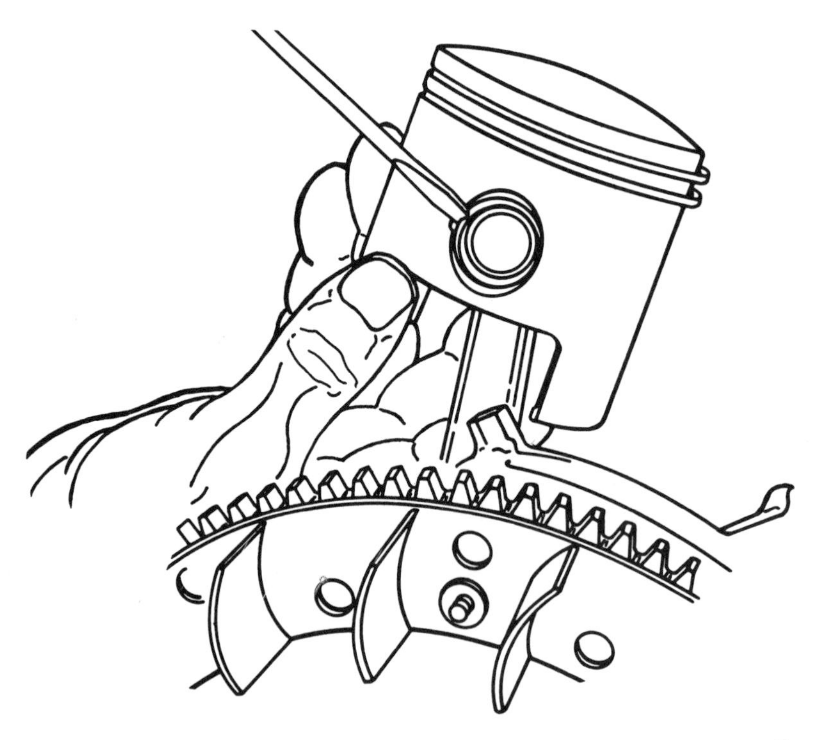

Fig. 2-31. The piston pin retainer rings are either pried out or are pulled out with needlenose pliers.

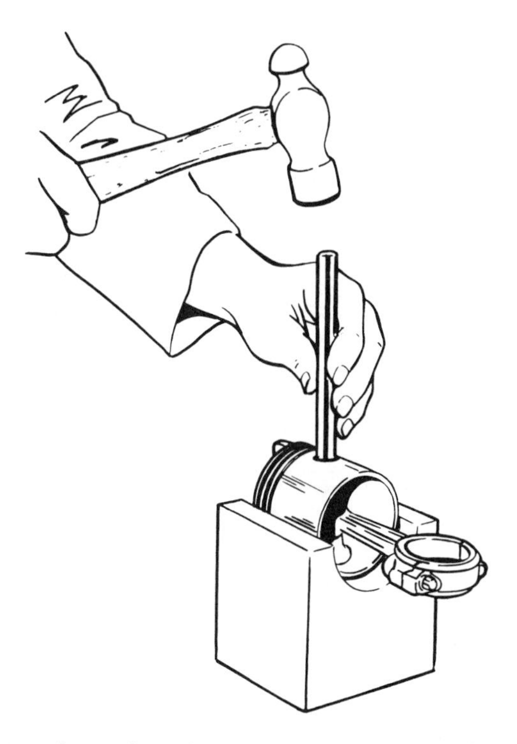

FIG. 2-32. The piston pin is driven from the piston. Use care not to damage the piston or pin.

4. Clean the carbon from the piston ring grooves. A special tool can be used or a broken ring can be filed flat (Fig. 2-33) and used to clean the grooves. Clean the piston in solvent.

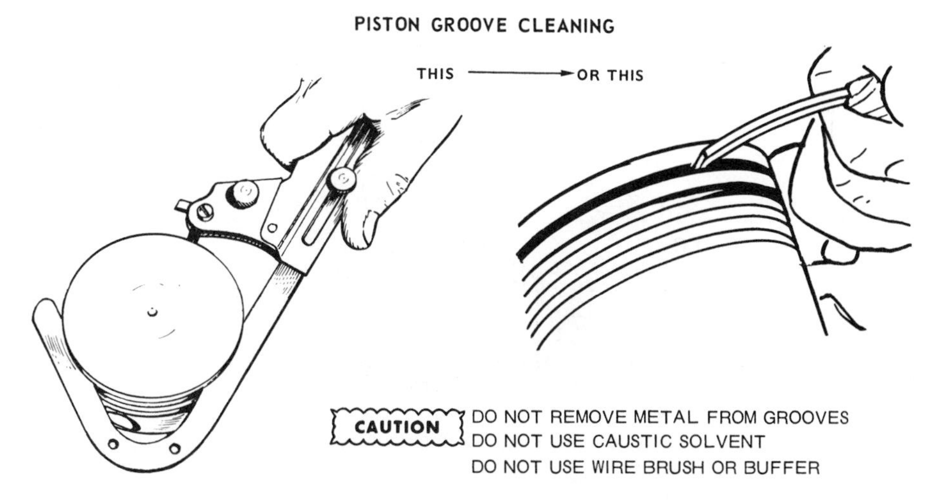

FIG. 2-33. Piston ring grooves can be cleaned with a special tool or a broken ring with an end filed flat.

5. Using a micrometer, measure the skirt diameter at a point 90 degrees from the piston pin bore (Fig. 2-34) at the lands, and down on the skirt. Compare the measurements with the manufacturers' specifications.

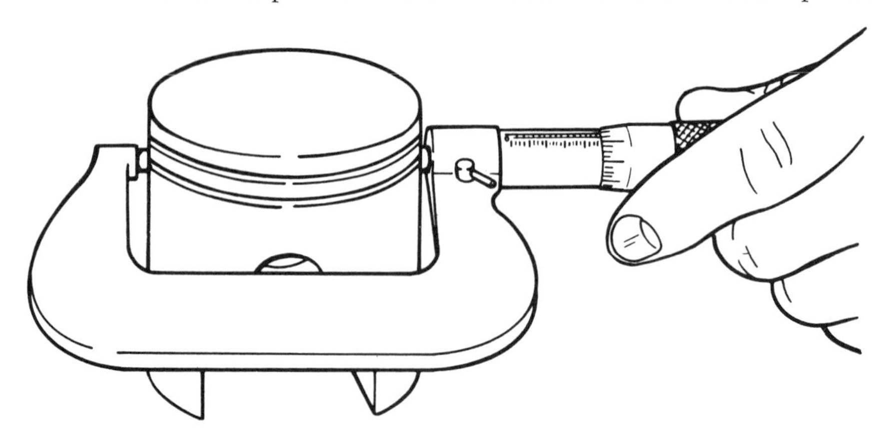

Fig. 2-34. Measure the piston at the locations recommended by the manufacturer.

6. Clean the rings in solvent, inspect for cracks, scoring, and roughness, and then check the fit of the rings in the cylinder as follows. Check the gap between the ends first; the gap is for expansion of the ring because of the heat. If the gap is too little, scoring occurs; if the gap is too large, there is leakage with loss of compression and oil consumption. To check the gap, place the piston ring midway down into the cylinder; use the piston to push the ring down. Check the end gap with a feeler gauge (Fig. 2-35) for proper gap as recommended by the manufacturer. If the

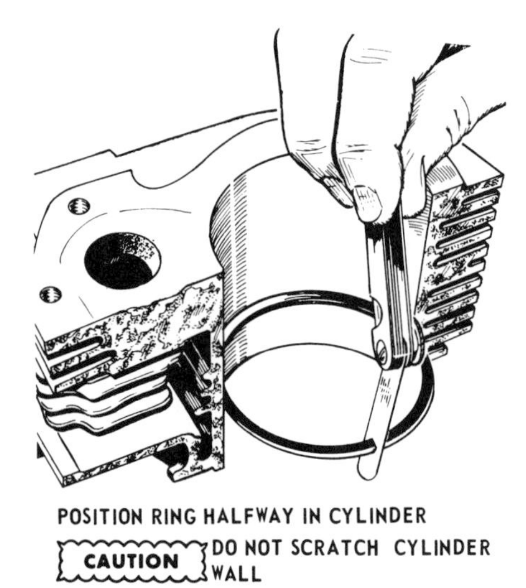

Fig. 2-35. This cutaway drawing shows the proper method of checking the end gap of a piston ring. The piston ring is halfway down the cylinder.

gap is too wide, replace the ring; if the gap is too narrow, the gap can be filed slightly and then rechecked.

7. Inspect the cylinder for scoring, pitting, and cracks. Repair or replace as required. Refer to Section 2-24.
8. After the gap is checked, measure the piston ring side clearance—the space between the ring and the groove. Hold the ring as shown in Figure 2-36 and measure the clearance. If the space is wider than allowed by the manufacturer's specification, replace the ring.

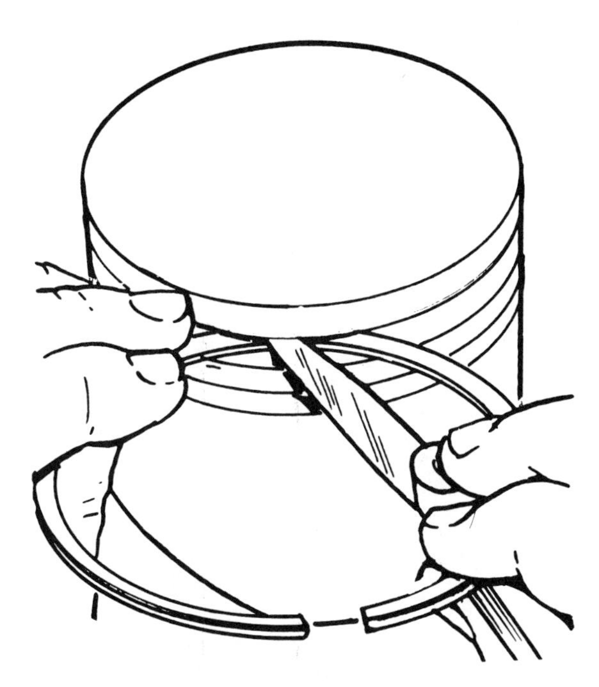

FIG. 2-36. Check the space between the ring and the groove with a feeler gauge.

9. Measure the diameter of the piston pin and the inside diameter of the piston pin bushing with a micrometer. If within the specification limit, the pin may be reused; if it is not, replace the pin with an oversize pin. Replacement with an oversize pin will also necessitate reboring the connecting rod bushing. New bushings may also be purchased.
10. Replace the piston pin, connecting rod, bearing, and retainer ring(s), as applicable, in the piston. Drive the pin in with the special tool or heat the piston with a propane torch and then insert the pin. Wear heavy fire resistant gloves.
11. Using a piston ring expander, reinstall the rings onto the piston grooves. Be sure to install the correct ring in each groove and be sure that the top of the ring is toward the top of the piston. Handle the rings carefully—they break easily. Stagger the ring gaps around the piston to

minimize compression leakage. If there are two rings, place the gaps 180 degrees apart; three rings, 120 degrees apart.

12. Oil the piston rings, piston, crankpin, and cylinder wall.
13. Carefully place a ring compressor (Fig. 2-37) over the piston and piston rings. Tighten the compressor and then insert the piston into the cylinder. Lower the piston skirt until the piston ring compressor rests on the cylinder. Use a hammer handle or other wood rod to drive the piston from the ring compressor into the cylinder.

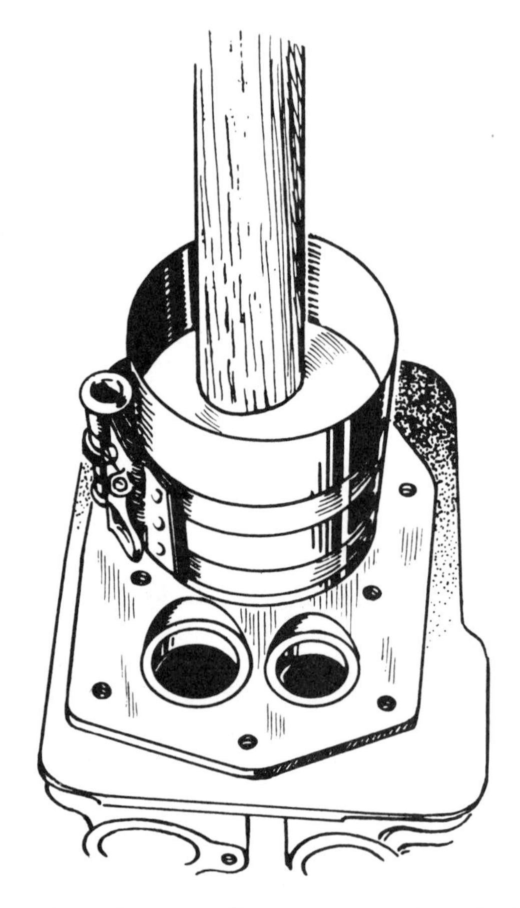

FIG. 2-37. A ring compressor is used to reinstall the piston into the cylinder.

14. Mate the connecting rod (Section 2-22) to the crankpin. Replace the rod bearing cap onto the connecting rod; match the assembly marks. Replace the oil dipper or slinger and bolts. Torque the bolts to the value specified by the manufacturer.

QUICK REFERENCE CHART 2-3

PISTON, PISTON RING, AND PISTON PIN MAINTENANCE

1. Remove cylinder head, crankcase base and/or end plate, and the connecting rod from the crankpin.
2. Remove piston rings.
3. Remove piston pin.
4. Clean carbon from piston ring grooves.
5. Measure piston diameter.
6. Measure ring end gap.
7. Inspect cylinder (Section 2-24).
8. Measure piston ring side clearance.
9. Measure piston pin (wrist pin) diameter and inside diameter of the piston pin bushing.
10. Replace piston pin, connecting rod, bearings, and retainer ring(s).
11. Reinstall piston rings into piston grooves.
12. Oil piston rings, piston, crankpin, and cylinder wall.
13. Using ring compressor, install piston into cylinder.
14. Attach connecting rod to crankpin.

2-24. CYLINDER MAINTENANCE

After the piston is removed, the cylinder is inspected for cracks, scoring, stripped bolt hole threads, and broken cooling fins. If sufficient cooling fins are broken off to prevent adequate air cooling, or if the cylinder is cracked, it must be replaced. Scored walls can be rehoned and stripped bolt hole threads can be corrected. The cylinder bore is also checked for proper dimension and out of round. Proceed as follows:

1. Remove the cylinder head base plate, end plate, piston, and connecting rod (Sections 2-21, 2-22, 2-23). Inspect the cylinder for cracks, scoring, stripped threads, and broken cooling fins. Replace the cylinder, if required.
2. Measure the cylinder bore with an inside micrometer (Fig. 2-38) or with a telescopic gauge and outside micrometer (Fig. 2-40). Measure the bore at two points 90 degrees apart; check the measurements against the manufacturer's specifications. If the cylinder is oversized, scored, or out of round, rebore by honing. A rehoned bore requires an oversized piston.

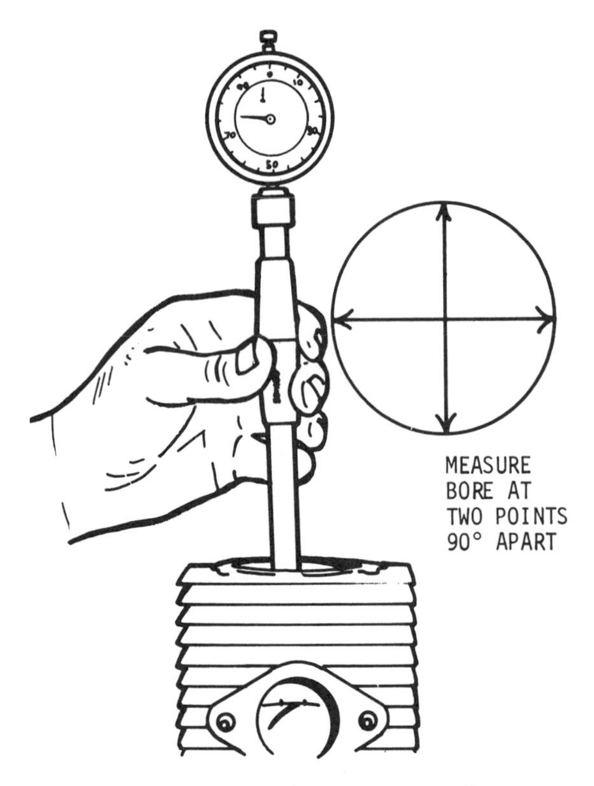

FIG. 2-38. An inside micrometer with a dial readout gauge is used to measure the cylinder bore.

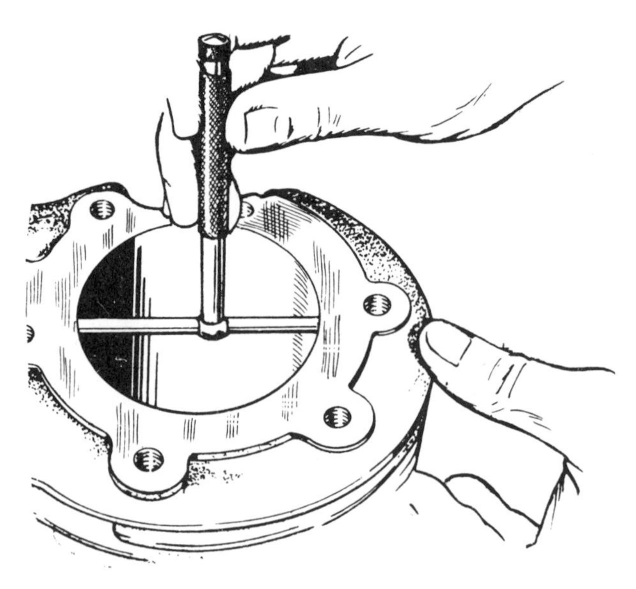

FIG. 2-39. A telescopic gauge and an outside micrometer can also be used to measure cylinder bore. Measure at least at two points 90 degrees apart.

Pistons are available from some manufacturers in sizes that are oversize by 0.010, 0.020, and 0.030 inch.

3. Cylinders are honed with a special honing tool (Fig. 2-40) held and turned by a drill press. Place and fix the cylinder block in place on the drill press stand. The honing tool is rotated at about 450 to 600 rev/min. Use a set of coarse stones first and hone to within 0.002 inch of the desired size before changing to a fine set of honing stones for the final honing. Follow the honing tool manufacturer's directions and the engine manufacturer's directions. It is desirable to have a crosshatched finish on the cylinder walls as shown in Figure 2-40 when the honing is completed. To accomplish this, the hone must be moved up and down at the correct speed as the hone rotates at the correct speed.

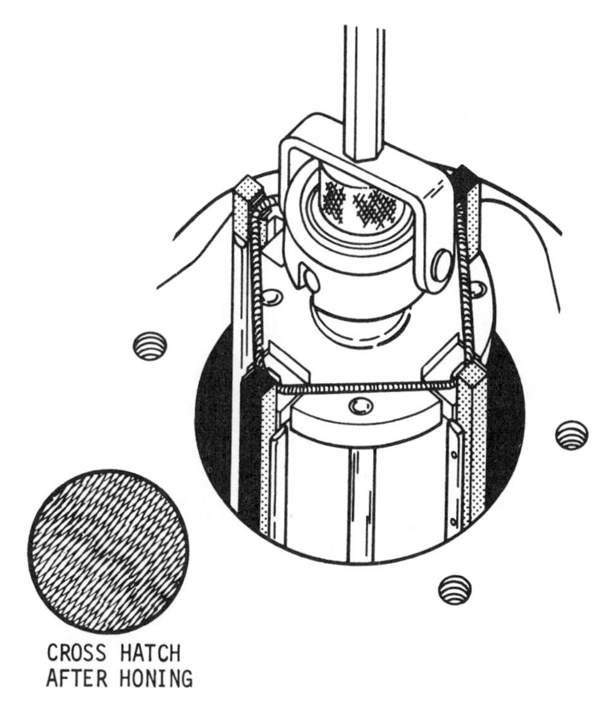

FIG. 2-40. Cylinder walls are honed to enlarge the cylinder; this is desirable if the cylinder bore is too small, is out of round, or if the cylinder walls are scored.

4. After honing, clean the cylinder with soap and water to remove all grit, metal particles, and oil.
5. Stripped bolt holes are repaired with Helicoils. Drill out the old threads and tap the hole with a Helicoil tap. Install a new Helicoil insert with the correct threads for mating with the original bolt(s).
6. Check the dimensions of the cylinder. Install the oversize piston and the other parts to reassemble the engine.

QUICK REFERENCE CHART 2-4

CYLINDER MAINTENANCE

1. Remove cylinder head, base plate, end plate, piston, and connecting rod (Sections 2-21, 2-22, and 2-23). Inspect.
2. Measure cylinder bore.
3. Rehone as required.
4. Clean cylinder after rehoning.
5. Repair stripped bolt hole threads with Helicoils.
6. Check dimensions of cylinder.
7. Reinstall parts and reassemble engine.

2-25. CAMSHAFT MAINTENANCE

Camshaft maintenance (four-cycle engines only) consists of cleaning, inspection, and measurement of camshaft dimensions. Proceed as follows:

1. Remove the base plate or the end cover.
2. Rotate the crankshaft until the timing punch marks on the crankshaft gear and the camshaft gear align (Fig. 2-41). Note the alignment; the camshaft cam *lobes* will not be in a position to press against the valve lifters.

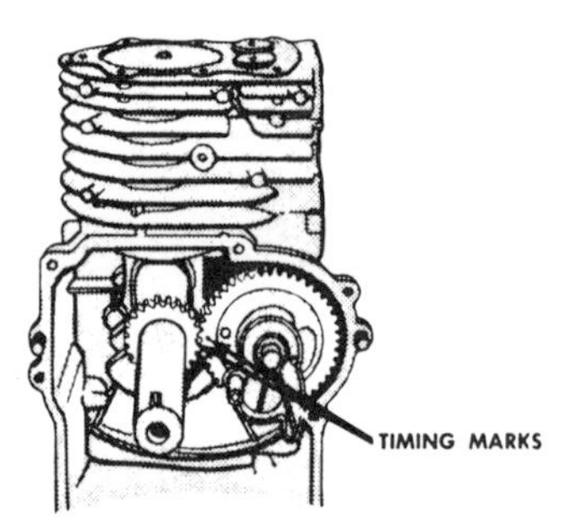

FIG. 2-41. When the camshaft is reassembled into the crankcase, the timing marks on the camshaft gear and the crankshaft gear must align.

3. Pull the camshaft out.
4. Clean the camshaft with solvent; be sure to clean out any oil ports.
5. Inspect the gear teeth and lobes for wear and nicks.
6. Measure the camshaft dimensions with a micrometer. Compare the measurements with the manufacturer's specifications; replace if required.

7. If any bearings are used, check them for wear. Replace if required.
8. After cleaning, inspection, and replacement, if applicable, are performed, reassemble the parts. Place the valve lifters into the cylinder block; ensure that the proper lifter is returned to the proper position. Install the camshaft into the block and align the timing marks (Fig. 2-41). Ensure that the valve lifters are correctly positioned (Fig. 2-42).

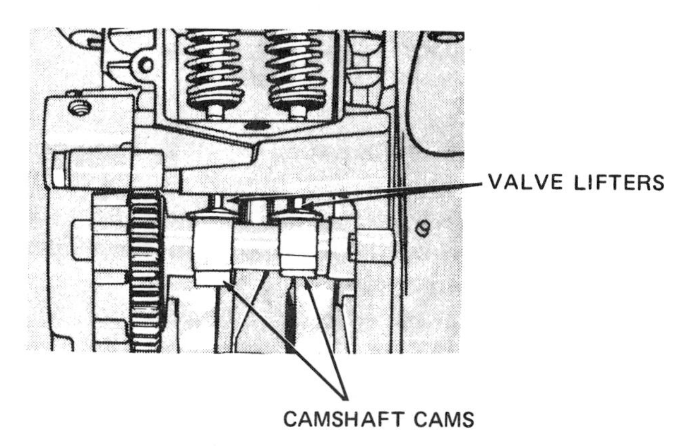

FIG. 2-42. When the crankshaft is positioned for the combustion stroke, the lobes of the camshaft are off of the valve lifters and the camshaft timing mark can be aligned with the crankshaft timing mark.

9. Align the oil pump, governor drive gear, ignition centrifugal advance mechanism, and the automatic compression release, if used on the engine.
10. Oil all parts.

QUICK REFERENCE CHART 2-5

CAMSHAFT MAINTENANCE

1. Remove base plate or end cover.
2. Rotate crankshaft until timing punch marks align.
3. Pull camshaft out.
4. Clean camshaft with solvent.
5. Inspect gear teeth and cam lobes for wear and nicks.
6. Measure camshaft dimensions and compare with specifications.
7. Check any bearings for wear.
8. Reassemble. Ensure that gear timing marks on camshaft gear and crankshaft gear are aligned.
9. Align other parts.
10. Oil all parts.

2-26. VALVE MAINTENANCE

The valves used to admit the air-fuel mixture into the combustion chamber and to let the exhaust out of the combustion chamber of four-cycle engines can develop several problems including sticking, cracks, pits, carbon fouling, and worn valve stems and guides. Valve maintenance includes maintenance of the valves, springs, seats, guides, and lifters.

VALVES

Valve maintenance consists of cleaning, inspection, measurement, and refacing. Proceed as follows:

1. Remove the cylinder head (Section 2-21) and the valve chamber access cover.
2. Using a valve spring compressor (Fig. 2-43), lift the end of a valve spring. Use needlenose pliers to hold and remove the retaining pin, collar, or keeper. If a valve spring compressor is not available, a screwdriver can be used, but it is much more difficult.

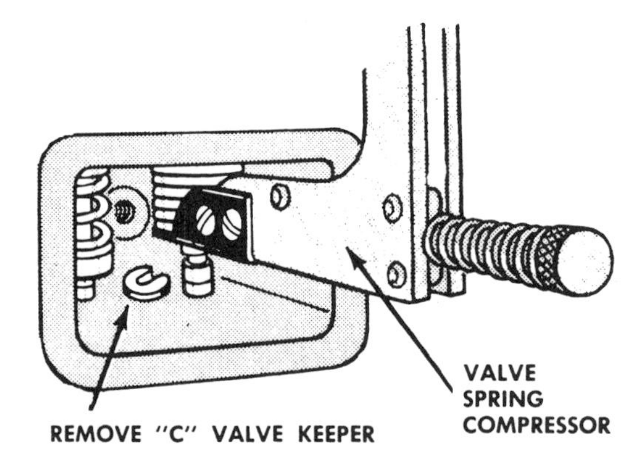

Fig. 2-43. A valve spring compressor is used to remove the valve. Compress the spring and then remove the retainer which can be a pin, collar, or keeper.

3. Remove the valve from the top of the cylinder and remove the valve spring from the access hole. In like manner, remove the other valve and spring. Tag the spring that was used with the exhaust valve—it is usually a heavier spring and must be returned to the correct valve.
4. Clean the valves, valve ports, and guides by scraping or wire brushing the carbon off. Then clean them with solvent.
5. Inspect each valve for bad burns, cracks, warpage, and bent stems. Measure the stem diameter with a micrometer and compare the measurement with the specification. If any of the aforementioned conditions exist or if

the stem is worn beyond the allowed tolerance, replace the valve with a new one.

6. If the valve has only a slightly burned or pitted face, or if a slight groove is worn on the face, the valve can be refaced using either a hand-operated or a motorized refacing tool. Follow the refacing tool and the engine manufacturer's instructions. The face is usually refaced at an angle of 45 degrees. The remaining margin usually has to be greater than 1/32 inch; if the margin is less than 1/32 inch, the valve is replaced.

VALVE SPRINGS

Valve springs are checked for breaks, bends, ends that are not square, and for correct tension if a tension reading machine is available. Proceed as follows:

1. Inspect the springs for breaks, bends, or distortion; if found, replace with new springs.
2. Place the springs on a flat surface next to a square. If the ends are not square, replace the springs.
3. If a spring tension reading machine is available, check the tension of the springs. Compare the results of the tension test with the engine manufacturer's specification for the spring tension; replace as required.

VALVE SEATS

If the valve seats are grooved, pitted, or worn, the seats need reconditioning. The reconditioning can be done by hand with a valve seat reamer or with a valve seat resurfacer that is powered by a hand drill or a drill press. Valve seats are of two types: either cut directly into the cylinder block or the seat is a separate part that is pressed into the block. Reconditioning requires the use of special tools and the manufacturer's procedure; a general guideline follows. To recondition a valve seat, proceed as follows:

1. Use the cutter (reamer or resurfacer) to cut away all of the oxidized metal until new and solid metal is exposed. Be sure to apply steady pressure directly downward (Fig. 2-44) to minimize the possibility of not having the seat true to the guide. The valve seat is cut to the angle specified by the manufacturer—approximately 43 to 45 degrees—and to the specified seat width—approximately 1/32 to 3/64 inch.
2. If the seat width of Step 1 cannot be met, a reamer or resurfacer with a more acute angle is used.
3. A fine grade of lapping compound is now used to remove the burrs caused by reaming/resurfacing; this assures a satisfactory seal between the valve face and the valve seat (refer to the next subsection on valve lapping).
4. Clean off all lapping compound, metal residue, and other foreign matter with solvent. Dry.

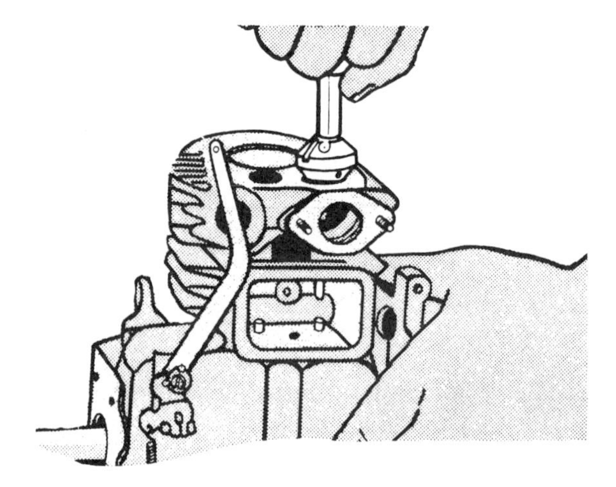

FIG. 2-44. Valve seats are reworked whenever a valve is refaced or replaced.

Install a new valve seat insert as follows:

1. On blocks that have valve seat inserts, first remove the insert by using a valve seat puller. Turn the puller bolt until the seat is drawn out. If a puller is not available, the seat can be removed with a hammer, chisel, and punch, but exercise extreme care not to cut into the block or to distort the block. On blocks that do not originally have valve seat inserts, a *pocket* has to be cut using a cutter and pilot assembly.
2. Use a cutter and cut the block to a specified depth for the insert and for holding the insert in place (Fig. 2-45). A *pilot* that is a part of the cutter must be tight in the valve guide to prevent the cutter from cutting oversize. Take periodic measurements so that the cutter does not go too deep.

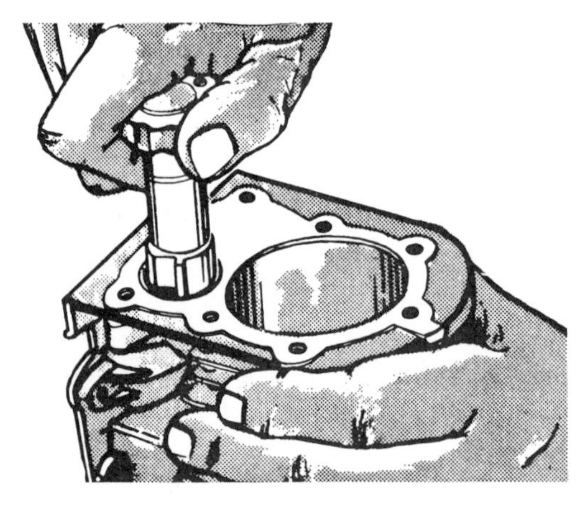

FIG. 2-45. A cutter is used to cut the cylinder block to the specified depth to receive the valve seat insert.

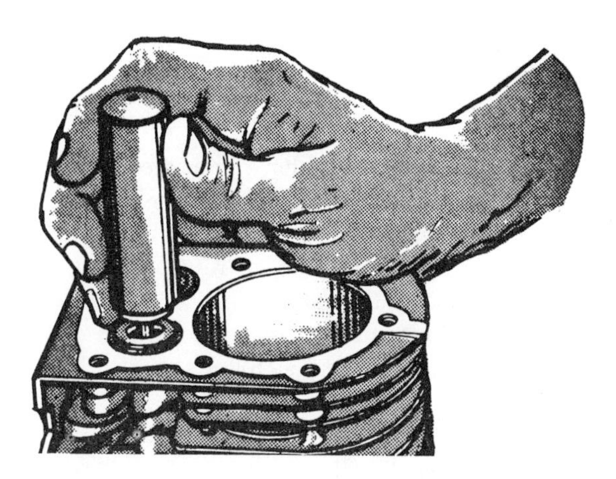

Fig. 2-46. A driver is used to install the valve seat insert into the cylinder block.

3. Clean the bore with compressed air to remove all metal chips.
4. Chill the valve seat insert to contract the metal before installing it.
5. Install the valve seat insert using a driver (Fig. 2-46). Install the insert so that the 45 degree bevel is up toward the driver; this aids in moving metal over the insert to hold it in place. Do not drive on a solid object when installing the insert as this will distort the block—you can hold the block against your body.

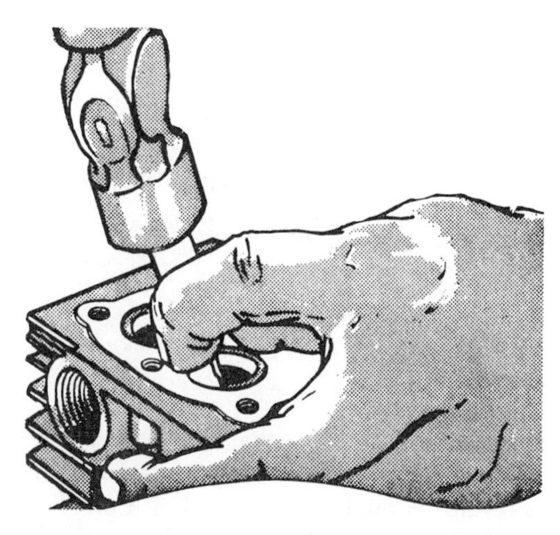

Fig. 2-47. The metal is peened around the valve seat insert to hold it in place.

6. After the insert is in place, the metal is peened over the top edge of the insert (Fig. 2-47). Move the metal toward the insert.

7. When the valve seat inserts have been installed, recondition the seats as previously described.

VALVE LAPPING

After the valve faces and seats have been reconditioned, the process of lapping is performed to remove grinding marks and burrs from the valve face and seat thus assuring a good seal of the valve into the seat. Two grades of lapping compound—coarse and fine—are used. Two types of valve grinders used for lapping are a suction cup type that sticks to the valve head and is rotated by a round handle in the hand palms and a grinder that is similar to a hand drill that oscillates back and forth. Lap the valve face and seat as follows:

1. Apply coarse lapping compound around the valve face (do not get any compound on the stem). Insert the valve into the block. Bear down slightly on the valve grinder (Fig. 2-48) and rotate the grinder in the palms of the hands.

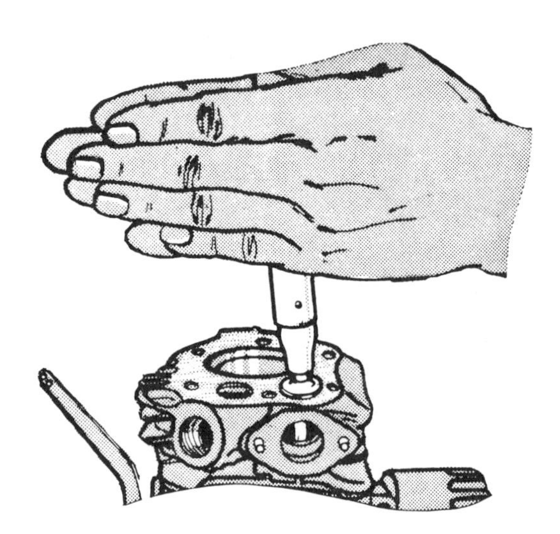

FIG. 2-48. The valve and seat are lapped by placing lapping compound on the valve face. The valve is then rotated with a valve grinder tool.

2. Rotate the valve to a position at 90°, then 180°, 270°, 360°, 90°, 180°, 270°, and 360° and rotate the grinder in the hand palms at each position. This results in producing a dull finish on the valve face.

3. Wipe the valve and the seat clean. Inspect for an even grind; if uneven, use the coarse compound again. If even, repeat with the fine lapping compound.

VALVE GUIDES

The clearance between the valve stem and the valve guide is checked by measuring the diameter of the valve guide. The result is compared with the engine specification and corrections made where necessary. Proceed as follows:

1. Using a small hole gauge and a micrometer, measure the diameter of the valve guides. Measure at the top and at the bottom of the guide and remeasure at 90 degree points to check for out of round. Compare the measurements with the specifications.
2. When the valve stem to valve guide clearance is over the maximum service clearance, one of four procedures may be performed: the valve guide can be made oversize, the guide can be replaced, a bushing can be inserted, or the guide can be knurled. These alternative procedures are discussed in Steps A to D.

   A. The valve guide can be bored or reamed to a standard oversize. New valves with oversize stems are used.
   B. The valve guide can be replaced. Press or drive the old guide out from the top of the cylinder. Press a new guide in from the top of the cylinder. Ream the guide to the proper diameter if it is not already at proper diameter.
   C. Ream the worn valve guides for installation of a bushing. Press the bushing into the valve guide. Ream the bushing to accept the standard size valve stem.
   D. A knurling tool is used to bring the valve guides (it can be used on the tappet guides also) back to a smaller size than the original bore. The bore is then reamed to the standard size. If the knurling tool is used, the valve seats should be reworked after knurling the guides to be sure that the seat is true to the guide.

VALVE LIFTERS (TAPPETS)

The valve lifters are inspected for score marks, burrs, and wear on the tappet. The stem is measured with a micrometer and the clearance between the lifter stem and the valve stem is checked for proper gap. Proceed as follows:

1. Remove the camshaft (Section 2-25) and pull the valve lifters (tappets) out; tag each lifter so that it is later returned to the same tappet guide.
2. Inspect the lifter faces for roughness and wear. Inspect the stem and stem end for wear. Replace the lifter if necessary.
3. Using a micrometer, measure the stem diameter and compare the measurement with the specification. If the clearance between the lifter stem and the guide in the block is excessive, the lifter guide hole can be reworked using a knurling tool to reduce the size of the hole. The hole is then reamed back to standard size.

4. To check the clearance between the valve lifters and the valve stems, reinstall the valve lifters into the correct guides and reinstall the camshaft (Section 2-25). Rotate the flywheel until the lifters are in the lowest position (the camshaft cam *lobes* are *not* against either lifter). Measure the clearance between the lifter stem and the valve stem with a feeler gauge (Fig. 2-49).

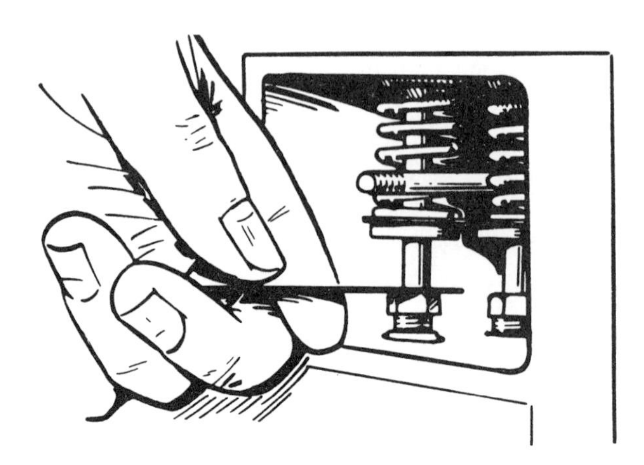

FIG. 2-49. Check the clearance between the valve lifter and the valve stem with a feeler gauge. Some lifters, as shown here, have an adjustment for setting the clearance; on others, the valve stem is ground.

5. Compare the measurement of Step 4 with the manufacturer's specification. If the clearance is too little, remove the valve and grind the valve stem slightly—square and flat. Replace and check again. Some engines have an adjustable screw on the lifter (Figs. 2-19 and 2-49) that is adjusted for proper clearance between the lifter and the valves; make the necessary adjustment instead of regrinding the valve stem.

INSTALLATION OF VALVES

Reinstall the valves, springs, and keepers as follows:

1. Place the valves in the valve guides.
2. Place the correct springs on the correct valves.
3. Using a valve spring compressor, compress the valve spring and insert the keeper, retainer, or pin. Release the spring and make sure that it is properly seated. Do the same for the other valve.
4. Install the valve spring cover and the cylinder head (Section 2-21). Reassemble the other parts of the engine.

QUICK REFERENCE CHART 2-6

VALVE MAINTENANCE

1. Inspect and reface valves, as required.
2. Inspect and replace valve springs, as required.
3. Inspect and recondition valve seats, or replace with valve seat inserts, as required.
4. Lap the valve faces and seats.
5. Inspect, measure, and recondition the valve guides, as required.
6. Measure and correct the clearance between the valve lifter stem and the valve stem, as required.
7. Reassemble.

2-27. CRANKSHAFT MAINTENANCE

The crankshaft is inspected visually for damage and is measured for correct journal size, taper, crankpin size, and out of round. If the crankshaft is bent, it must be replaced. Proceed as follows:

1. Remove the shroud.
2. Remove the flywheel (Section 5-8).
3. Remove the engine base plate.
4. Disconnect the connecting rod; note the position of the rod bearing cap (Section 2-22).
5. Remove the main bearing plate.
6. Remove the crankshaft and inspect it for: scored or damaged bearing surfaces; bent, cracked, or broken crankshaft; damaged keyways; damaged flywheel taper; and damaged or stripped threads. If any of these defects are found, the crankshaft should be replaced or reworked as subsequently discussed.
7. Measure the crankpin and journals with a micrometer—measure all the way around and end to end. If the crankpin or journals are worn, are out of round by 0.001 inch or more, are flat on one side, or are tapered, they can be reground for undersize bearings. The other alternative, if regrinding machinery is not available, is to replace the crankshaft.
8. Threads on the crankshaft are cleaned up with a die.
9. Reassemble the parts removed. Align the crankshaft and camshaft timing marks on four-cycle engines.

QUICK REFERENCE CHART 2-7

CRANKSHAFT MAINTENANCE

1. Remove engine parts so that crankshaft can be removed. This includes connecting rod lower end (and camshaft on four-cycle engines).
2. Remove and inspect crankshaft.
3. Measure crankpin and journals with micrometer. Replace or regrind.
4. Clean up threads with a die.
5. Reassemble.

2-28. MAIN BEARING MAINTENANCE

The main bearings can be bushings, ball bearings, or tapered roller bearings. Maintenance includes inspection, measurement, and replacement, if required. Proceed as follows:

1. Remove the crankshaft (Section 2-27).
2. Clean the bearings with a solvent.
3. Visually inspect the bearings for damage and measure for wear.
4. Relubricate the bearings and rotate to check for damage.
5. If necessary to replace the bearings, remove and reinstall new bearings. If the bearings are on the crankshaft, pull them off with a puller or press them off with an arbor press; they may also be punched out with a special tool. If the bearings are in the crankcase, place the crankcase on a hotplate at 400 degrees Fahrenheit. Tap the bearing lightly so that it drops out. Drop in the new bearing. Be sure to wear heavy gloves.
6. When bushings are replaced, they must be reamed to the proper diameter. Clean the metal cuttings away. Check the oil passages to make sure they are not obstructed.

QUICK REFERENCE CHART 2-8

MAIN BEARING MAINTENANCE

1. Remove crankshaft.
2. Clean bearings.
3. Visually inspect and measure bearings for wear.
4. Relubricate bearings and check for damage.
5. Remove old bearings and reinstall new bearings, as required.
6. Ream bushings to proper diameter.

2-29. OIL SEAL MAINTENANCE

The oil seals are of rubber or leather and are used at both main bearings to prevent oil leaks from the crankcase. The seals are replaced whenever the crankshaft is replaced because the seals are usually damaged by crankshaft removal.

The seals are sometimes held in place by a retainer and a snap ring. Pry the snap ring out of the groove.

The sharp edge of the leather or rubber seal is placed toward the inside of the engine. It is pushed in flush with the hub or left to protrude $1/16$ inch.

2-30. REED VALVE MAINTENANCE

Reed valves are used in two-cycle engines to allow the air-fuel mixture to flow into the crankcase. If the valves are not operating properly, an engine malfunction will occur.

Clean the reed valve assembly with a solvent; be careful of the reeds. Inspect the reed valve assembly for broken, bent, or distorted reed valves, a damaged or distorted valve seat, or a bent or broken reed valve stop. If any of these are noted, replace the defective part. Check for excessive clearance between the reeds and the reed plate with a feeler gauge (Fig. 2-50). The maximum clearance allowed is approximately 0.015 inch (see the engine specification).

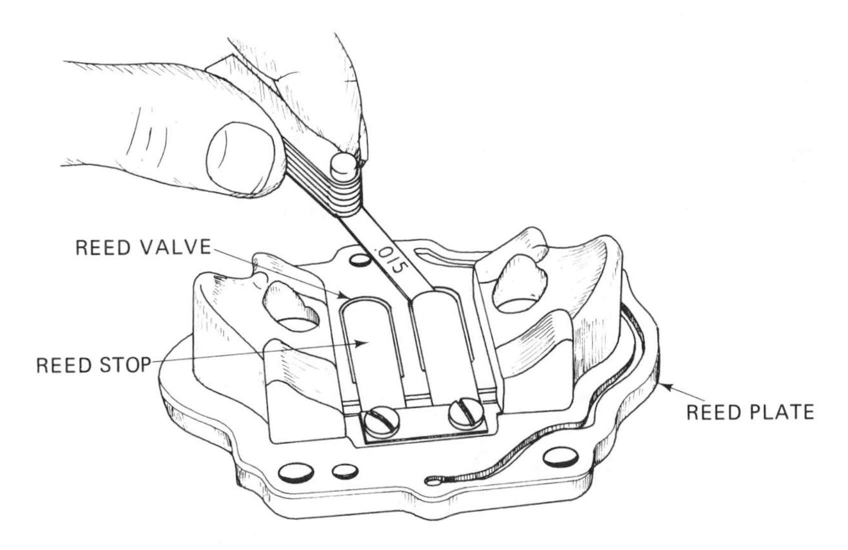

Fig. 2-50. Check the reed valves for distortion, breakage, and correct clearance.

2-31. COOLING SYSTEM MAINTENANCE

Maintenance of an air-cooling system consists of keeping the air cooling fins on the cylinder head, cylinder block, and flywheel free of dirt or other foreign matter and of keeping the engine shroud in place. Refer to Section 3-8 for periodic maintenance procedures. Maintenance of water-cooling systems includes flushing the system with clean water, making water (thermostat) temperature checks, and adding anti-freeze to prevent cooling system damage in below freezing temperatures.

Engine water-cooling systems are flushed every 1000 hours of operation to remove foreign matter. Open the drain cock at the bottom of the radiator, engine drain plug(s), and the top radiator cap. When all water has run out, connect a source of fresh water into the top of the radiator and flush until the water running out is clean. For extremely dirty water-cooling systems, the system can be back flushed under pressure from a hose. Follow the manufacturer's directions. Salt water is flushed from outboard engines and outboard engines are drained as described in Sections 3-5 and 3-4, respectively.

Engine temperatures of outboard engines are checked with a heat-sensitive stick similar to a crayon. The stick melts on contact with a surface at or above a specific range. Two sticks that melt at different temperatures are used—the lower melting stick indicates the operating temperature and the high temperature melting stick is to indicate that the engine temperature did *not* reach that temperature (for example, sticks melting at 125° and 163°F are used).

The temperature is best checked when the engine is operating on a boat. If this is not possible, run the motor in a test tank for at least 5 minutes at a maximum speed specified by the manufacturer—about 3000 rev/min. Using the two melting sticks, mark the surface to be checked; the marks will appear dull and chalky. When the surface temperature of the marked surface reaches the melting point of the stick, the mark will melt becoming liquid and glossy in appearance.

If the low temperature melting stick mark does not melt after a reasonable time, the thermostat is stuck open and the engine is running too cool. If the high temperature melting stick mark melts, the cooling system is not functioning correctly; the engine is overheating. Check for a stuck thermostat, restrictions in the cooling system, worn water pump, or a leaky water system.

3

Operation and Periodic Maintenance Procedures

The two-stroke cycle engine has a power stroke and a compression stroke with the actions of exhaust and intake occurring between the strokes. Oil is mixed with the fuel and the mixture is transferred from the carburetion system into the crankcase and then through internal ports to the combustion chamber. The oil in the fuel lubricates the engine.

The four-stroke cycle engine has four strokes: power, exhaust, intake, and compression. Fuel and air are mixed in the carburetor and flow through an intake valve into the combustion chamber where they are compressed and burned. Exhaust gases flow from the combustion chamber through an exhaust valve to the muffler. Oil is placed into the crankcase; it is not mixed with the fuel. The oil in the crankcase is splashed or pumped onto the moving engine parts to lubricate them.

The two- and four-cycle engines being studied fall into the classifications of: gasoline used as the driving force, reciprocating or rotary, internal-combustion, air- or water-cooled, and the crankshafts are principally located vertically, horizontally, or in any position such as in a chain saw. The engines, when new, operate efficiently and trouble-free providing power to help man to accomplish work or to further enjoy his leisure time. Efficient and trouble-free operation will not last indefinitely, but it can be extended significantly if the engine is cared for properly. This means operating the engine correctly and providing adequate maintenance—cleaning, lubrication, checks, tune-ups—to the engine on a scheduled periodic basis. Unfortunately, most engine owners usually don't operate their engines correctly nor do they provide sufficient, if any, periodic maintenance to the engines. This results in premature engine failures resulting in costly repair bills.

This chapter provides the correct procedures for operating an engine and for performing the periodic maintenance procedures necessary to extend the life of the small gasoline engine. There are many types and models of small gasoline engines—to explain specific operation and periodic maintenance procedures for each is impossible. This chapter provides the necessary detail for adequate periodic maintenance, but it is recommended that the owner's manual or operator's instruction book for the specific engine being operated and maintained be consulted as the most thorough and accurate information. Appendix H provides illustrated parts breakdowns of some typical small gasoline engines. Refer to the figures in Appendix H to identify *typical* parts mentioned in this chapter.

This chapter presents: engine starting procedures; engine operating procedures; engine shutoff procedures; winter or long duration storage; saltwater operation; returning the engine to service after storage; periodic maintenance and minor engine tune-ups; cleaning the engine; cleaning carburetor air cleaners; cleaning fuel filters (strainers); cleaning fuel tank vent caps; cleaning crankcase breathers; checking oil level and changing oil in the four-cycle engine; lubrication of two-cycle engines; lubrication of nonengine parts; adjusting belt tensions; checking, cleaning and regapping spark plugs; simple compression checks; and checking the battery. By knowing how to perform these procedures and by actually performing the procedures in the practical exercises at the end of this chapter, you will learn to extend the life of small gasoline engines. If you are a shop owner or repairman, this information is valuable for your tune-up work. If you pass along the starting, operating, and shutoff hints to your customers, you will earn future customers for both repairs and sales.

3-1. ENGINE STARTING PROCEDURES

There are more complaints about the hard starting of small gasoline engines than there are complaints about anything else. Perhaps this is because the correct procedures for starting an engine are not followed. Start a small gasoline engine as follows:

1. Place the engine on level ground.
2. On water-cooled engines, check that the water level is correct; fill as required. Also check for the proper oil level; fill as required (Sections 3-13 and 3-14).

CAUTION

Make sure that the air intake, or baffle, is cleared of all obstructions including weeds, water, dirt, and snow. Shrouds should always be in place because they direct air over the engine fins to cool the engine.

3. Fill the engine fuel tank with clean gasoline, or if it is a two-cycle engine, fill with the proper mixture of gasoline and oil (Section 3-14). Open the fuel shutoff valve and the gas cap vent shutoff (as applicable).
4. If the engine has a rope starter, put your foot on the engine to hold it still.
5. If it is a larger engine on a piece of equipment with brakes, apply the brakes.
6. Close the choke (close choke fully in cold weather; little or no choking is required with a warm engine).
7. If the engine has a priming button, prime the engine by pressing the button several times before attempting to start the engine. Choking is normally not required with a primer.
8. Crank the engine a few times to get gasoline into the carburetor. (To crank the engine means to turn the engine through its operating cycle. This can be accomplished in different ways depending upon the type of starter; pull the rope, release the spring crank handle, kick the starter, or turn the key and let the electric starter crank the engine.)
9. Set the throttle to the recommended starting position. If it is a single choke/throttle control, set the control to choke. When the engine starts, move the control to throttle.
10. Turn on the ignition and crank the engine. If it is a rope starter, pull the rope until compression is felt; then let the rope retract. Give a good pull and don't allow the rope to snap back.
11. If an electrical starter is on the engine, close the starter switch and crank until the engine starts. Release the starter switch as soon as the engine starts. If you have cranked for 10 to 15 seconds and the engine has not started, stop cranking. Avoid long cranking periods because they can damage the starter. Allow the starter to cool before attempting again. Use a retractable starter if the battery is low or dead.
12. If the engine doesn't start, open the choke part way and try again. The engine may have been flooded from being too rich with gasoline.
13. Once the engine is running, open the choke. Let the engine idle approximately two minutes to warm the engine before setting it to full. Don't run a cold engine at full power and don't load it down.
14. If the engine is an outboard engine, check that the water-cooling system is pumping water. This is detected by observing a small amount of water coming through "telltale" holes.

QUICK REFERENCE CHART 3-1

ENGINE STARTING PROCEDURES

1. Place on level ground.
2. Check and fill water and oil to correct levels.
3. Fill with clean fuel.
4. Open fuel shutoff valve and gas cap vent shutoff valve.
5. Clean away any obstructions.
6. Close choke.
7. Prime engine.
8. By whatever means available, crank engine over a few times to get fuel into carburetor.
9. Set throttle to recommended starting position.
10. Turn ignition on and crank the engine. If engine doesn't start, open choke part way and try again.
11. Once running, open choke and let engine warm up two minutes before applying full power or load.

3-2. ENGINE OPERATING PROCEDURES

The life of any small gasoline engine can be extended if it is operated correctly. There are two dangers of incorrectly operating small gasoline engines: *overspeeding* the engine and *overloading* the engine. Overspeeding can cause the engine to blow up or parts can fly off because of the high speed. Overloading of the engine results in overheating which causes rapid wear of the parts. Likewise, you should not operate the small gasoline engine for excessive periods of time at idling or slow speed operation, because this can result in crankcase flooding, carbon accumulation in the cylinder head and spark plug fouling.

To prevent overspeeding, do not disconnect or tamper with the governor. The governor is a speed regulating device that is covered in Chapter 7. To prevent overloading the engine, use it at a slightly slower operating rate, such as cutting the grass a little bit slower and cutting a narrower path.

New engines should be broken in properly. Initially adjust the carburetor of two-cycle engines for a fairly rich mixture and run the engine at this mixture for the first 10 hours of operation. On new four-cycle engines, adjust the throttle to minimum speed. Let the engine operate for 30 minutes without loading it. Change the oil after the first few hours of operation.

QUICK REFERENCE CHART 3-2

ENGINE OPERATING PROCEDURES

1. Operate engine at recommended speed. Do not overspeed the engine.
2. Do not overload the engine.
3. Avoid long periods of engine speeds at idle or slow speed of operation.

3-3. ENGINE SHUTOFF PROCEDURES

At first thought it seems like there would be no specific procedures for shutting an engine off. However by correctly shutting the engine off, the life of the engine will be increased. Shut off small gasoline engines as follows:

1. Remove the load from the engine.
2. Set the engine to idle speed and let it continue to run for approximately two minutes (this cools the engine by gradually reducing the thermal stresses on the parts).
3. Turn the engine off by moving the ignition switch to off or by placing the shorting bar to the tip of the spark plug.

CAUTION

Never pour gasoline into the fuel tank of a hot small gasoline engine. Any spilled gasoline could be ignited by the hot engine.

4. Close the fuel tank/carburetor shutoff valve, if applicable. If the engine will be used again shortly, fill the tank with clean fuel (*after the engine has cooled*). This keeps condensation from accumulating in the fuel tank.
5. If the engine will not be used within a month, drain the fuel tank and then start the engine again and run it until the gasoline runs out of the line and the carburetor. This prevents the fuel from gumming the carburetor passages. If the engine won't be used within a month, refer to Section 3-4.

QUICK REFERENCE CHART 3-3

ENGINE SHUTOFF PROCEDURES

1. Remove load.
2. Set engine to idle speed and run for two minutes.
3. Turn engine off.
4. Close fuel shutoff valve.

3-4. WINTER OR LONG DURATION STORAGE

If the engine will not be used for a period exceeding approximately one month, or if the engine is to be stored for the winter, special procedures are performed to protect it during the nonuse period. Stored gasoline becomes gummy and can cause the carburetor and ports within the small gasoline engine to become clogged; thus it is necessary to remove all gasoline from the engine and lubricate the engine to protect the parts during the period of nonuse. Prepare an engine for storage as follows:

1. Drain the fuel tank.
2. Start the engine and run it until it runs out of fuel; this clears the carburetor and lines of all fuel.
3. Drain the oil from the crankcase (Section 3-13). Some manufacturers recommend refilling the crankcase and leaving the oil in the crankcase during storage.
4. Remove the spark plug.
5. Pour one tablespoon of heavy engine oil into the combustion chamber.
6. Turn the engine over through several cycles to distribute the oil. Replace the spark plug.

NOTE

Some manufacturers recommend adding oil while the engine is running. To do this, first, be sure that the engine is in a dust-free area. Set the engine to a slow speed and remove the air cleaner element. Pour a tablespoonful of oil into the air intake. The oil will go through the carburetor and into the cylinder. When a blue smoke comes out of the exhaust, turn the engine off. Replace the air cleaner element. Remove all fuel from the fuel tank. Run the engine until it runs out of fuel.

7. Thoroughly clean the outside of the engine (Section 3-8) and dry it thoroughly.
8. Wrap the engine in an old blanket and store it in a warm, dry place. If possible, keep the engine indoors.

Outboard motors must have the water removed from within them to prevent the water from freezing. If the water is left in a water-cooled engine, it could cause the water jacket or other parts to crack if the water freezes. To remove the water, set the speed control to stop (to prevent accidental starting). Turn the engine over by hand several times.

QUICK REFERENCE CHART 3-4

STORAGE OVER LONG DURATIONS

1. Drain fuel tank.
2. Run fuel out of engine.
3. Drain crankcase.
4. Remove spark plug.
5. Pour one tablespoonful of heavy oil into combustion chamber.
6. Turn engine over manually.
7. Replace spark plug.
8. Thoroughly clean and dry engine.
9. Wrap engine for storage.

3-5. SALTWATER OPERATION

The salt in saltwater can eventually corrode the internal parts of outboard engines, although the parts of outboard engines have a saltwater corrosion material applied to them. It is always best to cycle fresh water through the cooling system after use and before storing. This is most easily accomplished by placing the engine so that the lower part of the drive shaft is in a tank of fresh water. Start the engine and let it run for several minutes after the engine has reached operating temperature as the fresh water is cycled into and out of the engine water cooling system.

3-6. RETURNING ENGINE TO SERVICE AFTER STORAGE

It is easy to return an engine to service after storage if the procedures for storing the engine (Section 3-4) were accomplished. To restore the engine to service, proceed as follows:

1. With four-cycle engines, fill the crankcase with new oil of the type recommended by the engine manufacturer (unless you did this at the time of storage).
2. Fill the fuel tank with *new* fuel. On two-cycle engines, mix the gasoline with the proper amount and type of oil. Use a metal can to thoroughly mix the oil and gasoline before putting the mixture into the fuel tank (Section 3-14).
3. Clean the air cleaner (Section 3-9).
4. Refill the bowl with oil, if an oil bath type air cleaner is used (Section 3-9).
5. Check the spark plug for damage. Clean and regap the plug as required (Section 3-17).
6. Perform the engine starting procedures (Section 3-1).

QUICK REFERENCE CHART 3-5

RETURNING ENGINE TO SERVICE AFTER STORAGE

1. Fill crankcase with new oil (four-cycle engine only).
2. Fill fuel tank with new fuel (mix fuel and oil for the two-cycle engines).
3. Clean air cleaner (and add oil to bowl of oil bath type).
4. Clean and regap spark plug.
5. Start engine.

3-7. PERIODIC MAINTENANCE AND MINOR ENGINE TUNE-UPS

Periodic maintenance is performed during normal engine use to keep the engine operating at maximum efficiency and to prevent wear of engine parts. The periodic maintenance is concentrated in the following areas: cleaning the engine; cleaning carburetor air cleaners; cleaning fuel filters; cleaning fuel tank vent caps; cleaning crankcase breathers; checking oil level and changing oil in the four-cycle engine; lubrication of two-cycle engines; lubrication of nonengine parts; adjusting belt tensions; checking, cleaning, and regapping spark plugs; performing a simple engine compression check; and checking the battery. Though this list seems lengthy, not all of these procedures are performed at one time, nor does it take much time to perform the procedures. By performing them, however, you will save large repair bills, or the possibility of complete engine failure. Table 3-1 lists the periodic maintenance requirements and indicates how often each procedure should be accomplished.

Table 3-1
PERIODIC MAINTENANCE CHART

* *Period—Hours of Operation*	*Task*	*Refer to Section*
Add oil to gasoline each time you refill.	Lubricate two-cycle engine.	3-14
Each 3 hours.	Check oil level (four-cycle engine only).	3-13
As required—perhaps after every use as in removing grass clippings from lawn mower.	Clean the engine.	3-8
Each 10 hours.**	Clean carburetor air cleaner.	3-9
Each 10 hours.	Check tightness of all hardware. Check sharpness of blades, teeth, etc. Check for cracks or breaks in all parts.	3-7
Each 10 hours.	Lubricate nonengine parts.	3-15
Each 10 hours.	Adjust belt tensions.	3-16
Each 25 hours.	Clean fuel filter.	3-10
Each 25 hours.	Clean fuel tank vent cap.	3-11
Each 25 hours.	Clean crankcase breather (four-cycle engine only).	3-12
Each 25 hours.***	Change oil (four-cycle engine only).	3-13
Each 50 to 100 hours.	Check, clean, and regap spark plug.	3-17
Each 50 to 100 hours.	Check compression.	3-18
Fall of year; before long periods of nonuse.	Put engine into storage.	3-4
Spring of year; after long periods of nonuse.	Return engine to service.	3-6
Upon removal of outboard engine from saltwater.	Rinse in fresh water.	3-5
Monthly.****	Check battery electrolyte level. Clean battery terminals. Check electrolyte level bimonthly.	3-19

* Each of these time periods should be reduced for engine operation in dusty environments such as tiller or garden tractor operation.

** Clean carburetor air cleaner each 3 to 5 hours in dusty operating environment.

*** Change crankcase oil after the first 2 to 5 hours of engine operation; then every 25 hours.

**** Check bi-weekly in hot weather.

One of the most important areas in periodic maintenance is inspection. You should inspect the engine after each use for the following:

1. Loose parts. Retighten as required.
2. Cracked or broken parts. Replace as required.
3. Frayed or dried out belts. Replace as required.
4. Blades, teeth, etc., for sharpness. Replace or sharpen as required.

The majority of engines brought into a repair shop should be given a minor engine tune-up. The do-it-yourself homeowner, repairman can also perform this tune-up when required. A minor engine tune-up consists of:

1. Cleaning and regapping, or replacing and regapping spark plug (Section 3-17).
2. Tightening spark plug and cylinder head bolts to the torque value specified by the manufacturer (Sections 3-17 and 2-21).
3. Testing compression (Section 3-18).
4. Cleaning carburetor air cleaner (Section 3-9).
5. Adjusting carburetor (Section 7-7).
6. Cleaning fuel tank, lines, and filter (Sections 3-10 and 3-11).
7. Adjusting governor speed (Section 7-9).

Major tune-ups, minor overhaul and major overhaul are covered in Section 9-8. They are not considered periodic maintenance.

3-8. CLEANING THE ENGINE

Dirt is one of the biggest causes of failure in air-cooled small gasoline engines. Dirt collects on the outside of the engine and prevents heat from radiating from the cooling fins. Hence the cylinder and other engine parts get too hot. The heat causes the oil to breakdown causing additional damage from excessive wear because of the lack of lubricating oil.

You can use soap and water, a degreasing compound, or other solvents to clean an engine, but you must be careful. Do not clean a hot engine with water because when the cold water hits the hot engine, it can crack the cylinder. Do not clean a hot engine with a degreaser either because some degreasers are flammable. Therefore never clean a hot engine.

To clean the engine with soap and water or a degreaser, use a stiff brush; do not use any metal implements such as a screwdriver, scraper, or a putty knife to clean engine parts because these implements will scratch the surfaces causing dirt to collect easily. Cleaning of the engine consists of cleaning the shroud, cylinder, cylinder head fins, flywheel vanes, muffler, and the exhaust ports of two-cycle engines.

NOTE

Do not clean an engine with water under pressure because the pressure can force moisture into parts to cause eventual, if not immediate, problems.

CLEANING THE SHROUD

The shroud is a cover that directs the flow of air from the fan (on the flywheel) through the cooling fins to remove heat from the cylinder head and the cylinder block (see Figs. H-1 to H-3). Clean the shroud as follows:

1. If necessary, remove parts that interfere with the removal of the shroud. These parts may include the air cleaner, muffler, spark plug wire, governor, or spring.
2. Remove the shroud.
3. Straighten bends and repair other damage to the shroud. Replace the shroud with a new one if necessary.
4. Scrape dirt and other material off of the shroud with a wooden stick or a plastic scraper. Brush the shroud with a stiff brush and water. Brush or wipe the shroud with degreaser or other solvent to remove oil.
5. Likewise, clean the air intake screen.

CLEANING CYLINDER FINS, CYLINDER HEAD FINS, AND FLYWHEEL VANES

The cylinder fins, cylinder head fins, and flywheel vanes must be kept clean to permit air circulation that results in maximum heat radiation and hence cooling of the engine. Do not let dirt or other deposits collect on the fins as this reduces the air cooling capacity of the system. Clean as follows:

1. Use a 1/4 inch wood dowel that is pointed in a pencil sharpener to scrape away dirt and other foreign matter.
2. Remove additional dirt and foreign material with a solvent or degreaser applied to a rag.
3. Check for cracks and leaks. Repair as required.

CLEANING MUFFLER AND EXHAUST PORTS OF TWO-CYCLE ENGINES

Carbon accumulates at the exhaust ports of two-cycle engines. This accumulation reduces engine performance and therefore must be removed. The carbon also causes overheating that results in damage to engine parts. Clean as follows:

1. Remove the muffler and the gasket.
2. Inspect the exhaust ports for an accumulation of carbon. If carbon is present, proceed to Step 3; if no carbon is present, proceed to Step 7.
3. Rotate the crankshaft until the piston covers the ports. This prevents the carbon from falling back into the cylinder.
4. Position the engine so that the exhaust ports are pointed down so that the carbon will fall out.
5. Scrape off the carbon with a wooden dowel or plastic scraper. Do not use a metal tool.

6. Use a brush to clean out the remainder of the dirt and carbon in the ports. If air pressure is available, blow out the dirt.

CAUTION

Be sure the piston is still blocking the ports before using air pressure. Wear safety glasses or a face shield to prevent the blown carbon from getting into your eyes.

7. Reinstall the muffler, a new gasket, and attaching hardware. The muffler should be the same distance from the exhaust outlet as when removed.

QUICK REFERENCE CHART 3-6

CLEANING THE ENGINE

1. Remove the shroud.
2. Straighten any bends, repair damage and then clean the shroud.
3. Clean the air intake screen.
4. Clean the cylinder fins, cylinder head fins, and flywheel vanes.
5. On two-cycle engines, remove muffler. Rotate crankshaft until piston covers exhaust ports and then clean exhaust ports.
6. Reinstall parts removed.

3-9. CLEANING CARBURETOR AIR CLEANERS

Air cleaners (Fig. 3-1) are used to trap dust and other foreign material from the air before the air is drawn into the carburetor. There are three types of carburetor

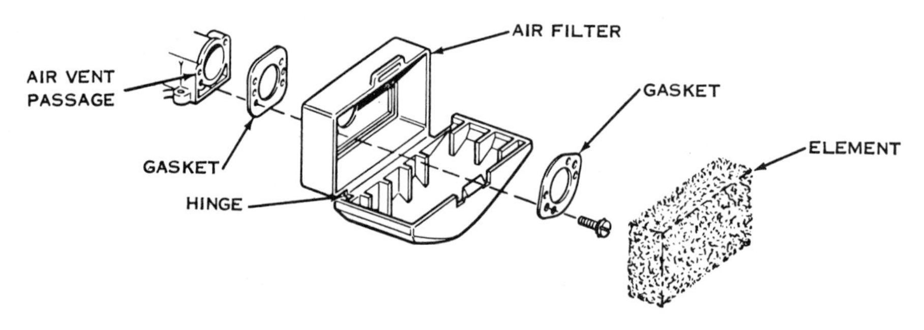

Fig. 3-1. Air cleaners (filters) trap dust from the air being drawn into the carburetor. There are three types of filters: dry filter, oiled filter, and oil bath. The filter shown is an oiled filter.

air cleaners used on small gasoline engines: dry filter, oiled filter, and oil bath air cleaners.

CLEANING DRY FILTER AIR CLEANERS

In a dry filter air cleaner, air passes through extremely tiny pores in the filter element, but dust is trapped. The filtering element is made of paper, felt, fiber, foil, moss, hair, or a metal cartridge. Clean the dry filter air cleaner as follows:

1. Remove the filter cover and filter element. Be careful not to damage the filter in any way. Replace damaged filters immediately—even a pin hole in the filter can let in sufficient dust to eventually ruin the engine.
2. Cover the carburetor intake with a piece of cloth or plastic to prevent dirt from blowing into the open intake.
3. Remove the filter and tap the bottom of the filter lightly on a flat surface to knock out the dirt. If the filter is paper, felt, a fiber hollow element, or a metal cartridge, tap the filter to remove the dirt. Do not wash these types of dry filters unless the manufacturer specifically recommends it. If the filter is fiber or moss, blow compressed air through it from the inside of the filter toward the outside (Fig. 3-2). Wash the filter in water. Do not use an oily solution because it can clog the filter. If the filter is foil, moss, or a hair element, it can be washed in a solvent and then dried before replacing it.

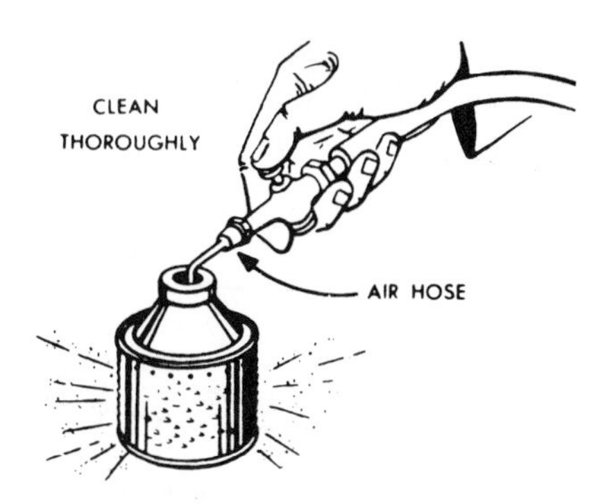

FIG. 3-2. Fiber or moss filters are cleaned by blowing compressed air through them from the inside of the filter toward the outside of the filter.

4. If the filter is worn out or excessively clogged, replace it with a new one. To determine if the filter is bad enough to need replacement, select a dust free location and run the engine first without the filter and then with the filter. Any noticeable drop in engine speed and output indicates that filter replacement is required.
5. Check that the gasket between the air cleaner and the carburetor is in good condition—a poor seal will admit dusty air. Replace the gasket, as required.
6. Reinstall the filter element and cover.

CLEANING OILED FILTER AIR CLEANERS

In the oiled filter type of air cleaner, the filter element is a metal mesh or a foam pad that is dampened with oil. Air passes through the mesh or pad and dust clings to the oil. Clean as follows:

1. Remove the cover (it may simply snap on) on the filter element.
2. Cover the carburetor intake with a piece of cloth or plastic to prevent dirt from blowing into the open intake.
3. Wash the filter element in kerosene or in detergent and water (Fig. 3-3). Foam pads should be repeatedly squeezed in the solution until clean. Swish metal mesh filters through the solution and rub with a stiff metal bristled brush until clean (Fig. 3-4).

FIG. 3-3. Wash foam elements in a detergent and water. Squeeze the filter dry and apply a few drops of oil. Work the oil throughout the filter. Wring out the filter to remove excess oil.

4. Dry the element. Wring out a foam pad and then press the pad between dry cloths. Swish a metal mesh filter in the air or blow it dry with compressed air.
5. Oil the filter. Saturate foam pads with oil and work the oil throughout the pad; then wring out the pad to remove excess oil. Dip a metal mesh filter in oil; let the excess oil drip off.
6. Reinstall the filter element and cover. Some foam pads have a coarse and a fine side; the coarse filter faces outward toward the incoming air. Some elements are designed so that the pad provides the air seal. Be sure it is

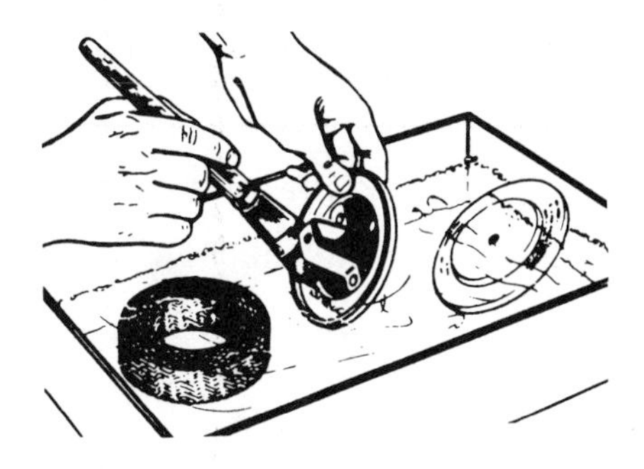

FIG. 3-4. Swish metal mesh filters through a solvent and rub with a stiff metal bristled brush. Dry the filter. Dip the filter in oil and let the excess drip off.

installed correctly. If there is a gasket, insure that it is in good condition; replace a damaged gasket.

CLEANING OIL BATH AIR CLEANERS

An oil bath air cleaner has an oil cup in the bottom of the cleaner. Air passes over the oil and picks the oil up as a fine mist. This mist is carried up to a metal mesh filter element. The dust is trapped on the oily surface and is washed down into the oil cup. Clean the oil bath air cleaner as follows:

1. Remove the filter (some filter elements unscrew).
2. Cover the carburetor intake with a piece of cloth or plastic to prevent dirt from blowing into the open intake.
3. Separate the filter parts. Pour out and discard the old oil.
4. Wash the oil cup, filter, and cap in kerosene or detergent and water. Swish the parts through the solution and rub the parts with a stiff bristle brush until clean (Fig. 3-5). Dry the parts.
5. Refill the oil cup to the mark (not above the mark) with the oil specified —usually SAE 30.
6. Check the gasket for damage; replace if damaged.
7. Reinstall the oil cup and filter element.

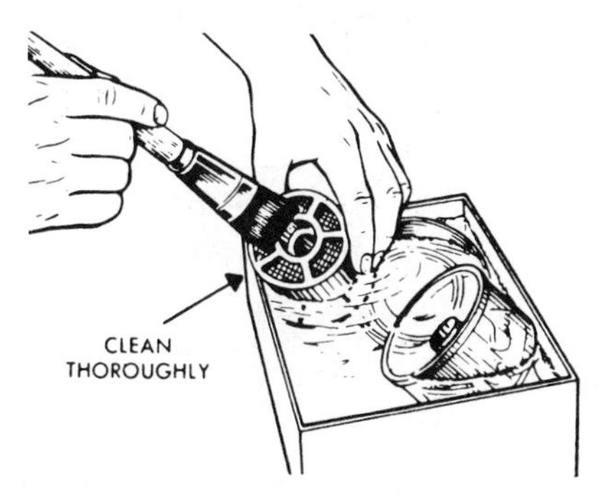

FIG. 3-5. Swish the parts of an oil bath air cleaner in solvent and brush the parts with a stiff bristle brush until clean. Dry. Refill the oil cup.

QUICK REFERENCE CHART 3-7

CLEANING CARBURETOR AIR CLEANERS

1. Locate air cleaner and determine type.
2. Remove filter cover and element.
3. Cover carburetor intake.
4. Clean element and all parts.
5. Dry parts.
6. Coat with or add oil (except for dry filters).
7. Reinstall all parts and attaching hardware.

3-10. CLEANING FUEL FILTERS (STRAINERS)

Fuel filters, also called strainers, are used to filter dirt and other foreign matter from the fuel before the fuel enters the carburetor. There are four types of fuel filters: one is fixed in the fuel tank, the second is a weighted filter at the end of a hose in a fuel tank, the third is a bowl type (Fig. 3-6) filter mounted externally to the fuel tank, and the fourth is a filter at the end of the fuel line at the fuel pump inlet (Fig. 3-7).

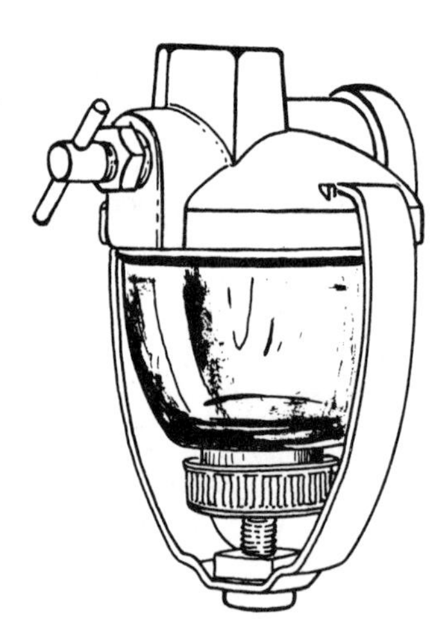

Fig. 3-6. The bowl type fuel filter is easily cleaned by removing the bowl, filter, and gasket. Swish the bowl and filter through a solvent, dry, and reinstall.

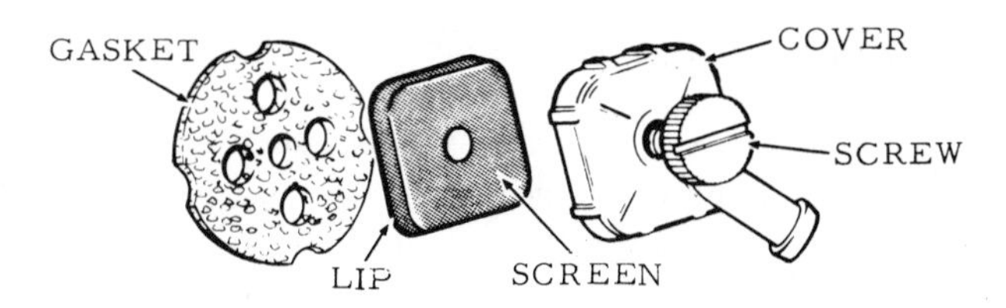

FIG. 3-7. This fuel filter is attached to the fuel line at the fuel pump inlet of an outboard engine.

CLEANING FUEL FILTERS THAT ARE FIXED IN THE FUEL TANK

Fuel filters that are fixed in a tank are located in a corner. The filter may or may not be removable (Fig. 3-8). If the filter is not removable, pour the fuel out of the tank and pour in a fresh supply. Swish the fuel around in the tank and then pour it out and discard it. If the filter is removable, pour out the old fuel and proceed as follows:

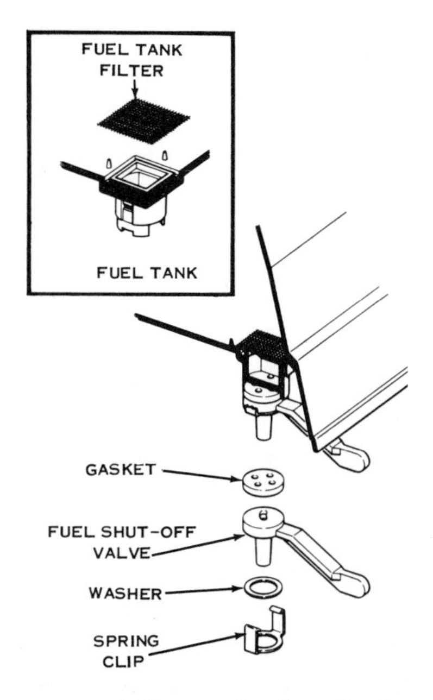

FIG. 3-8. Some fuel tanks have internal filters and a shutoff valve.

1. Remove the filter attaching hardware; this may necessitate removal of the fuel shutoff valve.
2. Remove the filter.
3. Swish the filter through a solvent until the filter is clean.
4. Swish the filter in the air to dry it.
5. Reinstall the filter and fuel shutoff valve.

CLEANING WEIGHTED FUEL FILTERS

Weighted fuel filters are used on engines such as a chain saw that must be operated in any position. The filter is attached to a flexible hose that is weighted to make the hose fall to the lowest part of the fuel tank. Clean a weighted fuel filter as follows:

1. Remove the fuel tank vent cap.
2. Gently pull the hose until the weighted filter can be pulled through the tank neck. You may need to reach the hose with a piece of bent wire. Cover the tank neck with a piece of cloth or plastic to prevent dirt from entering.
3. Remove the filter from the hose and swish it through solvent until it is clean.
4. Dry the filter by swishing it in the air.
5. Replace the filter onto the weighted hose.
6. Place the weighted hose back in the tank and replace the fuel tank vent cap.

CLEANING BOWL TYPE FUEL FILTERS

A bowl type fuel filter (Fig. 3-6) is easily recognized as a glass bowl with fuel in it. Fuel flows through the inlet port, passes into the bowl, then passes through the filter element and the outlet to the carburetor. The filter and the bowl are cleaned as follows:

1. Close the fuel shutoff valve on the filter.
2. Remove the bowl, filter and gasket.
3. Open the fuel shutoff valve and let a cup of fuel run out. This will clean out the fuel line. Discard the fuel.
4. Swish the filter and bowl through a solvent until clean. Dry the filter and bowl by swishing them in the air.
5. Check the gasket. If it is damaged, replace it.
6. Reinstall the filter and bowl with attaching hardware or bail. Before attaching the hardware securely, open the fuel shutoff valve and fill the bowl. This prevents air from getting trapped in the line.

CLEANING FUEL FILTER ATTACHED TO FUEL PUMP

This type of fuel filter (Fig. 3-7) is used with outboard engines. The filter is attached to the fuel line with a hose clamp and to the fuel pump with a thumb screw. Clean the filter as follows:

1. Loosen the thumb screw and remove the cover, filter screen, and gasket.
2. Swish the filter in solvent until clean.
3. Dry the filter with compressed air or by swishing it in the air.

4. Inspect the gasket. If damaged, replace it with a new one.
5. Reinstall the gasket, filter screen, cover, and thumb screw.

QUICK REFERENCE CHART 3-8

CLEANING FUEL FILTER

1. Remove filter (and bowl).
2. Swish parts in solvent until clean.
3. Dry.
4. Check gasket and replace, if damaged.
5. Reinstall filter and hardware.

3-11. CLEANING FUEL TANK VENT CAPS

The fuel tank vent cap has a tiny hole in it. Atmospheric pressure (gravity) causes fuel to flow from the fuel tank to the carburetor. If the fuel tank vent cap hole is clogged, there is no flow because a vacuum is created in the fuel tank. This can cause the engine to run a few minutes and then die; the vacuum can then decrease and the engine can be restarted. Again it will die. Some vent caps, as shown in Figure 3-9 include an air screen filter. Clean a vent cap as follows:

1. Remove the vent cap from the fuel tank. Cover the tank neck with a piece of cloth or plastic to prevent dirt from entering.
2. Disassemble, if applicable, and soak the vent cap in solvent. Then swish the vent cap in the solvent and use a small brush to remove all foreign matter.

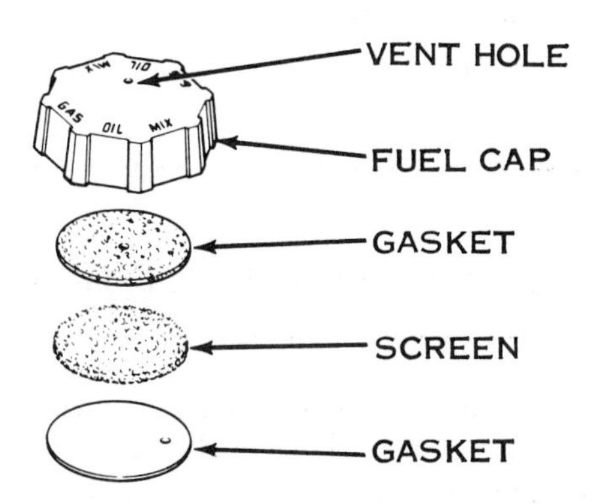

FIG. 3-9. When fuel tank vent plugs become clogged, a partial vacuum occurs in the fuel tank causing the engine to stop. Some vent caps do not include a screen filter.

3. Swish the vent cap (and screen) in the air to dry it and then replace it onto the fuel tank neck.

3-12. CLEANING CRANKCASE BREATHERS

A crankcase breather (Figs. 2-7 and 2-8) needs to be cleaned to insure that the blow-by gases that pass from the combustion chamber of a four-cycle engine past the piston and rings into the crankcase are removed. Without the proper venting of these blow-by gases, the engine parts will become corroded, and the pressure can cause the oil to leak from the seals. Clean the crankcase breather—reed valve and filter element—as follows:

1. Remove the attaching hardware, cover, gasket, filter element, reed and plate. Be sure to notice the disassembly order so that reassembly will be easy. Note any oil drain holes in the breather; the breather will be installed with the drain hole down so that any oil splashed onto the breather from the lubrication system will be returned to the crankcase.
2. Clean the filter element and other parts by swishing them in solvent. Dry by swishing in the air.
3. Check the gaskets for damage; replace if necessary.
4. Reinstall all parts. If there is an oil drain hole, install the drain hole of the breather toward the bottom of the crankcase.

QUICK REFERENCE CHART 3-9

CLEANING CRANKCASE BREATHER

1. Noting sequence and direction of removal, remove all hardware and breather parts.
2. Clean filter element and other parts in solvent.
3. Check gaskets. Replace if necessary.
4. Reinstall all parts.

3-13. CHECKING OIL LEVEL AND CHANGING OIL IN THE FOUR-CYCLE ENGINE

The oil level in a four-cycle engine must be maintained at the proper level so that the engine parts are properly lubricated. Failure to adequately lubricate the parts will result in rapid wear with eventual engine failure. Check the oil level every three hours of operation and after each use of the engine. The oil level is checked in different models by use of a dipstick, an oil minder sight tube, or by checking that the oil is level with the fill plug hole. Always place the engine on a level surface to check the oil level. Proceed as follows:

NOTE

Some manufacturers specify that the dipstick plug is *not* to be screwed into the threads to check the oil level; others specify to screw the plug into the threads. See the manufacturer's recommendations.

1. If an oil dipstick (Fig. 3-10) is incorporated in the engine, remove the dipstick and wipe it clean with a cloth. Reinstall the dipstick completely into the engine. Remove it again and inspect the oil level. If it is at the full mark, no additional oil is required. If it is below the full mark, add sufficient oil to bring it to the proper level. Refer to Step 4.

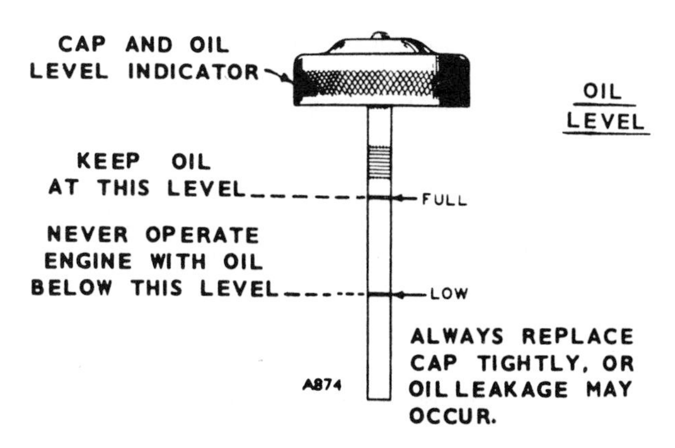

FIG. 3-10. A dipstick is used to determine the level of the oil in the engine.

2. If an oil minder sight tube is incorporated into the engine, squeeze the plunger (bellows) several times. If the plastic sight tube fills with oil, the level is correct. If it does not fill, add oil to the engine and check the level again. Refer to Step 4.
3. Some engines have a slanted fill neck (Fig. 3-11) with the base of the filler plug level with the correct oil level in the engine. Remove the filler plug from the fill neck and inspect that the oil is level with the neck. If it is not, fill the crankcase with oil. Refer to Step 4.
4. If oil is needed, fill the engine with the oil recommended by the manufacturer. Fill to the *mark* or until the oil overflows as described in Step 3. *Do not overfill;* excessive oil causes smoking from the engine. Reinstall the oil fill cap.

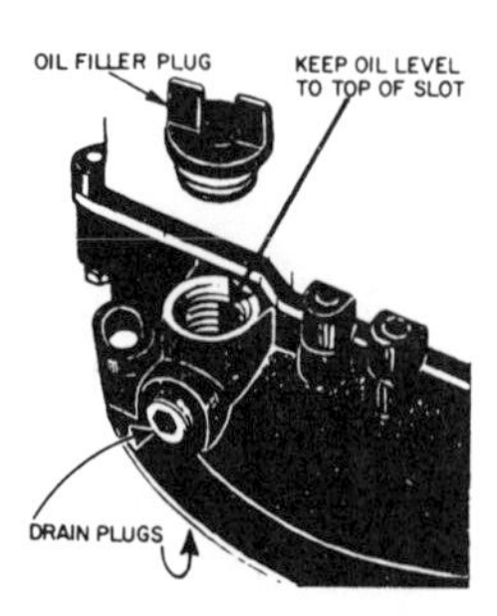

FIG. 3-11. For engines that have a slanted fill neck, add oil into the crankcase until the oil is level with the neck.

Change the oil as often as recommended by the manufacturer—at least every 25 hours of operating time. On new engines, the oil is often changed after the first 2 to 5 hours (depending upon the manufacturer's recommendations). Too soon cannot be early enough to change the oil of a new engine. This initial oil change gets rid of metal particles that may have been ground off as a result of initial operation. If the engine is operated in a dusty environment, such as the engine in a tiller or garden tractor, change the oil more frequently. Change the oil as follows:

1. With the engine hot, but turned off, locate and remove the oil drain plug.
2. Position the engine so that the drain hole is to the bottom for complete drainage. Discard the oil.
3. If the drain plug is not the same plug as the fill plug, reinstall the drain plug.
4. Refill the crankcase with the oil recommended by the manufacturer. Refer to Chapter 8.
5. Reinstall the oil fill cap.

QUICK REFERENCE CHART 3-10

CHANGING OIL IN THE FOUR-CYCLE ENGINE

1. Run engine to warm up the oil (and sludge).
2. Remove oil drain plug.
3. Position engine with drain hole down. Drain and discard oil.
4. Reinstall drain plug.
5. Fill crankcase with oil recommended by the manufacturer.

3-14. LUBRICATION OF TWO-CYCLE ENGINES

In the two-cycle engine, the oil is mixed with the gasoline at the proper ratio as recommended by the manufacturer. The carburetion system combines air with the fuel mixture—gasoline and oil—and sends this mixture to the crankcase as a mist. During the cycle of operation, the gas-oil-air mixture is within the crankcase lubricating the crankcase parts and is then forwarded as a mist through the intake port to the combustion chamber for lubrication and ignition. It is important that the proper ratio of oil and gasoline is mixed according to the directions of the manufacturer. Too much oil causes incomplete combustion and rapid build-up of carbon that fouls spark plugs, becomes heavy on the piston head, and clogs ports. Too little oil will deprive the engine of proper lubrication causing overheating, scuffing, scoring, and eventual engine seizure. Oil is never added to gasoline for four-cycle engines.

Mix oil and gasoline for a two-cycle engine as follows (Fig. 3-12):

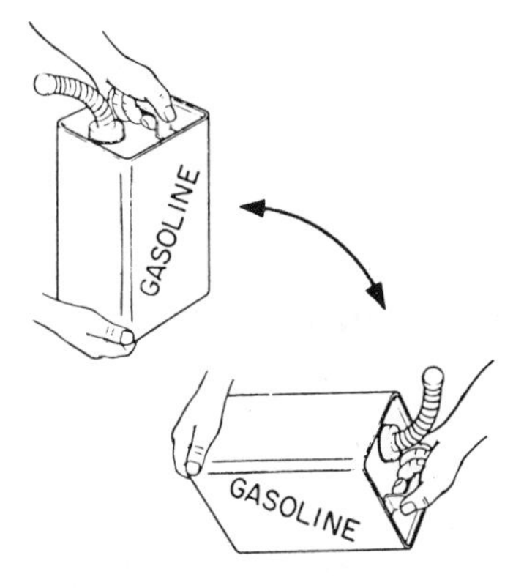

FIG. 3-12. Always mix gasoline and oil for two-cycle engines in a metal container. Never mix in a glass bottle.

1. Use a clean mixing can—*never use a glass bottle.*
2. Pour in three-fourths of the amount of gasoline required.
3. Add the proper amount of oil.
4. Mix thoroughly by vigorously shaking the can.
5. Add the remainder of the gasoline.
6. Mix thoroughly by vigorously shaking the can.

It is important to follow the mixing procedure described to thoroughly mix the fuel and oil. If the oil is simply poured into a full can of fuel and is then shaken, the oil may not mix thoroughly and, instead of mixing, will stratify. Mix the oil in three-fourths of the fuel; add the rest of the fuel and mix again for every refill.

One manufacturer of engines suggests the use of the following fuel mixture: regular leaded gasoline or premium grade gasoline (92 octane minimum) is mixed with SAE 30 or SAE 40 two-cycle air-cooled type engine oil in a gasoline to oil ratio of 20 to 1 for the first fill of a new engine. Thereafter use a 40 to 1 ratio. Low lead or nonleaded gasoline is not approved for two-cycle engine operation. The use of a premium grade gasoline may be especially beneficial in warm weather operation to help prevent detonation or an after-run condition. Gasoline to oil ratios for standard gallon size containers are given in Table 3-2.

Table 3-2
SUGGESTED GASOLINE TO OIL MIXES FOR STANDARD GALLON SIZE CONTAINERS

Gasoline	*1 Gallon*	*2 Gallons*	*3 Gallons*	*4 Gallons*	*5 Gallons*	*6 Gallons*
Oil (U.S. std. measure)	3.25 Ounces	6.5 Ounces	9.5 Ounces	13 Ounces	16 Ounces	19 Ounces
Oil (imperial measure)	4 Ounces	8 Ounces	12 Ounces	16 Ounces	20 Ounces	24 Ounces

Outboard engines have one of two types of tanks: either a portable tank or a built-in tank. With portable tanks, pour the lubricating oil into the tank and then add the gasoline at the proper ratio recommended by the engine manufacturer. Replace the filler cap. Mix the fuel by tipping the tank as shown in Figure 3-13. With the built-in tank, use a large funnel with a fine mesh strainer (100 mesh or finer). Slowly pour the oil with the gasoline into the tank as shown in Figure 3-14.

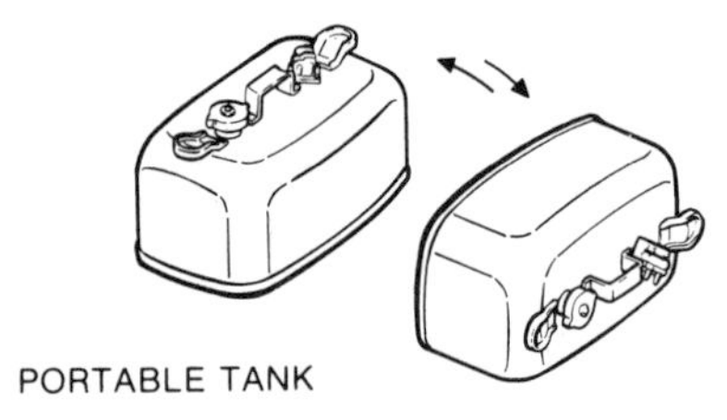

FIG. 3-13. Fill the container three quarters full with gasoline. Add the proper amount of oil and mix by tilting the tank. Then add the remainder of the gasoline and again mix thoroughly by tilting.

FIG. 3-14. To mix gasoline and oil in a built-in tank, use a large funnel with a fine mesh strainer. Slowly pour the oil with the gasoline into the tank.

Other manufacturers recommend other ratios of gasoline to oil. You should refer to the specifications of the manufacturer of the engine being used. Do not use automotive oils and do not use additives in the oils. Also do not use two-cycle outboard motor oil in two-cycle engines that are not used as outboards; air-cooled engine operation encompasses more widely varying speeds and much higher combustion chamber temperatures than are experienced with water-cooled engines. The best procedure is to use the oil recommended by the manufacturer.

3-15. LUBRICATION OF NONENGINE PARTS

The engine is associated with other parts that are either used to transport the engine and machine or the engine drives other parts to perform some type of work. Consult the manufacturer's specification as to lubrication of these parts; do not overlook them and only be concerned with the engine. Lubricate the following types of parts as applicable: linkage, gear reduction units, transmissions, chains, wheels, shafts, axles, and other machinery.

3-16. ADJUSTING BELT TENSIONS

The pulleys and belts that are driven by the engine should be checked for proper tension. Pulleys must also be maintained in perfect alignment to minimize belt drag and resulting wear (Fig. 3-15). Check the manufacturer's literature when doing any work that requires removal or adjustment of the pulleys. Removal or adjustment usually includes the loosening of a setscrew, repositioning of the pulley, and the retightening of the setscrew. Figure 3-16 illustrates a loose drive belt. The two left hand figures illustrate a belt that is too long or loose and a belt where excessive speed is required for the belt to engage. The right hand figure shows that the belt is too loose and is sagging. Belt tension can be increased by increasing the distance between the centers of the pulleys.

FIG. 3-15. Belt pulleys must be maintained in perfect alignment to minimize belt drag and resulting wear.

3-17. CHECKING, CLEANING, AND REGAPPING SPARK PLUGS

Spark plugs should be cleaned and regapped every 50 to 100 hours of operation. The spark plug provides "telltale" signs of an engine's condition of operation (Section 5-7). Spark plugs can foul easily from carbon, oil, gas and lead. Also the electrodes burn and the insulators chip. A normal or good spark plug has tan or gray

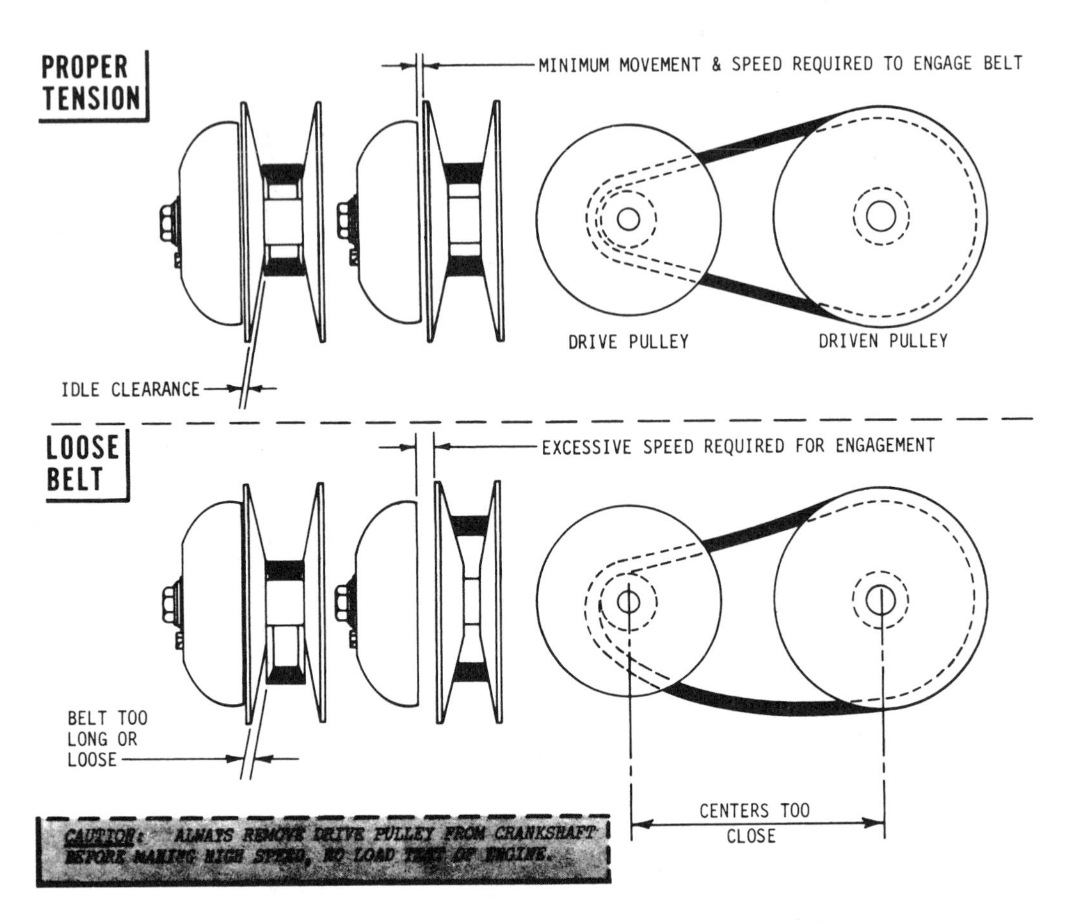

FIG. 3-16. Belts must be of proper tension to prevent slippage on the pulleys; too much tension can put stress on the belt and on the bearings of pulley axles.

deposits and has no more than 0.005 inch increase in the original gap between the electrodes. A good plug can be cleaned and reinstalled in the engine. Worn plugs having tan or gray deposits and wear of 0.008 to 0.010 inch should be thrown away. Other colors on the spark plug indicate trouble. Refer to Section 5-7. Clean spark plugs as follows:

1. Remove the spark plug and gasket from the engine and cover the port with a piece of cloth or plastic to prevent dirt from entering the cylinder.
2. Check the plug for cracks, breaks in the insulator, badly burned electrodes with pits or thinly worn electrodes, or other damage. If any of these signs are evident, discard the plug.

3. Wire brush the shell and threads with a hand- or motor-powered wire brush (Fig. 3-17). Don't brush the insulator or the electrodes. If the threads are damaged, replace the plug with a new one—properly gapped.

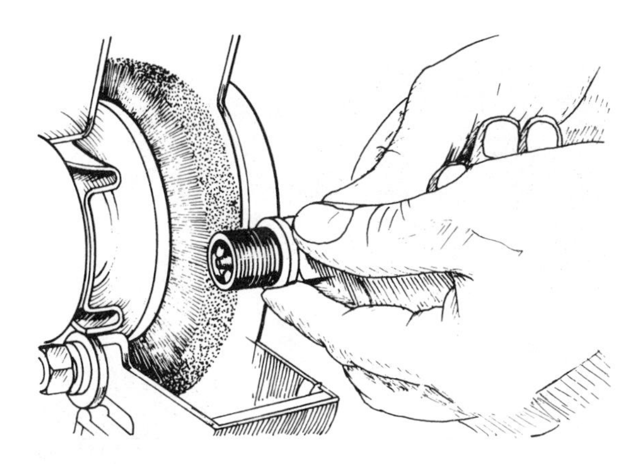

FIG. 3-17. The threads of a spark plug can be cleaned with a hand or motor powered wire brush.

4. Wipe the entire plug with a rag dampened with solvent. Be sure the terminal is free of all oil.
5. File the oxide and scale from the sparking surfaces of the electrode using several strokes of a point file (Fig. 3-18). An abrasive blast machine may also be used, if it is available, and if it is recommended by the specific engine manufacturer. Some manufacturers void the warranty if the plugs are sandblasted.

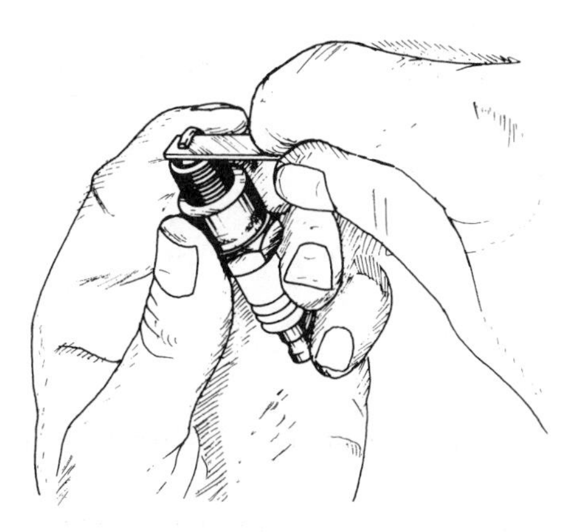

FIG. 3-18. Oxide and scale are filed from the sparking surfaces of the spark plug electrodes with a point file.

6. Regap the electrodes to the gap specified by the engine manufacturer by bending the side electrode. Never bend the center electrode because you can easily damage the plug. Check for proper gap with a round wire gap gauge (Fig. 3-19).

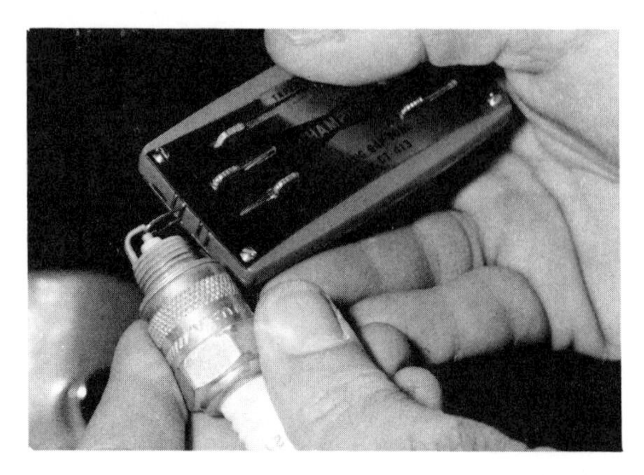

FIG. 3-19. A plain flat feeler gauge cannot accurately measure the true width of the spark gap. Use a round wire gauge as shown. To change the spark gap, bend the side electrode (never bend the center electrode).

7. Using steel wool, clean off the copper gasket. Then wipe the gasket with a rag dampened with solvent. Check that the plug seat in the cylinder head is clean and free of obstructions. Clean, if required (Fig. 3-20).

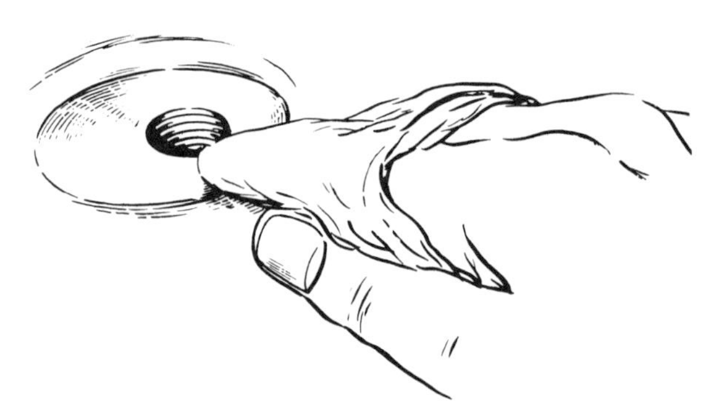

FIG. 3-20. Clean the spark plug seat with a lint free cloth dampened with solvent.

8. Reinstall the gasket and spark plug. Tighten the plug hand tight; then tighten the plug with a torque wrench to the torque value specified by the engine manufacturer. If a torque wrench is unavailable, tighten the plug 1/4 turn beyond hand tight.

NOTE

Replace spark plugs only with recommended new plugs.

QUICK REFERENCE CHART 3-11

CHECKING, CLEANING, AND REGAPPING SPARK PLUGS

1. Remove plug and gasket.
2. Cover plug port.
3. Check plug for cracks, breaks in insulator, burned or worn electrodes, and other damage. Discard, if required.
4. Wire brush shell threads. Inspect threads.
5. Wipe plug with rag dampened with solvent.
6. File oxide and scale from electrodes.
7. Regap electrodes (bend side electrode).
8. Clean gasket.
9. Reinstall plug and torque to correct valve.

3-18. SIMPLE COMPRESSION CHECK

This test is used to make a simple check of the engine compression. If there is little compression, the engine will have little power. If you determine from this test that there is little compression, then you should refer to Chapter 2 for engine maintenance procedures of the piston, piston rings, and cylinder. Make a simple compression check as follows:

1. Ensure that the on-off switch is off and/or that the spark plug wire is disconnected.
2. Pull the engine through several cycles with the starter. Feel the compression as you pull the engine through. If the engine spins easily, it has little, if any, compression and needs additional servicing. Refer to Chapter 2.
3. If the engine resists the pull of the starter, it has compression. If you spin an engine with good compression fast enough, you can hear the carburetor sucking air and the noise of kickback.
4. Also listen for any scraping or squeeking internal noises that could indicate piston and cylinder scoring. If you hear these noises, refer to Chapter 2.

3-19. CHECKING THE BATTERY

CAUTION

Use extreme care to avoid spilling or splashing electrolyte. The sulphuric acid in the electrolyte can cause body burns and destroy clothing. If the electrolyte is splashed on the body or clothing, it should be neutralized immediately with a solution of baking soda and water. Then flush the area with clean water.

Electrolyte splashed in the eyes is extremely dangerous. Always wear safety glasses when working with batteries to protect the eyes. If electrolyte is splashed in the eye, force the eye open and flood it with cool clear water for five minutes and call a doctor immediately. Do not put any medication or eye drops in the eye until advised by a doctor. If electrolyte is spilled or splashed on the painted surfaces of the engine or surrounding machinery, neutralize it immediately with a solution of baking soda and flush it with clean water.

An explosive gas mixture forms in each cell of a battery when the battery is being charged. Part of this gas escapes through the holes in the vent plugs and may form an explosive atmosphere around the battery if the ventilation is poor. Sparks or flames can ignite this gas causing an internal explosion which may shatter the battery.

The following precautions should be observed to prevent an explosion. Do not smoke near batteries being charged or which have recently been charged. Do not break live circuits at the terminals of batteries, because a spark usually occurs at the point where the live circuit is broken; this spark could cause an explosion.

Periodic maintenance of a battery includes: addition of water to the electrolyte level; cleaning of the terminals and cable connectors; checking the battery for cracks and leaks; checking the specific gravity of the electrolyte; and recharging a discharged battery. The procedures for each of these maintenance tasks are discussed in this section.

CHECKING ELECTROLYTE LEVEL

The battery electrolyte level should be checked monthly, but more often in the summer because of the increased temperature. Observe the electrolyte by removing the vent cap and viewing down into the vent well. The well has a split down each side of the vent. The bottom of the split causes the surface of the electrolyte to appear distorted when it makes contact with the split. The level of electrolyte is correct when the distortion first appears (Fig. 3-21). If the level of the electrolyte is below the split, add either distilled water or colorless, odorless, drinking water until the level rises to the vent well splits and appears distorted. Do not overfill, as this shortens the electrolyte life and may cause the sulphuric acid to run out and cause damage or corrosion to the engine or machinery parts. Likewise, do not let the electrolyte level get below the tops of the plates because this will cause the plates to dry out and become useless. If water is added in freezing temperatures, charge the battery to full charge immediately. Excessive use of additional water indicates that the battery is being overcharged. This could be caused by too high a voltage regulator setting or too high operating temperatures.

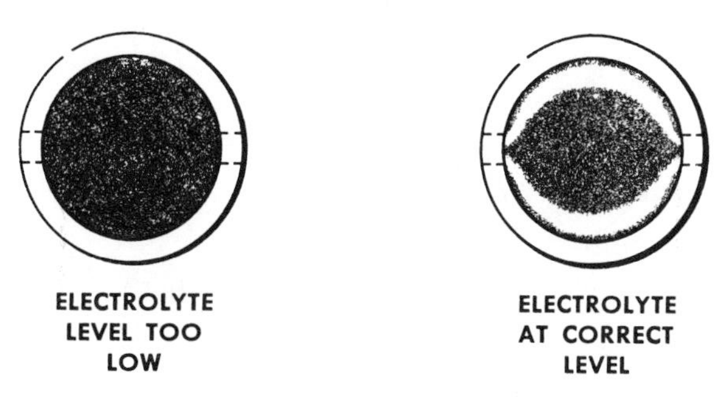

FIG. 3-21. Add water to the battery electrolyte until the electrolyte level appears distorted in the vent well. Do not overfill the battery.

CLEANING

Check the battery periodically for cracks, leakage of electrolyte, dirt, and corrosion. Wash the top of the battery with diluted ammonia or baking soda solution to neutralize the acid followed by a thorough flushing with clean water. Do not let these solutions get into the vent wells of the battery. Check the electrical cables, connectors, and battery posts for breaks, cleanliness, lack of corrosion, and tightness. If corrosion exists, disconnect the terminals and clean the terminals and clamps with a baking soda solution and a wire brush. After cleaning, replace the connectors, secure them, and then apply a thin coating of petroleum jelly on the posts and clamps to retard corrosion. Be careful in removing clamps so that you do not damage the battery posts. Inexpensive special tools for removing the clamps and for cleaning the clamps and posts are available and are recommended.

The cable connectors used on sealed terminal batteries secure the connection and make the connection tight and clean. Tighten the stud with a torque wrench to 70 inch pounds. Do not coat the terminals with grease or petroleum. Hold-down bolts prevent the battery from moving in its case and should be torqued to 60 to 80 inch pounds for bottom recess hold-downs and 40 to 60 inch pounds for top bar and frame hold-downs.

CHECKING SPECIFIC GRAVITY

The specific gravity reading of the electrolyte of a battery indicates the charge/discharge condition of the battery. The specific gravity reading is taken with an instrument known as a *hydrometer* (Fig. 3-22) which consists of a glass tube with a rubber bulb at the upper end and a hydrometer float inside the glass tube. A small amount of electrolyte is drawn into the tube from one of the cells in the battery. This causes the hydrometer to float and indicates a specific gravity reading. The hydrometer scale reads from 1000 to 1300 and the readings indicate the specific gravity and hence the battery charge/discharge condition as follows:

above 1270	fully charged
1200	half charged
1100	discharged
1000	water

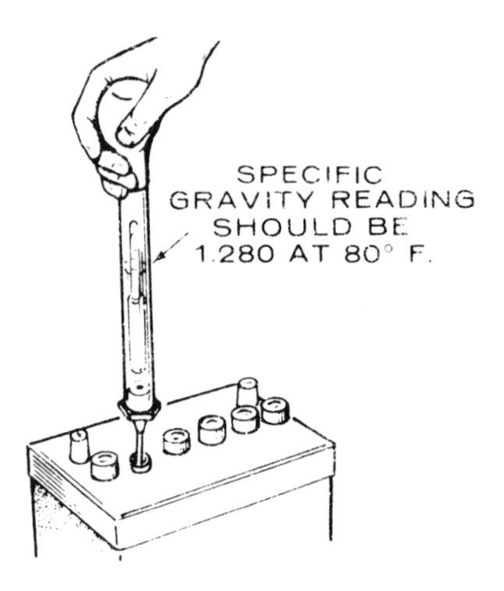

Fig. 3-22. The specific gravity reading of the electrolyte indicates the charged condition of a battery.

These specific gravity readings are at approximately 80° Fahrenheit and must be corrected for other temperatures. For every 10° above 80° Fahrenheit, a factor of 4 points is *added* to the reading; for every 10° below 80°, 4 points is *subtracted* from the reading. Thus if a specific gravity reading is 1250 at 50°, the actual specific gravity is 1238. Note that a hydrometer reading is not accurate if water has been recently added because the water has not mixed with the electrolyte.

BATTERY CHARGING

Battery charging consists of applying an electrical current at a charge rate in amperes for a period of time in hours. Thus a 5 ampere charge for a period of 7 hours would be a 35 ampere-hour charge to the battery. The charging rates should be between 3 and 50 amperes with the lower rates recommended over the higher rates. Specific charging rates or times cannot be specified in general for a battery because the size or electrical capacity in ampere-hours of the battery is unknown, the temperature of the electrolyte is unknown, the presently charged state (as fully charged or half charged) is unknown, and the battery age and condition are unknown. Thus the following information applies generally to any battery charging situation. Any large battery may be charged at any rate above 3 amperes as long as spewing of the electrolyte due to bubbling of the gas does not occur, and as long as the temperature of the electrolyte does not exceed 125° Fahrenheit. If the electrolyte spews out, or if the temperature exceeds 125°, the charging rate of amperes must be reduced or temporarily terminated to prevent damaging the battery. Small batteries such as used on motorcycles should be charged at a 0.5 ampere rate; a higher rate may heat and damage the battery.

The battery is fully charged after two hours at a low charging rate when all cells are gasing freely, but are not spewing electrolyte, and when no change has occurred in the specific gravity reading. The fully charged specific gravity is 1260 to 1280 corrected for the electrolyte temperature and with the electrolyte level filled to the split ring. If, after prolonged charging, a specific gravity of at least 1230 cannot be reached on all cells, the battery will not perform like new, but it may continue to operate satisfactorily as it has in the past. Variation in readings of over 30 points in any two cells indicates a questionable battery. Refer to Table 3-3 for a battery charging time guide.

BATTERY TESTING PROCEDURE

A battery testing procedure is used to determine whether a battery is good and useable, requires recharging, or is worn out and should be replaced. The accuracy of the testing is dependent upon variables that include temperature, specific gravity, age of the battery, etc. Therefore, an accurate test requires a number of steps. Appendix G illustrates a battery testing procedure provided by the Delco-Remy Battery Company. Refer to it for making battery tests.

Another battery voltage test can be performed by using the engine load instead of a discharge tester for resistance. Disconnect and ground the spark plug leads. Connect a voltmeter across the battery terminals. Crank the engine for 15 seconds. If the voltage reading is 9.6 volts or higher (for 12-volt batteries) at the end of 15 seconds, the battery has good output capacity.

QUICK REFERENCE CHART 3-12

BATTERY CHECKS

1. Observe all cautions—refer to Section 3-19.
2. Check electrolyte level. Add water as required.
3. Clean battery and terminals.
4. Check specific gravity.
5. If required, charge battery.

Table 3-3

BATTERY CHARGING GUIDE

(6-Volt and 12-Volt Batteries)

Recommended Rate* and Time for Fully Discharged Condition

Twenty Hour Rating	5 Amperes	10 Amperes	20 Amperes	30 Amperes	40 Amperes	50 Amperes
50 Ampere-Hours or less	10 Hours	5 Hours	2½ Hours	2 Hours		
Above 50 To 75 Ampere-Hours	15 Hours	7½ Hours	3¼ Hours	2½ Hours	2 Hours	1½ Hours
Above 75 To 100 Ampere-Hours	20 Hours	10 Hours	5 Hours	3 Hours	2½ Hours	2 Hours
Above 100 To 150 Ampere-Hours	30 Hours	15 Hours	7½ Hours	5 Hours	3½ Hours	3 Hours
Above 150 Ampere-Hours		20 Hours	10 Hours	6½ Hours	5 Hours	4 Hours

*Initial rate for constant voltage taper rate charger

To avoid damage charging rate must be reduced or temporarily halted if:

1. Electrolyte temperature exceeds 125°F.
2. Violent gassing or spewing of electrolyte occurs.

Battery is fully charged when over a two hour period at a low charging rate in amperes all cells are gassing freely and no change in specific gravity occurs. **For the most satisfactory charging, the lower charging rates in amperes are recommended.**

Full charge specific gravity is 1.260-1.280 corrected for temperature with electrolyte level at split ring.

4

Basic Electricity and Electrical Components

Small gasoline engines use electricity in the ignition and starting systems. In order for you to troubleshoot and repair these systems, it is necessary that you understand their operation. This necessitates learning some fundamentals of magnetism and electricity, and how different electrical components are used. This chapter describes atoms, electricity, electrical units, magnetism and induced voltage, electrical components, continuity, and test equipment used to check electrical components.

4-1. ATOMS

All natural matter is made up of 105 fundamental constituents called *elements*. These elements exist in the free state (that is they are not combined with anything else) as substances such as iron, oxygen, and copper. The atom is the smallest unit which retains the characteristics of the *original* element. Combinations of atoms result in *molecules*. A molecule is the smallest unit of any *compound*. A compound consists of chemical units of two or more elements—you are familiar with some of these compounds such as water (hydrogen and oxygen), salt (sodium and chlorene), and rust (iron and oxygen).

Atoms are extremely tiny units of matter. There are literally billions of atoms in a single speck of dust. The atom is further divided into two main parts: a positively charged *nucleus,* and a cloud of negatively charged *electrons* that surround the nucleus. The nucleus or center of the atom is made of *protons* and *neutrons* (Fig. 4-1). A proton is charged positively; the neutron consists of an electron and a proton bound together. The neutron has no charge. The negatively-charged electrons whirl in orbits, or rings, around the positively-charged nucelus. There is an electron for each proton; the atom, therefore, has no charge in its natural state.

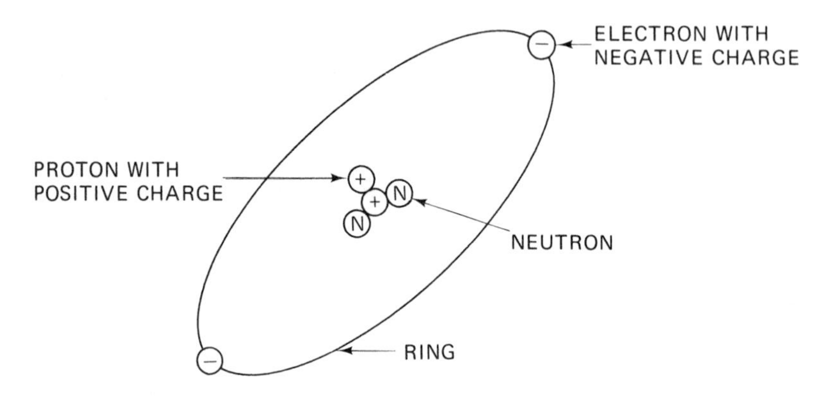

FIG. 4-1. An atom is the smallest unit which still retains the characteristics of the original element. The atom is made up of a positively charged nucleus and a cloud of negatively charged electrons that surround the nucleus.

The atoms of different elements vary with respect to the charge on the positive nucleus and the number of electrons revolving around the charge. Hydrogen, for example, has a net charge of one on the nucleus and one orbital electron, whereas copper has a net charge of 29 on the nucleus and 29 orbital electrons. The number of orbital electrons is called the *atomic number* of the element.

Electrons are grouped in rings around the nucleus. One or more of the rings in each element (except inert gases) are not completely filled with electrons. If the uncompleted ring is nearly empty, the element is *metallic* in character; it is mostly metallic when there is only one electron in the outer-most orbit. Examples of metallic elements are silver, copper, and aluminum.

If the incomplete ring lacks only one or two electrons, the element is usually nonmetallic. Atoms having outer rings approximately half filled with electrons exhibit both metallic and nonmetallic characteristics and are referred to as *semiconductors*. Examples of semiconductor materials are carbon, germanium, and arsenic.

In the metallic elements, the electrons in the outer rings are held togethei rather loosely. Consequently there is a continuous movement of these electrons between one atom and another. These electrons that move about are called *free* electrons. The ability of these free electrons to drift from one atom to another makes the flow of electric current (the flow of electrons) possible.

An element that has many free electrons is called a good *conductor;* the electrons can easily move (conduct) from one atom to the next. If there are few free electrons, the element is a *nonconductor*; electrons do not move freely (do not conduct) from one atom to the next. Nonconductors that are virtually free of electrons are insulators; wood, plastic, mica, rubber, fiber, and porcelain. They are good insulators because they are made up of elements having virtually no free electrons in their rings. Insulators cover conductors so that one conductor does not touch another conductor causing conduction between the two. That is why wires (copper, aluminum) are covered with a rubber or plastic insulating material.

Electrons are negatively charged and protons are positively charged. Like the poles of magnets in which opposite poles attract each other and like poles repel each other, opposite charges attract each other and like charges repel each other. That is, electrons repel electrons and protons repel protons, but electrons and protons attract one another. Atoms stay together because the electrons and protons attract each other; but the electrons and protons stay apart because of the centrifugal force of the electrons spinning in their orbits around the nucleus. Electrons do not cling together because their charges repel and, likewise, protons do not cling together because they are of like-charge and repel each other. The neutrons tend to hold the protons together in the nucleus of the atom.

4-2. ELECTRICITY

Free electrons move about constantly and haphazardly in the atoms of a conductor such as a copper wire. A drift of electrons (an electric current) is caused only when there is a difference in electrical "pressure" or "potential" between the ends of the wire. This potential difference can be produced by connecting a source of electrical potential to the ends of the wire.

One way of producing this difference of potential is to connect a battery to the wire. A battery (Section 4-5) has a deficiency of electrons at the positive terminal and an excess of electrons at the negative terminal. When a wire is connected between the terminals, the deficiency at the positive terminal attracts the free electrons from the wire. The wire, in turn, seeks free electrons from the negative terminal. This attraction, or seeking of electrons, continues until neutral—that is, until there are as many electrons at the negative terminal of the battery as at the positive terminal of the battery.

A potential difference is the result of a difference in the number of electrons between the two points in question. The force or pressure due to a potential difference is termed the electromotive force. It is expressed in units called *volts.*

An electric current is the flow of electrons along a conductor due to the application of an electromotive force. This flow is in addition to the irregular movements of electrons between the rings of atoms. Electrons do not actually flow along a conductor from one end to the other. Each free electron travels only a short distance and bumps the next electron which travels a short distance and bumps into another electron. The electrons are actually moving from the ring of one atom to the ring of another atom and so on.

The flow of current, as defined in the electron theory, is the flow of negatively charged electrons from one point to another point that has a *more* positive charge.

4-3. ELECTRICAL UNITS

Three electrical units with which you need to become familiar are the ampere, the ohm, and the volt. The *ampere* is a fundamental unit of current flow, or the rate of flow of electricity. The ampere and the flow of current can be compared to the amount of flow of water in a given time, or the rate of flow of water. One ampere is the quantity of electricity equivalent to one *coulomb* passing a given point per second (a coulomb is 6.28 quintillion electrons).

The *ohm* is a measurement of the resistance offered to the flow of electricity. Resistance is the property of a substance that resists the flow of electrons. It can be compared to the flow of water which is regulated by the size of the pipe and the smoothness of the inside of the pipe. Every substance has resistance, including such materials as copper and aluminum. Wire is made of copper or aluminum, and you will find that the wire used to carry heavy current is of a thick diameter in order to reduce it's resistance to the flow of electrons. Thus the thicker the wire, the easier is its ability to carry heavy currents. An *ohm* of resistance is equal to the amount of opposition offered by a conductor to the flow of one ampere of current when a pressure of one volt is applied across its terminals.

The *volt* is the electromotive force that will produce a current of one ampere through a resistance of one ohm. That is, a volt is equivalent to the electric pressure required to force one ampere of current through a resistance of one ohm. The electromotive force or pressure can be likened to a water system in which the pump pressure causes the flow of current.

These units—the ampere, ohm, and volt—have a relationship with one another. This is expressed in Ohm's law, which is a relationship between the electromotive force (voltage), the flow of current (amperes), and the resistance that impedes the flow of current (ohms).

$$I = \frac{E}{R} \qquad (4\text{-}1)$$

where $I =$ current in amperes

$E =$ electromotive force in volts

$R =$ resistance in ohms

This formula can be manipulated by algebra to solve for any unknown. Thus:

$$R = \frac{E}{I} \tag{4-2}$$

$$E = IR \tag{4-3}$$

4-4. MAGNETISM AND INDUCED VOLTAGE

At first you may wonder why you need to study magnetism in a small gasoline engine course, but magnetism is a means of generating an electric current and it is used in the magneto-ignition systems of small gasoline engines.

No doubt you are already familiar with the basics of magnetism—that a magnet has a north and a south pole. A magnet is made of an iron material that can become *permanently* magnetized; that is, once magnetized, the material remains magnetized even after the magnetizing force is removed. Other metals, such as soft iron, will magnetize, but do not remain permanently magnetized. When the magnetizing force is removed, the magnetized soft iron loses its strength. A piece of metal can be magnetized by stroking the material with another permanent magnet, or by passing an electric current through or very near the material. When a material is magnetized, the molecules tend to line up as if there were billions of tiny magnets inside of the material. Each tiny magnet has a north and a south pole and they line up one behind the other, resulting in a large magnet having a north and a south pole. You are also familiar with the fact that like-poles repel and unlike-poles attract. Thus when the north end of two magnets approach each other, they repel, but when a north and a south pole approach each other, the two magnets attract one another and cling together.

Magnets in the shape of a bar or a horseshoe have a magnetic field that surrounds the magnet and attracts other magnetic materials such as iron or steel to them. Another magnetic field the same as that surrounding a magnet also surrounds a piece of wire through which there is a flow of electrons. Without potential, or voltage, there is no magnetic field surrounding a wire because the electrons do not all move in one direction.

The magnetic field surrounding a conductor is increased considerably in strength by winding the wire into a coil. The magnetic field surrounding each turn of the wire combines with the fields of adjacent turns and forms a total field of increased strength. The field is further strengthened when the coil is wrapped around a bar of metallic material.

When a flow of electrons first begins by closing a switch letting a direct current flow through the wire, the magnetic field builds up. When the switch is opened, the field rapidly collapses. If another coil is placed near the first, a flow of electrons can be induced into the second coil by opening and closing the switch in the circuit of the first coil. This is known as electromagnetic induction. When one coil is near another, a varying flow of electrons in the first produces a varying magnetic field which cuts the turns of the other coil and induces a flow of electrons in the second.

When a permanent magnet is passed by or through a coil of wire, the magnetic field of the magnet causes a flow of electrons into the coil. This is the principal upon which the magneto-ignition system of small gasoline engines works. A permanent magnet is cast into the flywheel. As the flywheel rotates, the permanent magnet passes by a coil (the ignition coil) inducing a flow of electrons. Thus, it can be seen that magnetism can produce a flow of electrons (a current) and conversely a flow of electrons can produce a magnetic field. Generators, alternators, regulators and ignition coils each contain conductors that produce electromagnetism.

4-5. ELECTRICAL COMPONENTS

Electrical components are used in the ignition and starting systems of small gasoline engines. In order to repair these systems, it is necessary that you become familiar with the operation and use of the components in the systems. The electrical components used include: condenser (capacitor), coil, solenoid, breaker points, spark plug high-tension lead, spark plug, resistor, battery, transistors, diodes, and rectifiers.

CONDENSER (CAPACITOR)

A condenser, which is called a capacitor in the electronics field, is a device capable of storing electrical energy. A condenser is made of two metallic plates (or foils) separated from each other by a layer of insulating material (Fig. 4-2). When a source of direct current is momentarily placed across the condenser plates through its two leads (sometimes the metal case is one of the leads), the plates become charged; there is an excess of electrons on one plate and a deficiency of electrons on the other plate. The plate having the excess of electrons is said to

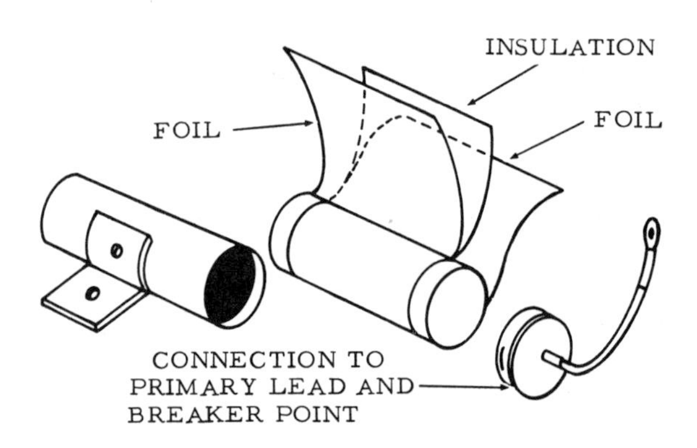

FIG. 4-2. A condenser stores electrical energy. It consists of two metal plates (foil) separated by an insulator.

be negatively charged, and the plate having a deficiency of electrons is said to be positively charged. If the condenser plates become joined by means of a wire and a switch, they will discharge and electrons will flow until a neutralized state is reached whereby each plate of the condenser will have the same amount of electrons.

When the condenser is charged, it stores potential energy as electrostatic energy. This electrostatic, or potential energy, is capable of doing work when the charge is released through an external circuit. The unit of capacitance is the *farad.* This unit is too large for normal use; the capacitance of a condenser used in the ignition systems of small gasoline engines is within the range of 0.1 to 0.5 microfarad (a microfarad is one-millionth of a farad). It is connected across the breaker points to prevent electric current from arcing when the points are opened to initiate ignition. (Ignition is discussed in detail in Chapter 5).

COIL

A *coil* is simply a coil of wound wire. Its purpose is to induce a flow of electrons when a permanent magnet is passed by. This is the principle of the magneto-ignition system discussed in Chapter 5 in which a permanent magnet, attached to the flywheel, passes a coil on each revolution of the flywheel and induces a flow of current. A coil core made up of a number of thin, soft iron laminations strengthens the magnetic field.

The coil in the ignition system (Fig. 4-3) has two wire wrappings that surround the center leg of an E-shaped laminated iron core. One wrapping is called the *primary* and consists of relatively few turns (about 200) of heavy wire. The second wrapping is the *secondary* and consists of thousands of turns of thin wire. When the magnet on the flywheel passes the coil, a current is induced into the primary coil generating another magnetic field. When the flow of current is broken by opening the breaker points, the magnetic field surrounding the primary coil rapidly collapses and induces a high voltage into the secondary coil (which is connected to the spark plug).

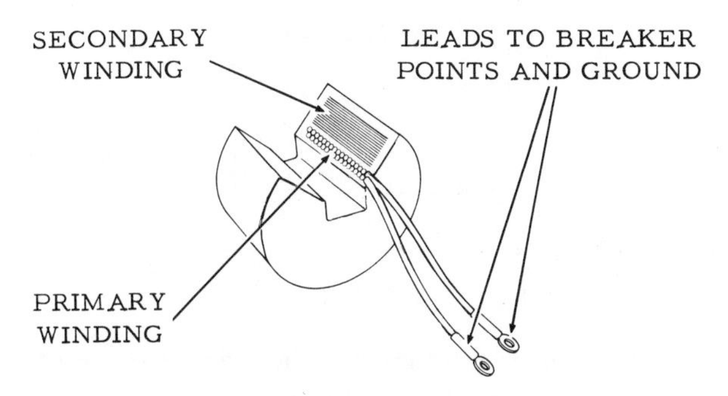

FIG. 4-3. A voltage is induced when a magnet is passed by a coil.

One end of each coil is grounded to the laminated core. The other end of the primary coil connects to the insulated and stationary breaker point. The other end of the secondary coil connects to the spark plug. (Refer to Chapter 5.) The voltage induced into the secondary coil is in proportion to the turns ratio of the primary to secondary coils. For example, if the turns ratio is 60:1 and the primary voltage is 170 volts, the secondary voltage is 10,200 volts which is more than ample to fire across the gap of the spark plug electrodes. The coils are molded to keep out moisture.

Coils have a value known as *inductance*. A circuit has an inductance of one *henry* when a current, changing at the rate of one ampere per second, induces an average of one volt.

SOLENOID

A *solenoid* is an electromagnetic helix. A system of equal circular currents flows in a uniform direction about a single straight or curved axis. An example of a solenoid used with a gasoline engine is a solenoid relay (Fig. 4-4). The relay acts as a switch to complete the starting motor circuit from the battery terminal to the starter cable (Fig. 4-5). The switch is activated when the starter switch is momentarily closed. A current flows through the solenoid creating an electromagnetic force that attracts a piece of soft iron called an *armature*. The armature has a movable electrical contact on it; when the armature moves because of the electromagnetic force overcoming an opposing spring force, the movable contact closes. This connects the battery to the starter through the solenoid contacts. When the starter switch is released, the current (or the electromagnetic force) in the coil is discontinued. The spring returns the armature and electrical contact to the normally open position breaking the path of current from the battery to the starter.

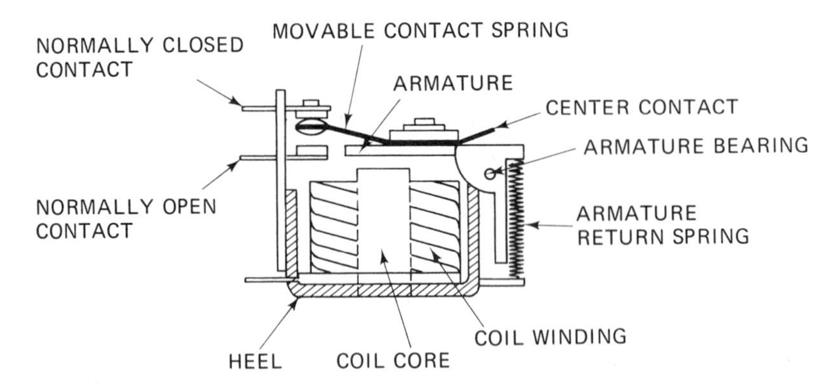

Fig. 4-4. When an electric current flows through the solenoid coil winding, a magnetic field is generated that pulls the armature to the coil. This action connects the normally open contact to the movable contact to complete a current path.

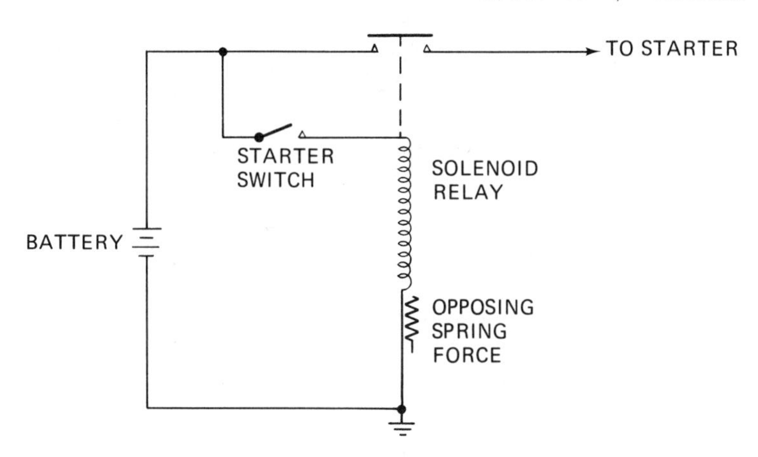

FIG. 4-5. When the starter switch is closed, the solenoid relay is energized. The battery is then connected through the solenoid relay to the electric starter.

BREAKER POINTS

Breaker points (Fig. 4-6) are used in magneto- and some battery-ignition systems (Chapter 5) to break the current flow in the primary of the ignition coil causing a collapsing magnetic field. This collapsing field induces a voltage into the secondary coil that causes the spark (from the spark plug) for engine ignition.

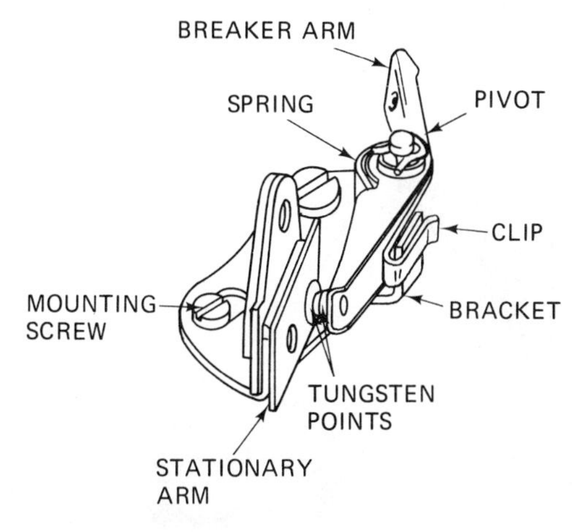

FIG. 4-6. Breaker points are used in magneto-ignition systems to interrupt the current flow in the primary of the ignition coil.

The breaker point assembly consists of a stationary arm, a breaker arm on a pivot, a spring, and mounting holes and screws. The stationary and movable breaker arms each have a tungsten contact (the *point*). The movable arm is spring-loaded. A cam moves against a rubbing block on the arm, moving the arm to and fro to open and close the breaker points. The points are usually opened when the lobe, or high point of a cam, rides against the breaker arm rubbing block. In addition to a cam, some breaker points are opened and closed by means of a cam and a push rod, plunger, or a trip lever. The points, which open from approximately 0.016 to 0.020 inch, open and close between 800 to 4500 or more times per minute depending upon engine speed. A slotted mounting hole or an eccentric-head adjusting screw can be adjusted to position the breaker arm in relation to the cam so that the proper maximum point gap can be set. The gap space is critical for proper engine ignition; the gap dimension is measured with a blade-type feeler gauge (Section 5-9).

The stationary arm of the breaker is sometimes mounted to a bracket (Fig. 4-6); in other designs, the stationary arm point is the tip of the condenser. The actuating cam on two-cycle engines is on the crankcase; on four-cycle engines, the cam is usually on the camshaft so that the points are opened on the power stroke, but not on the exhaust stroke. The shape of the cam depends upon the number of cylinders and the design of the ignition system.

SPARK PLUG HIGH-TENSION LEAD

The spark plug high-tension lead (Fig. 4-7) connects the secondary winding of the coil to the spark plug. The lead is heavily insulated to prevent the voltage passing through it from short-circuiting to ground if the lead touches the cylinder head. The cap of the lead should always be clean so that good electrical contact is made with the spark plug stud.

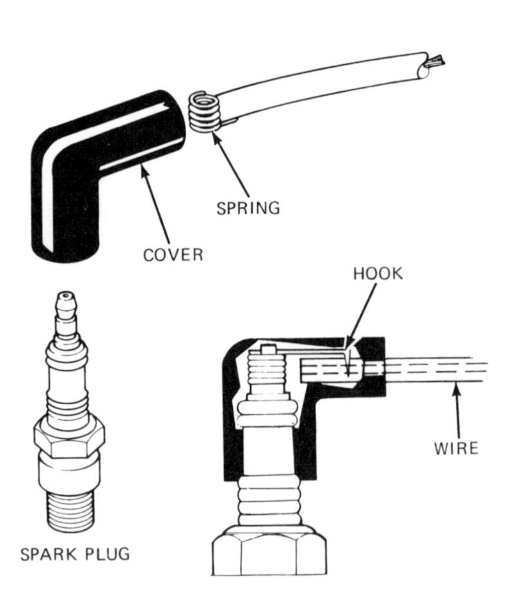

FIG. 4-7. The high-tension lead must be clean and fit tight to the spark plug. Check that the insulation is not cracked or broken.

SPARK PLUG

A spark plug (Fig. 4-8) is used to ignite the air-fuel mixture in the combustion chamber; the force of the combustion drives the piston down on the power stroke.

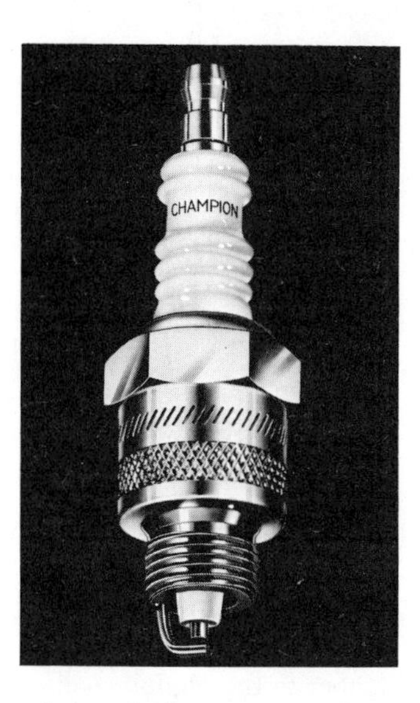

FIG. 4-8. The spark plug ignites the air-fuel mixture in the combustion chamber when a spark jumps between the plug electrodes.

Figure 4-9 illustrates the construction of the spark plug. The *stud* connects externally to the high-tension spark plug wire from the ignition coil secondary and internally to the center electrode; the stud and electrode carry the ignition spark voltage that jumps the air gap to the ground electrode for ignition. The center electrode is insulated from the shell by the ceramic insulator; the center electrode should never be bent in an attempt to set the spark gap because the ceramic insulator could be broken. The insulator dissipates the heat of combustion. The tip is designed to burn away fuel-wasting and power-wasting deposits.

The *reach* of a spark plug is the linear distance from the shell gasket seat to the end of the threaded portion of the shell. The reach is established by the cylinder head design so that the electrodes are positioned at the most satisfactory depth in the combustion chamber to ignite the air-fuel mixture. It is important to install the plug with the proper gasket and to torque the plug to the correct value so that the electrodes are positioned correctly for the most efficient engine ignition; if the electrode depth is raised or lowered due to the installation of either two gaskets or no gasket under the plug gasket seat, the temperature range at which the insulator nose stabilizes may change enough to cause overheating.

The difference between a *hot* and a *cold* plug (Fig. 4-10) is the difference in the temperature at which each type plug will operate inside a given combustion

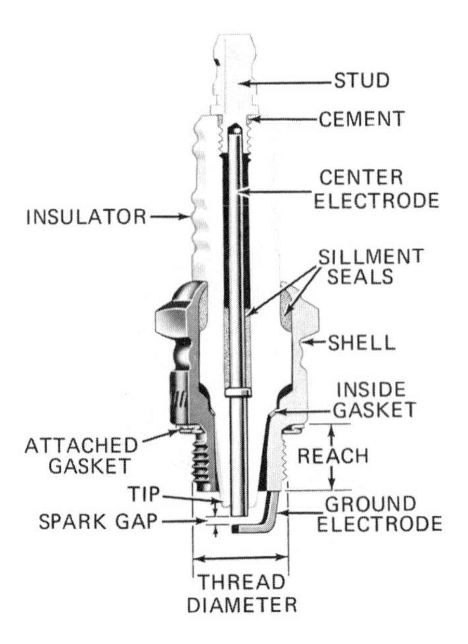

Fig. 4-9. When setting the spark plug gap, always bend the ground (side) electrode.

chamber; it is *not* a difference in the intensity of the spark. The heat range of a spark plug is a measure of the plug's ability to dissipate heat received from the engine combustion chamber to the engine cooling system. The rate of heat transfer is a product of spark plug design.

A spark plug in the correct heat range will operate satisfactorily in the engine

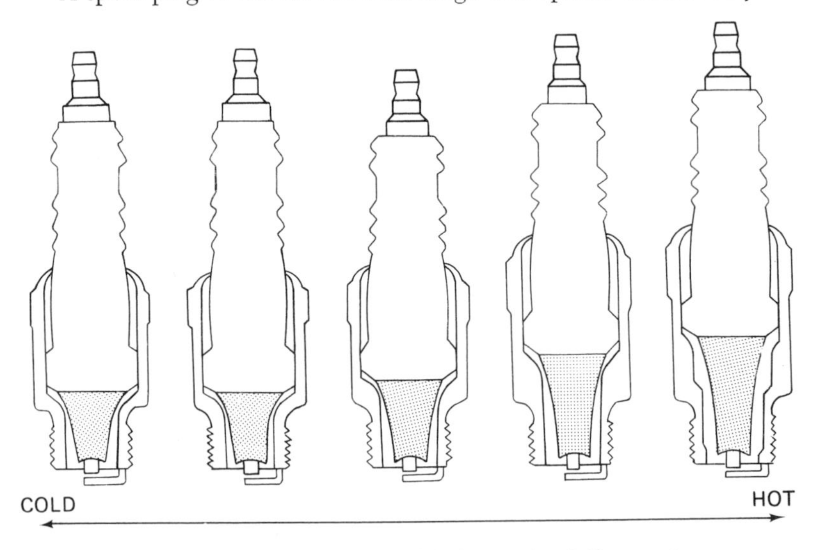

Fig. 4-10. The difference between a hot and a cold plug is the difference in temperature at which each type plug will operate inside a given combustion chamber.

for which it is recommended. This means the top limit of engine load should not cause overheating and preignition and the lowest limit should not encourage fouling from carbon or lead deposits.

When the air-fuel mixture is ignited in the combustion chamber, intense heat is developed and all parts exposed to the heat become hot, especially the firing tip of the spark plug insulator. To prevent the insulator tip from reaching a temperature that will cause preignition, the heat must be dissipated along the heat path to the cooling system (indicated by arrows in Fig. 4-11). Improper torque at installation, or the use of several or no gaskets on one plug will hinder the transfer of heat to the cooling system and cause the plug to operate at a higher temperature than its designed heat range.

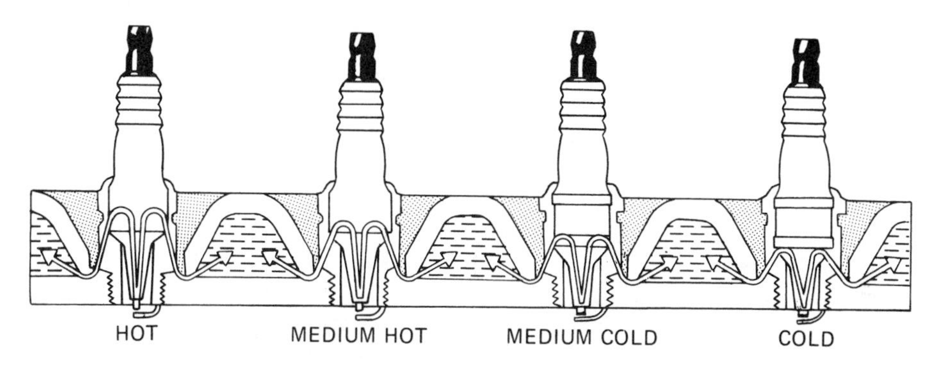

Fig. 4-11. Heat is dissipated along the heat path shown to the cooling system to prevent the insulator tip from reaching a temperature that will cause preignition.

When an engine is operated under varying conditions, it is sometimes recommended to change from one type of heat range plug to another. For example, when using an engine for trailriding, it is advisable to use a hot plug to eliminate the possibility of spark plug fouling. If the engine is run at high speeds for extended periods, a cold plug is advisable. If the engine is to be used in very heavy high-speed conditions, such as competition, then a very cold plug is advisable.

RESISTOR

A *resistor* is a device that impedes the flow of electrons through an electric circuit. Actually, all elements of an electric circuit offer resistance to the flow of electrons including; wire, coils, condensers, diodes, transistors, etc.

Resistors are either metallic or nonmetallic. Metallic resistors are made of wire or ribbon and are commonly called wire wound resistors. Carbon or graphite are used in making nonmetallic resistors. Resistors may be fixed, variable, adjustable, tapped, or automatic resistance control. The fixed resistors are of a fixed value that cannot be changed. Adjustable resistors are adjusted and set to a specific value and are then left alone. Variable, tapped, and automatic resistance control find few, if any, applications in small gasoline engines. The fixed and adjustable resistors are used in electronic-ignition systems (Chapter 5).

BATTERY

A battery (Fig. 4-12) converts chemical energy to electrical energy and provides direct current (DC) whenever a current-consuming device is connected to it. The battery is used in some ignition and starting systems of small gasoline engines. The electrical energy is produced by a chemical reaction between the materials in the plate and the electrolyte, which is a dilute solution of sulphuric acid (H_2SO_4). The amount of sulphuric acid to water in a fully charged battery is 36 percent acid to 64 percent water.

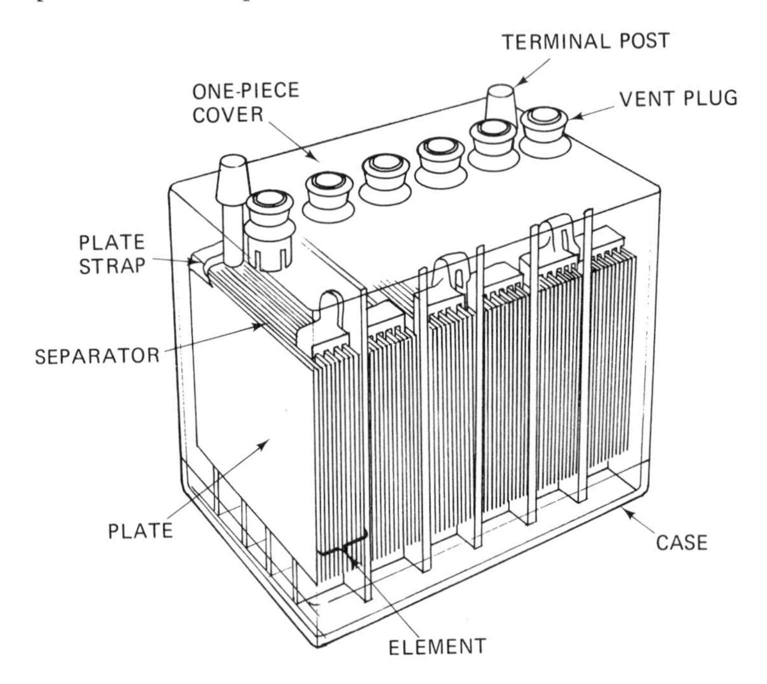

FIG. 4-12. A battery converts chemical energy into electrical energy.

Battery capacity is expressed in *ampere-hours*. The capacity is proportional to the active material on the plates. The ampere-hour rating is based on the number of amperes that a battery can deliver over a specified length of time. For example, a battery that can deliver 12.5 amperes for 8 hours is a 100 ampere-hour battery.

A battery is made up of a number of storage cells grouped together as a single source of direct current. A cell is a single element of a battery and has a voltage of approximately 2 volts. Thus, a 12-volt battery has 6 cells of 2 volts each.

A cell has two types of plates, one positive and the other negative. A plate consists of grids; the grids are flat, rectangular, lattice-like castings with fairly heavy borders and a mesh of horizontal and vertical wires. The grids are pasted with chemically active materials. The positive plates contain lead peroxide (PbO_2) and the negative plates contain sponge lead (Pb). Plate groups are then made by connecting a number of similar plates to a lead casting called a lead plate. The plate groups of opposite polarities are interlaced together; the negative and positive plates are alternated. Each plate in the interlaced groups is kept apart by plastic or rubber separators that permit rapid diffusion of the electrolyte to the plate area. The plate groups with their installed separators are called elements. When the element is all assembled, it is placed in a cell in the battery case.

Electrical energy is produced by the chemical reaction of the active materials of the dissimilar plates and the electrolyte. The amount of available electrical energy depends upon the active area, the weight of the material in the plates, and upon the quantity of sulphuric acid in the electrolyte. After most of the active materials have reacted, the battery cannot produce additional energy until the battery is recharged. Recharging is accomplished by using another source of direct current flowing in a direction opposite to that of the output battery current. Figure 4-13 illustrates the chemical compounds for a fully charged and a fully discharged battery. When a battery can no longer produce the desired voltage, it is discharged. Figure 4-14 illustrates chemical reactions during discharging and charging.

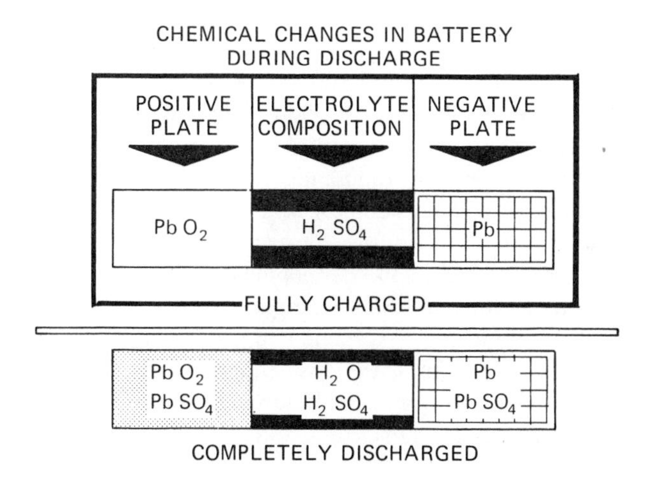

FIG. 4-13. In a fully charged battery, the positive plate is lead oxide and the negative plate is lead. The electrolyte is a solution of sulfuric acid and water.

In normal operation the battery gradually loses water from the cell due to dissociation of the water into hydrogen and oxygen gases which escape to the atmosphere through the vent caps. This rate of evaporation is increased by heat (such as the high temperatures of summer). The water must be replaced or the electrolyte will be overly concentrated, and the tops of the plates will be exposed

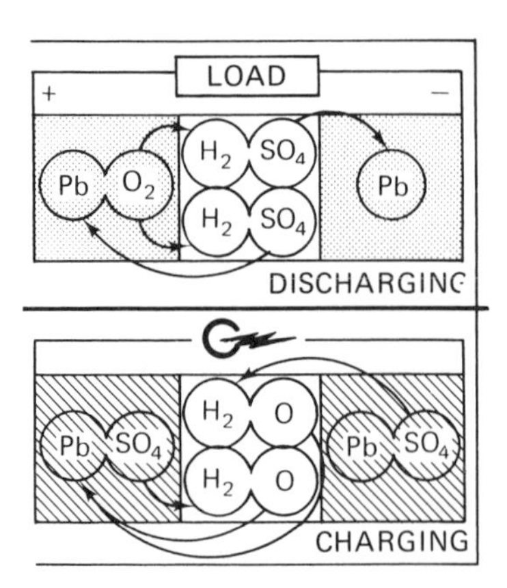

FIG. 4-14. When a battery is charging, the molecules combine to form the original materials of the plates.

to the air and will dry and harden. Lack of water results in a premature failure of the battery. Distilled water or colorless, odorless, drinking water should be added as required to keep the battery filled to the required level (Section 3-19).

As has been stated, sulphuric acid and water make up the electrolyte. Since sulphuric acid is heavier than water, the specific gravity is an indicator of the percentage of sulphuric acid in the electrolyte. Section 3-19 describes how to use a hydrometer to make specific gravity readings to check the condition of a battery. The specific gravity of a fully charged battery is 1.300; the specific gravity of a fully discharged battery is 1.100 and of water is 1.000. When the hydrometer reading of the specific gravity is below 1.225, the battery should be recharged. (In actuality, the hydrometer is numbered from 1,000 to 1,300; the numbers are expressed in terms of thousands rather than in decimals. An electrolyte reading may be 1250, for example.)

New design concepts in the construction of conventional lead acid battery construction has resulted in the development of rugged, rechargable, maintenance-free mini-batteries for applications ranging from those requiring a short burst of power to others requiring a sustained high voltage under load for a substantial length of time. The mini-battery is spillproof and is capable of withstanding severe vibrations and environmental extremes. Therefore the mini-battery has application in the ignition and starting systems of small gasoline engines (as well as for cordless power tools).

Cells are formed in a mini-battery from wall segments that telescope together and form the pocket for sandwiched plate separators and electrolyte. A special porous material immobilizes the electrolyte against the plates, thus insuring full contact with the plate surfaces. The plates are joined, the output leads and connector are attached, and the complete mini-battery is potted in an outer case. A lifetime supply of electrolyte is added during manufacture. No periodic maintenance is required. Some of the mini-batteries also provide a built-in overcharge protection circuit. This feature permits the use of a simple inexpensive charger.

The output voltage and the polarity of the output are identified on the case. The battery output is terminated in a polarized connector so that an incorrect connection is prevented.

The design of these high-energy mini-batteries permits installation and operation in any position without leakage. It also prevents damage from freezing even if the battery is discharged.

TRANSISTORS, DIODES AND RECTIFIERS

A *transistor* is a device used in electronic circuits to amplify a signal (as sound in a transistor radio), or to act as a variable switch (as in a transistorized ignition system of a small gasoline engine). The transistor is made of germanium or silicon with added impurities such as arsenic or antimony. The transistor consists of layers or regions of germanium (or silicon) and arsenic (or antimony) in a sandwich-like manner. The transistor replaces (in most applications) electron tubes and is more reliable, rugged, smaller, and produces less heat.

A transistor is made by joining what are known as N and P zones together. The N zone is a negative zone, because it has an excess of electrons. The P zone is a positive zone, because it has a deficiency of electrons. The zones are made within the transistor into NPN or PNP configurations.

A transistor has three terminals called the emitter, base, and collector. Electrons flow (are emitted) from the emitter, through the transistor and out of the collector terminal. The base provides a means of controlling the amount of electron flow through the transistor from emitter to collector. Thus, the base can turn on or can turn off the flow of electrons through the transistor. If it is necessary to replace a transistor in a transistorized ignition system, ensure that the transistor is correctly installed—the emitter, base and collector are connected to their respective connections.

Diodes are used to rectify, or change, alternating current to direct current. Diodes are made by simply joining a P and N zone. A diode has two terminals called the anode (+) and a cathode (—). Electrons flow through the diode in a direction of negative to positive—cathode to anode. The diodes are marked with a band (or in another manner) to indicate the cathode side. It is important that replacement diodes be placed back into the circuit with the cathode ends connected to the same points that the original diodes were connected to.

A rectifier is a device that changes alternating current into direct current. Alternating current is an electric current that reverses in direction at rapid regular intervals. During each interval, the current rises from zero to maximum, diminishes from maximum to zero, diminishes to a maximum negative value, and then rises back to zero (Fig. 4-15). Alternating currents (AC) are generated by alternators. A direct current is a current that flows in only one direction. Direct current generators actually generate an alternating current, but the *commutator* rectifies it to direct current. Batteries provide a source of direct current.

As previously mentioned, a diode is one device that is used as a rectifier. The diode allows electrons to flow easily in one direction while it opposes the flow of

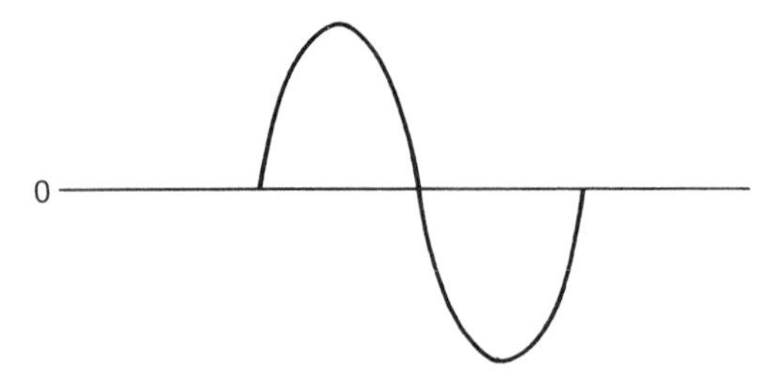

FIG. 4-15. In alternating current, the current rises from zero to maximum and then diminishes from maximum to zero, diminishes to a maximum negative value, and then rises back to zero.

electrons in the opposite direction. When an alternating voltage is applied to a diode, therefore, it allows electrons to flow in one direction (for example, during the positive swing of the alternating current), but opposes the flow in the opposite direction (the negative swing of the alternating current). Thus the diode rectifies or changes the input alternating current into a direct current.

Diode rectifiers are often wired into a bridge rectifier circuit as shown in Figure 4-16 to convert the alternating current output of an AC generator (Section 6-3) to direct current for charging an engine starting battery. When the voltage in the coil is of the polarity shown, the path of electron flow is through diode D1, to ground, to the negative terminal of the storage battery, from the positive terminal of the storage battery, through diode D3, and back to the coil. When the AC voltage alternates to the other half of the cycle, the polarities of the coil will be opposite of that shown in Figure 4-16. Electron flow now proceeds through D2, to ground, to the battery, through D4, and back to the coil. Thus, the bridge rectifier circuit of diodes D1 to D4 has allowed current flow in only one direction through each diode. This is a result of the property of diodes which was previously described that allows electrons to flow in only one direction because of low resistance. The alternating current input has been rectified, or changed, to a direct current to charge the storage battery.

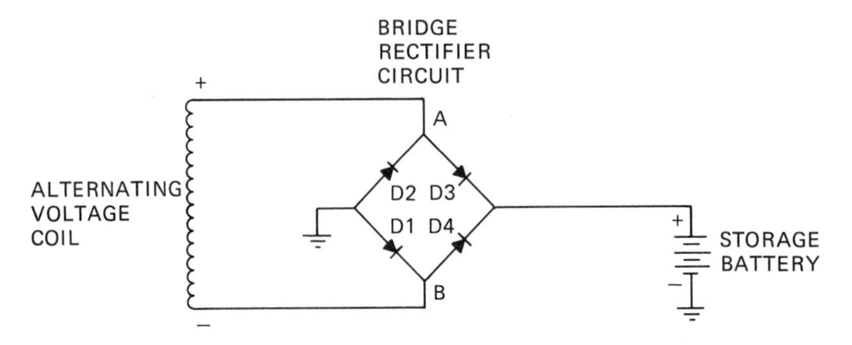

FIG. 4-16. Diodes are used in a bridge rectifier circuit to convert the alternating current output of an ac generator to direct current.

4-6. CONTINUITY

Continuity means that electrons are free to flow continuously from one point to another in an electrical wire or circuit. If there is an "open" in the circuit (such as a broken wire at some point) then there is no continuity; there is a break in the continuity. An ohmmeter (Section 4-7) is a test instrument used to check continuity. Other "jigs" such as a flashlight circuit can likewise be used to check continuity in small gasoline engine circuits.

4-7. TEST EQUIPMENT

There are several pieces of test equipment that are used to check electrical circuits in small gasoline engines. These are an ammeter, a voltmeter, and an ohmmeter. A multimeter combines the functions of all three of these meters into one combined test instrument. An inexpensive multimeter is a great aid in locating trouble in the ignition and starting systems of small gasoline engines.

An *ammeter,* or *ampere* meter, is used to measure current in an electric circuit. The meter must be inserted *into* the circuit to determine current flow because it measures the flow of electrons. A voltmeter is used to measure electrical voltage or pressure. The two leads of a voltmeter are placed *across* a particular circuit (such as across the terminals of a battery) to measure the electrical voltage. The ohmmeter measures resistance and is used to make continuity tests. Electric power is always turned off when the ohmeter is used.

4-8. SUMMARY

This chapter has built the foundation in electricity and magnetism that is needed for you to understand the operation of the ignition and starting systems discussed in Chapters 5 and 6.

5

Ignition Systems

Now that you understand the internal operations of the engine—the reciprocating motion of the piston, the movement of the connecting rod, the rotary motion of the crankshaft, the timing and movement of the valves, and the camshaft—it is necessary to learn how the spark is generated during each cycle to cause the combustion of the air-fuel mixture in the combustion chamber. Chapter 4 presented the fundamentals of an ignition system—magnetism and electricity; it also described many of the components used in ignition systems—magnets, condensers, coils, breaker points, spark plugs, high-tension leads, resistors, batteries, transistors, diodes, and rectifiers. This chapter ties the fundamentals of electricity and magnetism and the various electrical components into the various types of ignition systems.

The function of an ignition system is to ignite the air-fuel mixture in the combustion chamber to start the power stroke. A typical ignition system consists of a source of power (magneto or battery); a timer (to initiate the spark for ignition at the proper time during the cycle of operation); breaker points (to interrupt the current to the ignition coil); a condenser (to prevent arcing across the points); an induction coil (to generate an electromagnetic field); a spark plug for each cylinder (to provide spark between its electrodes); and a distributor cap (in multicylinder engines only to distribute the high voltage to the spark plug). Some ignition systems use a solid state module to replace some of the mechanical parts.

There are three types of ignition systems used in small gasoline engines: the magneto-ignition system, the solid-state ignition system (also called the capacitor discharge (CD) system or transistorized ignition) and the battery ignition system. Of the three, the magneto-ignition system is the most widely used in small gasoline engines. There is no need for a battery in many machines powered by small gasoline engines because there are no electric starters, lights, radios, or other electrical accessories; therefore, the self-contained magneto-system is the most practical and economical.

The new solid-state ignition system has some advantages over the magneto-ignition system. The major advantage is that solid-state electronic components are used instead of the mechanical breaker points, breaker cam, and a spark advance assembly. With the elimination of these mechanical parts, the solid state system is more reliable.

Some small engine-powered machines do use battery-ignition systems. As you will learn in this chapter, a battery is used in place of a magnetic system; thus electrochemical energy rather than magnetic energy is used to initiate the spark for air-fuel combustion. The battery-ignition system is usually used when the machine has lights or other electrical components on the machine.

5-1. MAGNETO-IGNITION SYSTEMS

The magneto-ignition system is a self-contained electrical spark generating system used solely for ignition purposes. The system generates a high voltage from 10,000 to 20,000 volts for ignition of the air-fuel mixture in the combustion chamber. This high voltage is generated by the interaction of a permanent magnet and a coil and the principle that when a magnetic field passes a coil, a current is generated (Section 4-4). The generated voltage is great enough to jump across the spark plug electrode air gap and ignite the mixture in the combustion chamber.

A magneto-ignition system (Fig. 5-1) consists of a *primary* and a *secondary* circuit. The primary circuit consists of breaker points that are actuated by a rotating breaker cam, a condenser, the primary winding of an ignition coil, and a shorting switch. The secondary circuit consists of the secondary winding of the ignition coil, the spark plug high-tension lead, and the spark plug.

Permanent magnets are built into the flywheel that revolves around or near the other ignition system components. The magnets revolve very close to the ignition coil so that on each revolution of the magnets, a magnetic field crosses the primary and secondary coils. This magnetic field induces a current (hence a low voltage) into the primary winding of the coil. This action takes place while the breaker points remain closed (Fig. 5-1) because of the cam position. When the cam lobe, or high point, revolves and opens the breaker points (Fig. 5-2), there is a rapid collapse of the magnetic field that was caused by the current in the primary coil of the ignition coil. This rapid collapse of the magnetic field induces a voltage into the secondary windings of the ignition coil. Since there are many turns of wire in the secondary coil, the voltage is added in series until a very high voltage is generated. This voltage is sent through the high-tension lead to the spark plug and is of suf-

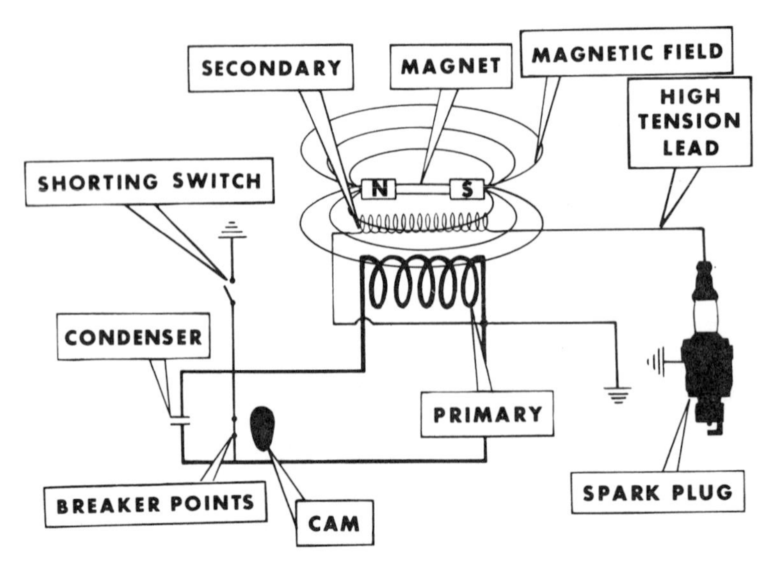

FIG. 5-1. The primary circuit of the magneto-ignition system consists of cam actuated breaker points, condenser, the primary winding of the ignition coil, and a shorting switch; the secondary of the ignition coil, the high-tension lead, and the spark plug make up the secondary circuit.

ficient power to cause a spark to arc from the center electrode to the ground electrode of the spark plug. When the points are open, the condenser acts to store the electricity that is in the primary circuit. The condenser thus prevents a voltage from arcing across the breaker points when they are opened by the cam. This prevents burning and pitting of the breaker points.

The permanent magnets that are cast into the flywheel are either ceramic or alnico. Both are powerful magnets that can quickly generate a magnetic field. An alnico magnet is made of aluminum, nickel and cobalt. The breaker points, condenser, ignition coil, and spark plug have been discussed in Section 4-5.

The locations of most of the magneto-ignition system parts are shown in a typical example in Figure 5-3. The ignition coil with its laminated core is placed in very close proximity to the flywheel (not shown). There is an air gap of approximately 0.010 inches between the flywheel and the ignition coil. The condenser, breaker points, and actuating cam are located behind or in another location close to the ignition coil and flywheel; in this example, these parts are located on a piece called the armature plate. The breaker points are opened by either a cam or a push rod. On two-cycle engines, the cam is usually on the crankshaft and thus, for one revolution of the crankshaft, the breaker points will be opened by the lobe of the cam. On most four-cycle engines, the cam is located on the camshaft, which has one revolution for each two revolutions of the crankshaft. Therefore, the spark to the spark plug is initiated only on the power stroke and not on any other stroke because the breaker points are not opened. There is a voltage of about 100 volts generated in the primary coil, which induces a voltage of 10,000 to 20,000 volts in the secondary coil at the time of the opening of the breaker points.

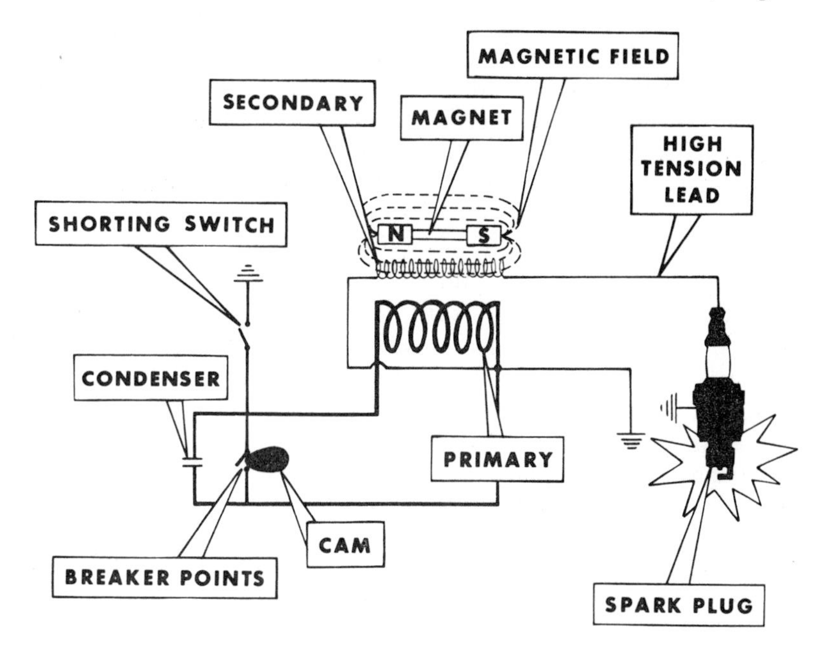

FIG. 5-2. When the breaker points are opened, there is a rapid collapse of the magnetic field which induces a voltage into the secondary windings of the ignition coil.

Switches are used in the magneto-ignition system to shut the engine off. One type of circuit (Fig. 5-1) has a switch which, when closed, connects the breaker points and condenser to ground. This short circuits the primary ignition circuit and stops the engine. On other engines, a blade of metal is located next to the spark plug; the metal blade switch is pressed against the spark plug to short-circuit the secondary circuit to ground, stopping the engine.

Sometimes an *external* magneto-system is used. The external magneto consists

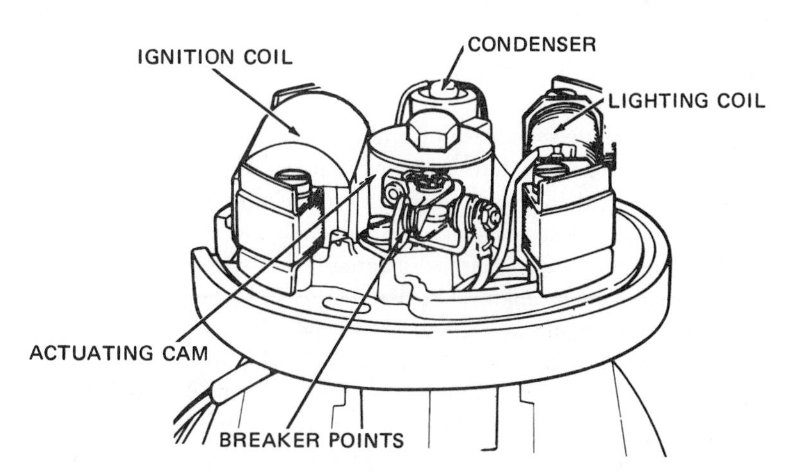

FIG. 5-3. The magneto-system parts are located in close proximity to the flywheel. The flywheel magnets also generate an ac voltage in the lighting coil for electric light power.

of a rotor driven through an impulse coupling. The rotor spins, developing a magnetic field within a laminated core. The primary and secondary coils are wound on the laminated core. The rotor has a north and a south side and when it revolves, it changes magnetic field direction in the core twice for each rotation. The magnetic field builds up in one direction, collapses slowly, and builds up in the opposite direction in a sinesoidal pattern (the pattern is the same as for the alternating current shown in Fig. 4-15). With the breaker points closed, the magnetic field is constantly increasing and decreasing in the sinesoidal pattern. When the ignition spark is to occur, the magnetic field is at its maximum strength and the breaker points are opened. The magnetic field collapses rapidly inducing a high voltage into the secondary coil and circuit to the spark plug. A condenser is used with an external magneto, as it is with the self-contained magneto-ignition system, to prevent the current from arcing across the breaker points.

Impulse coupling provides two functions to improve engine starting; it retards the ignition for better starting and it produces a higher voltage and a stronger spark at initial starting. When the ignition is cranked over by hand, the voltage to the spark plug is lower because the voltages produced by the magneto depend upon the speed of the magnetic lines cutting the primary coil. A stronger spark is generated if the magnetic lines are moved more rapidly through the iron core when the primary current is interrupted. The lines move faster and hence generate a higher voltage in the secondary. Impulse coupling causes the magnetic lines to move faster by use of a delayed spring action in conjunction with retractable pawls. Spring-tension builds up during a part of the revolution of coupling and then releases to spin the rotor quickly ahead. The rotor turns part way and then stops until spring-tension builds up again and is then flipped forward. After the engine starts, centrifugal force unlocks the impulse coupling making it inoperative. At the higher speeds, the rotor revolves in time with the engine.

5-2. SOLID-STATE IGNITION SYSTEMS

A solid-state ignition system is a broad term applied to any ignition system which uses electronic devices such as diodes, transistors, silicon-controlled rectifiers, or other semiconductors in place of one or more standard ignition components. Electronic devices are small, have no moving parts, no mechanical adjustments, are not subject to wear, deliver uniform performance throughout component life, and are hermetically sealed against dust, dirt, oil, fuel, and moisture.

The solid-state, or capacitor discharge (CD) system, is breakerless; an electronic device replaces the mechanical breaker points and related parts including the breaker cam, the spark advance assembly, etc. In the solid-state system, the permanent magnets in the flywheel are the only moving parts. Figures 5-4 through 5-9 illustrate the theory of operation of the solid-state ignition system. As the flywheel magnets (Fig. 5-4) pass the solid-state module laminations, a low alternating current is induced into the charge coil. The alternating current passes through a rectifier (Fig. 5-5) which converts the alternating current to a direct current. The direct current travels to the capacitor (condenser) (Fig. 5-6) where it is stored.

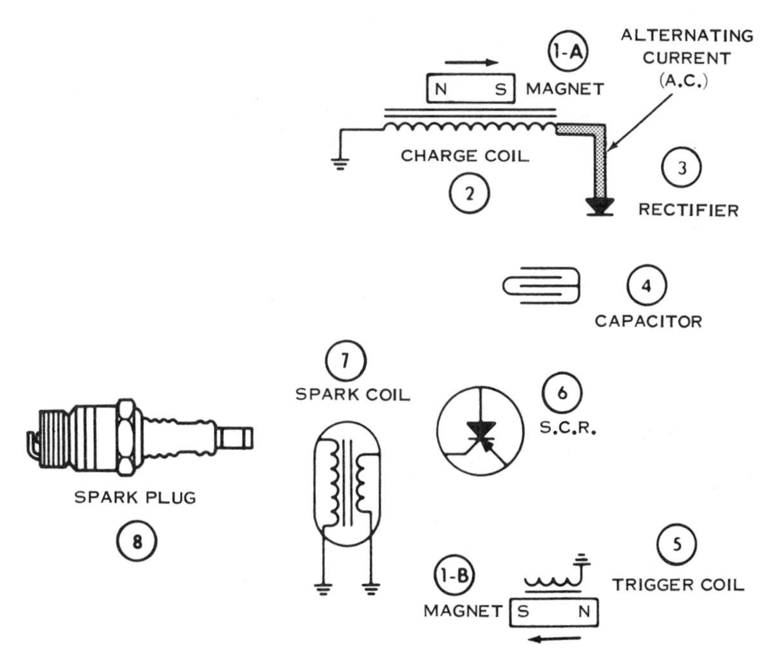

FIG. 5-4. In a solid-state ignition system, the flywheel magnets generate an ac voltage in the charge coil.

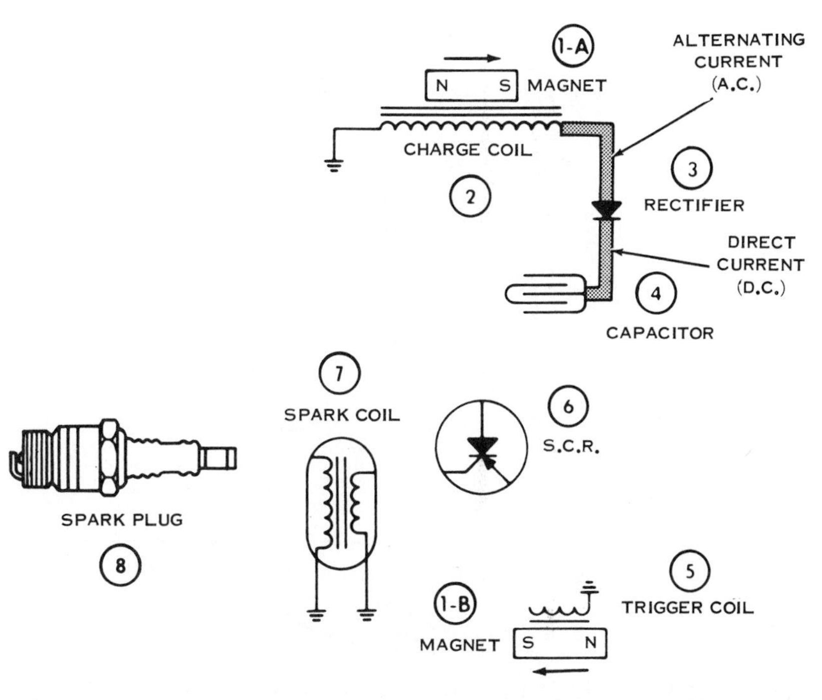

FIG. 5-5. The rectifier in the solid-state ignition system converts the ac to dc.

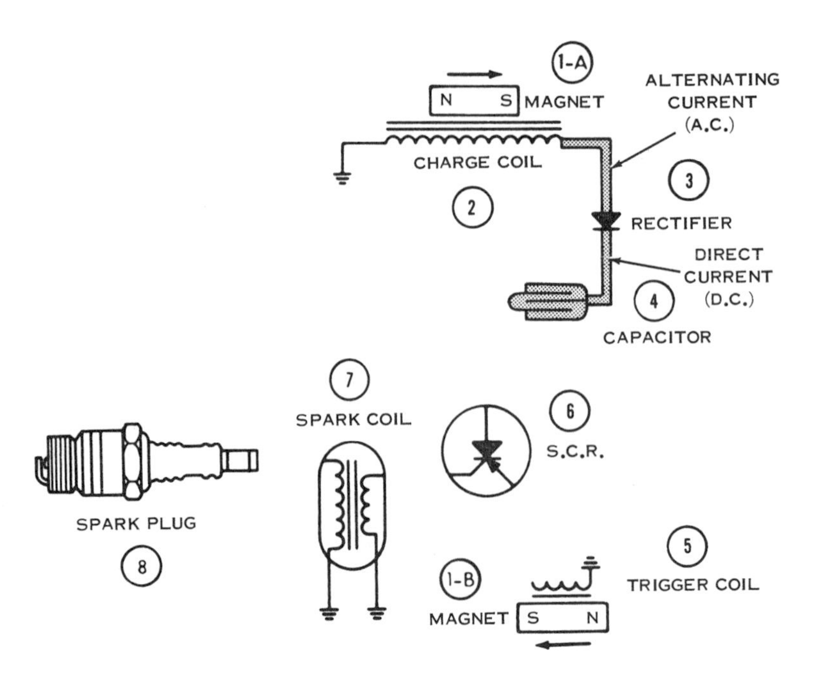

FIG. 5-6. The dc is stored in the capacitor (condenser).

The flywheel magnets rotate (Fig. 5-7) approximately 351 degrees until they pass laminations inducing a small electrical charge into the trigger coil. At starting

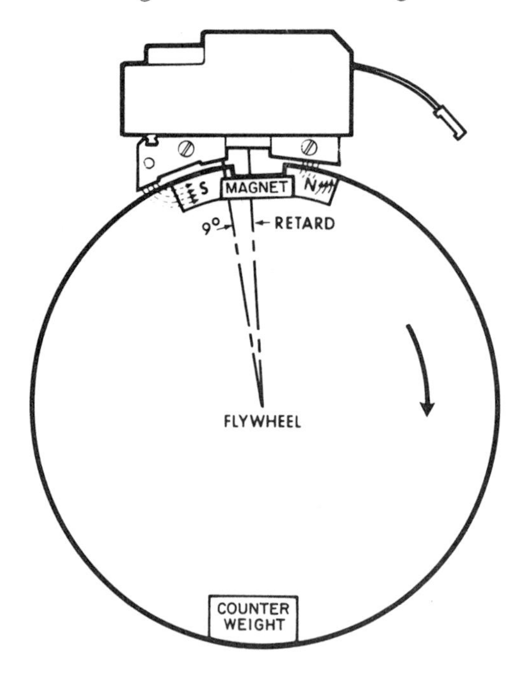

FIG. 5-7. The flywheel magnets generate a voltage in the trigger coil to fire the silicone controlled rectifier (SCR) at a retarded position (9°) for easy starting.

speeds, this charge at the trigger coil has sufficient magnitude to turn on the silicon-controlled rectifier which acts as a switch at a retarded position for easy starting. This is illustrated in Figure 5-7 as the 9-degree or retard-firing position. When the engine speeds up and reaches approximately 800 rev/min (Fig. 5-8), advanced

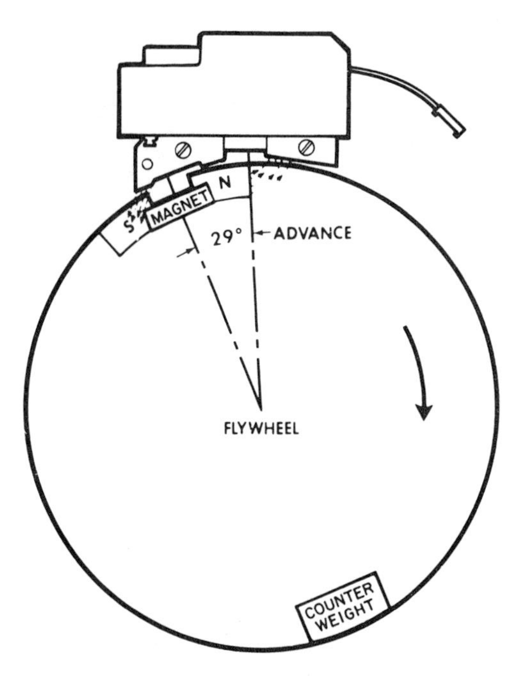

FIG. 5-8. When the engine speeds up, the trigger coil fires the SCR at the advanced firing position (29°).

firing begins. The flywheel magnets travel approximately 331 degrees at which time enough voltage is induced into the trigger coil to fire the silicon-controlled rectifier solid-state switch. This is the advanced firing position (retarded and advanced starting are further discussed in Section 5-4).

When the silicon-controlled rectifier is triggered (Fig. 5-9), approximately 300 volts stored in the capacitor pass through the silicon-controlled rectifier to the spark coil. The spark coil steps up the voltage instantaneously to approximately 30,000 volts. This voltage is sent to the spark plug where it is of sufficient voltage to jump the air gap between the spark plug center and ground electrodes to ignite the air-fuel mixture in the combustion chamber.

The solid-state ignition system consists of only three parts: the solid-state package, the spark plug, and the magnets in the flywheel. The solid-state package houses the charge coil, trigger coil, spark coil, rectifier, capacitor, and the silicon-controlled rectifier. The entire package is hermetically sealed so that moisture does not affect it. Thus, the only moving parts are the magnets in the flywheel. Maintenance requires only the determination of whether or not a spark is being produced. If there

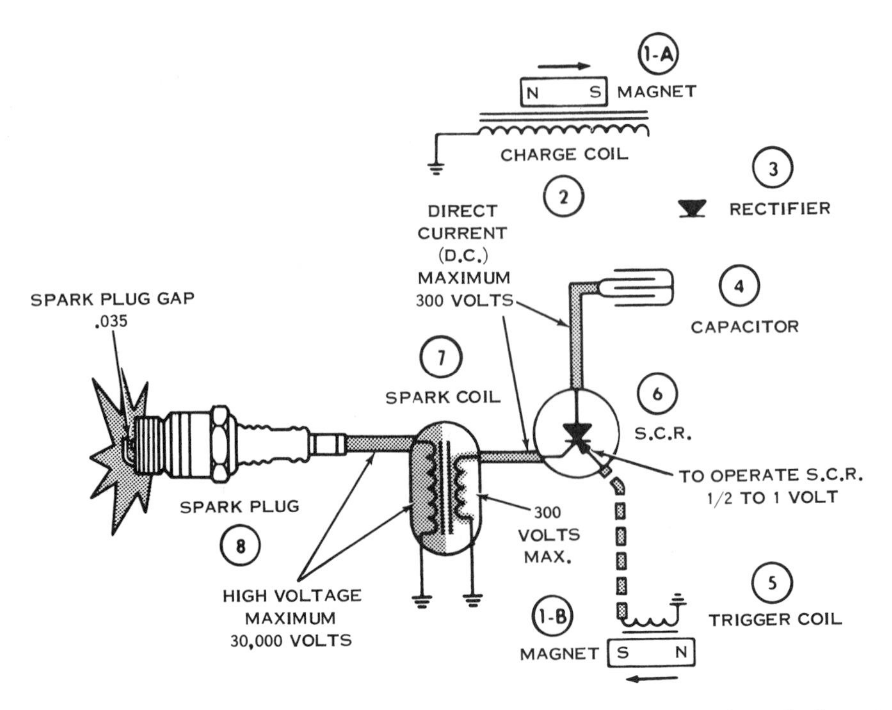

FIG. 5-9. When the SCR fires, the voltage stored in the capacitor passes through the primary of the spark coil. The high voltage induced into the secondary is discharged across the spark plug electrodes.

is an ignition switch, it is the most vulnerable part of the solid-state ignition system.

Instead of using magnets to induce voltage into the charge coil, a solid-state ignition system may instead use a battery to generate the current in the charge coil. A projection on the flywheel rotates past the trigger coil and halts the flow of current causing the magnetic field in the primary coil to collapse. This induces a high voltage in the secondary coil to fire the spark plug.

5-3. BATTERY-IGNITION SYSTEMS

The battery-ignition system (Fig. 5-10) is similar to the magneto-ignition system. In the battery-ignition system, a 12- or a 6-volt battery provides a current through the primary of the ignition coil; this replaces the magnets that induce the voltage into the primary coil in the magneto-ignition system. The battery-ignition system includes a battery, an off-on switch, an ignition coil, breaker points, an ignition cam, a condenser and a spark plug.

In operation, current flows through the primary of the ignition coil when the ignition switch is on and the breaker points are closed. At the time of ignition, a cam opens the breaker points and a voltage of approximately 100 volts is developed through the self-inductance of the primary winding of the coil. This collapsing

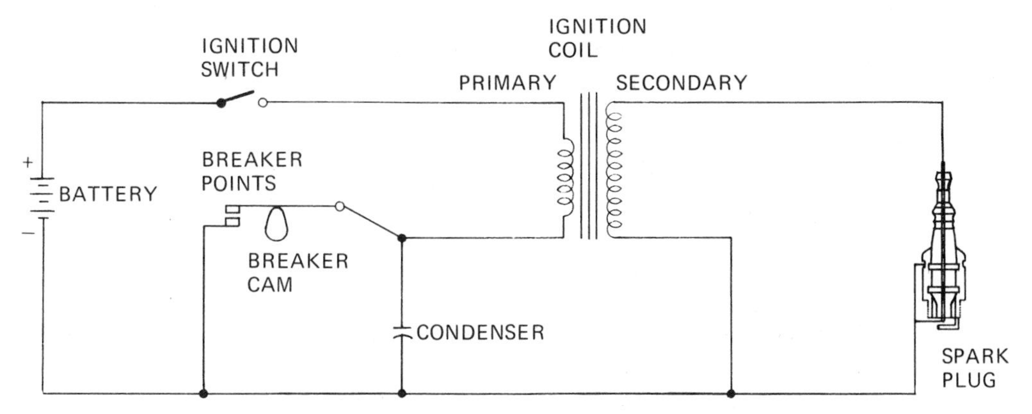

FIG. 5-10. A battery-ignition system is similar to a magneto-ignition system except that the current to the ignition coil primary is generated by a battery instead of magnets.

voltage induces a voltage into the secondary coil of approximately 20,000 volts which is sufficient to cause a spark to jump the air gap between the electrodes of the spark plug igniting the air-fuel mixture in the combustion chamber. The condenser protects the breaker points by providing a momentary place for the current in the primary to flow as the points begin to open; thus, the condenser protects the points from burning from a current arcing across them. The condenser also brings the current flow in the primary to a quick end. This hastens the collapse of the magnetic field in the primary. The battery-ignition system is turned off with the switch which breaks the battery current passing through the primary circuit. No magnetic field is therefore built up.

5-4. TIMING ADVANCE MECHANISMS

It is important for correct operation and maximum engine efficiency that the ignition spark be generated at the proper time during the engine cycle. If the spark occurs too early before the piston approaches top dead center, the combustion pressure will oppose the flywheel momentum and in extreme cases can cause the piston to rotate in the wrong direction causing a *kickback*. Similarly, if the spark occurs after top dead center, the full compression pressure will not be used to its fullest advantage.

You can visualize that as the piston moves up and down in the cylinder when the engine is starting and the crankshaft is turning at a slow speed, the spark should occur just prior to top dead center. However, if the engine is operating at a faster speed, then the momentum of the flywheel is driving the piston rapidly; the ignition spark must take place a number of degrees earlier before TDC so that combustion occurs and drives the piston down with maximum force. The timing advance mechanism provides a means of changing the ignition time for combustion. The ignition time is designated by the number of degrees of crankshaft rotation before top dead center.

Section 5-2 described the advance system in the solid-state ignition system. In that system, when the engine is running at starting speeds, the flywheel magnets induce a small electric charge into a trigger coil which fires the silicon-controlled rectifier that discharges the capacitor. This produces a current in the primary of the spark coil at a retarded firing position of 9 degrees. When the engine reaches a higher speed, advanced firing (that is, a greater number of degrees prior to TDC) occurs and a voltage is induced into the trigger coil at 29 degrees before top dead center. Thus in a solid-state ignition system, the timing advance is controlled electronically by the voltage induced into the trigger coil.

The timing advance mechanism in a magneto-ignition system operates mechanically. In addition to the permanent magnets in the flywheel, a magnet ring (Fig. 5-11) also includes a timing advance retard mechanism. This mechanism consists of a centrifugal flyweight lever, a spring, an ignition advance cam, and a retainer. The advance mechanism shifts from the retarded position into the fully advanced position (higher speed operation) with no intermediate positions. At starting speeds, the centrifugal flyweight lever is spring-held to the retard position to fire the spark plug when the piston is almost at top dead center. After the engine speed increases to a certain level, centrifugal force overcomes the force of the return spring allowing the centrifugal flyweight lever to shift into the advanced timing position. This in turn shifts the ignition cam. Hence the lobe of the cam is introduced earlier causing the breaker points to open earlier before the piston reaches TDC so that the full force of combustion is reached just after the piston starts downward on the

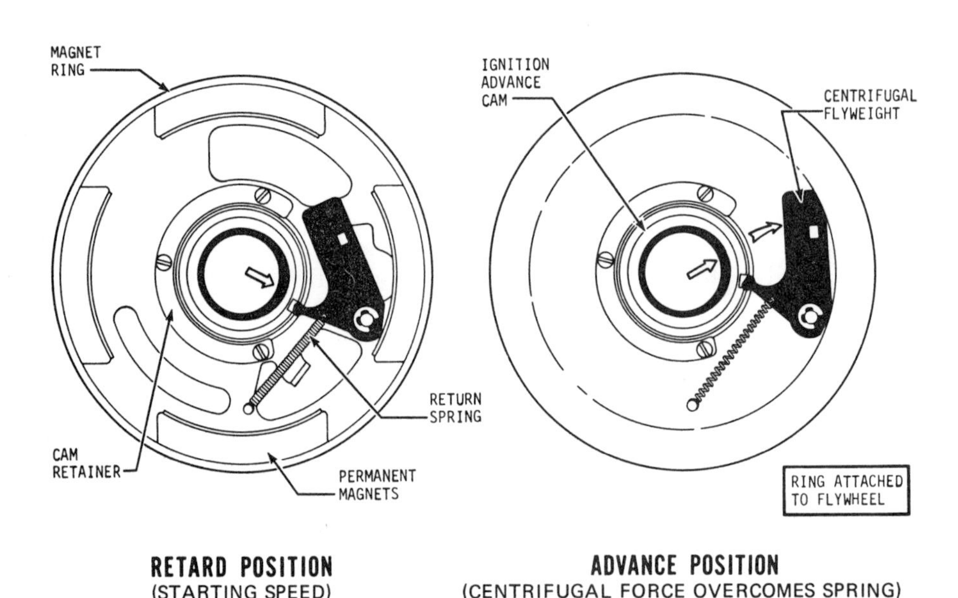

Fig. 5-11. Centrifugal force overcomes the spring force at higher speeds. The final result is that the breaker points open earlier before TDC so that the full force of combustion is reached just after the piston starts downward on the power stroke.

power stroke. The retard position is approximately 6 degrees of crankshaft rotation before the piston reaches TDC. When the engine speed has increased to about 1000 rev/min, the advanced position comes into operation causing the spark to occur at about 26 degrees of crankshaft rotation before TDC. At the higher speeds, the spark occurs earlier giving the mixture time to burn and to deliver its power to the piston. The maximum advance varies widely in different engines and may be as great as 45 degrees on very high-speed engines.

The preceding description has been automatic spark advance. Manual spark advance is accomplished by loosening the breaker point assembly and rotating it slightly to an advanced position. The assembly is secured in that new position.

5-5. MULTICYLINDER ENGINES

In addition to one-cylinder engines, there are also small gasoline engines having two, three, or four cylinders. A system is needed to distribute the electrical ignition spark to the spark plug of the correct cylinder at the correct time. This system is called the distribution system. The distribution system consists of a distributor that is usually housed in an assembly with a rotor, a distributor cap, driving cam, breaker points, and a condenser.

Instead of the secondary of the ignition coil being connected to the spark plug, the secondary coil is connected to the rotor in the distributor. The rotor revolves in a circle as driven by a driving gear. A spring-like contact on the end of the rotor contacts metal contacts on the distributor cap. Through wires from the cap, the electrical ignition voltage is sent to the proper spark plug at the proper time. The shape of the driving cam is dependent upon the number of cylinders in the engine. The cam spins and the cam lobes cause the breaker points to open at the correct time to send the high voltage through the rotor to the correct spark plug.

Another method of multicylinder ignition is the use of a separate magneto-system for each cylinder. This is commonly used in two-cylinder outboard engines.

5-6. IGNITION SYSTEM MAINTENANCE

Ignition system maintenance consists of periodic maintenance, troubleshooting, repairing, and adjusting. The periodic maintenance consists of checking, cleaning, and regapping spark plugs (Section 3-17) and checking the battery (Section 3-19) of battery-ignition systems.

The initial troubleshooting procedure is to test the overall system to determine whether there is a problem with the ignition system (Section 9-3). Once the trouble has been isolated to the ignition system, then procedures for checking each of the components are given in this chapter. Usually, there are several alternate methods of checking each of the parts of the ignition system. One method is simply to substitute the suspected bad part with a known good part. A second method is to check the parts with an ohmmeter. Finally there are special testers manufactured for the specific purpose of testing breaker points, condensers, and ignition coils. Three manufacturers of such testers are Merc-o-tronic, Graham, and Stevens. These testers (called ignition analyzers) check all functions of each of the parts of the ignition system. The testers check the parts without needless replacement of parts when they are only suspected as being bad. However, to test most of the parts, it is necessary to remove the parts from the engine. It is recommended that small shop owners look into and purchase one of these testers. However the homeowner or the occasional repairman can either substitute parts or use an ohmmeter (or multimeter) both inexpensively and adequately. The procedures for testing the ignition system parts are generally the same for each of the manufactured testers. General testing procedures using the testers are given in this text, but the repairman should thoroughly read and understand the instructions provided by the tester manufacturer.

CAUTION

All tests made using the Merc-o-tronic, Graham, Stevens, or similar ignition analyzer testers should be made on a wooden workbench. High voltages are available from the leads and probes coming from the testers. Contact of these leads with a metal bench could cause short circuits.

To avoid electrical shock from the testers, turn all power off when connecting the tester to the component being tested. Do not come into contact with the metal sections of the probes or leads.

Do not let the metal tips of the probes of the ignition analyzer touch each other.

Ignition system failures can be caused by one or more of the following:

1. Spark plug is worn or bad.
2. Spark plug gap is improperly set.
3. Terminal is missing from the spark plug high-tension lead.
4. Electrical lead is pulled out of the ignition coil or ignition coil is open.
5. Insulation on the high-tension lead wire or on other wires and terminals is cracked (the spark short circuits to ground).
6. Condenser is poor or bad or there is a bad ground connection.
7. Breaker points are burned or pitted.
8. Breaker point gap is incorrectly set.
9. Breaker point fiber is worn.
10. Electrical connections are poor.
11. Air gap between the flywheel magnets and the coil laminations are set incorrectly.
12. Flywheel magnets are weak or battery is dead.
13. Spark advance assembly is damaged or is installed incorrectly.

14. Solid-state module is faulty.
15. Ignition system timing is incorrect.

When troubleshooting magneto problems, make sure that the shutoff ground wire terminals do not touch the armature plates; if this happens, the ignition system will be permanently grounded. Also the shutoff switch may become inoperative if dirt and grime collect between the shutoff switch screw and the shutoff blade. The insulation on all wires should be examined.

There are no troubleshooting procedures for the solid-state module of a solid-state ignition system. The ignition system is tested very simply by using a test spark plug or a standard spark plug grounded to the engine to see if the system is producing a spark (Section 9-3). If there is no spark, the problem is either in the module, the ignition switch, the switch lead, the flywheel magnets, or an incorrect air gap. The ignition switch is the most vulnerable part of the solid-state ignition system because it can be affected by moisture; the hermetically sealed solid-state module is not affected by moisture or dust.

Once the cause of the trouble has been found, a repair can be made. Repairs on the ignition system are mostly replacements of the bad components; you cannot repair breaker points, condensers, spark plugs (except for cleaning and regapping), coils, broken wires, solid- state modules or components, or demagnetized magnets.

There are four adjustments that must be correctly made for efficient ignition system operation. These adjustments are spark plug gap (Section 3-17), breaker point gap (Section 5-9), air gap setting between the magnets on the flywheel and the soft core laminations on the ignition coil or solid-state module (Section 5-11), and the setting of ignition timing (Section 5-13).

5-7. SPARK PLUGS

Spark plugs are not only the easiest component to test in the ignition system, but they are also the most revealing component and give the maintenance technician a clue as to the malfunction in the engine. The spark plug is also one of the most probable causes of engine malfunctions. By observing the spark plug tip, the technician can determine the probable cause of engine failure. Table 5-1 lists normal spark plug conditions and five conditions that indicate causes of engine failure. Each spark plug description includes the condition of the plug, an analysis, and the maintenance procedures for correcting the condition.

Table 5-1
SPARK PLUG CONDITIONS, ANALYSIS, AND MAINTENANCE

NORMAL (Fig. 5-12)

Condition
A. Few combustion deposits present on plug
B. Electrodes not burned or eroded

Table 5-1 (Continued)

C. Insulator tip color: brown to light tan
D. Insulator dry (providing engine was not excessively choked prior to plug removal)

Analysis	*Maintenance*
A. Ignition and carburetor in good condition	A. Clean and reinstall plug or perform step B
B. Plug is correct heat range	B. Install a new plug of the same heat range

FIG. 5-12. A spark plug tip under normal conditions has few combustion deposits, the electrodes are not burned or eroded, the insulator tip color is brown to light tan, and the insulator is dry.

OXIDE FOULING (Fig. 5-13)

Condition

A. Electrodes not worn (but may be covered with deposits)
B. Insulator tip choked, splattered, or "peppered" with ash-like deposits
C. Deposits may be thrown against and adhere to side electrode in extreme cases
D. Deposits may wedge between electrodes, momentarily or permanently shorting out the plug

Analysis	*Maintenance*
A. Excessive combustion chamber deposits	A. Clean (Section 2-21)
B. Clogged exhaust ports or muffler	B. Clean (Section 3-8)
C. Use of nonrecommended oils	C. Use correct oil
D. Wrong fuel mix	D. Use correct mix
	E. Replace plug

FIG. 5-13. This spark plug tip exhibits oxide fouling.

Table 5-1 (Continued)

WET FOULING (Fig. 5-14)

Condition

A. Insulator tip black
B. Damp oily film over firing end
C. Carbon layer over entire nose
D. Electrodes not worn

Analysis	*Maintenance*
A. Idle speed too low	A. Adjust idle (Section 7-8)
B. Idle adjustment too rich	B. Adjust idle (Section 7-8)
C. Weak ignition outputs	C. Check ignition (Section 5-6)
D. Air filter badly clogged	D. Clean air filter (Section 3-9)
E. Wrong fuel mix (too much oil, wrong type oil)	E. Use correct mix and proper oil
F. Excessive idling (not shutting off engine)	F. Do not idle engine except for short periods
G. Plug too cold for type of work	G. Replace with hotter plug (Sec. 4-5)
H. High-speed carburetor and adjustment not set with engine under full load	H. Load engine and reset high-speed carburetor adjustment (Section 7-8)

FIG. 5-14. This plug tip exhibits wet fouling. Note the differences between an oxide-fouled tip and a wet-fouled tip.

Table 5-1 (Continued)

OVERHEATED (Fig. 5-15)

Condition	
A. Electrodes burned	
B. Insulator tip color: light gray or chalk white	
Analysis	*Maintenance*
A. Carbon clogged exhaust ports or muffler	A. Clean exhaust ports or muffler (Section 3-8)
B. Dirty or clogged cylinder fins	B. Clean cylinder fins (Section 3-8)
C. Lean carburetor setting	C. Readjust carburetor (Section 7-8)
D. Overloaded engine	D. Reduce overload
E. Spark plug too hot	E. Replace with a cooler plug (Section 4-5)

FIG. 5-15. An overheated plug has burned electrodes; the insulator tip is light gray or chalky white.

WORN-OUT (Fig. 5-16)

Condition	
Electrodes worn away by corrosive gases and sparks	
Analysis	*Maintenance*
Plugs require more voltage to fire	Replace with plugs of same heat range

FIG. 5-16. Electrodes eventually wear away from corrosive gases and sparks.

Table 5-1 (Continued)

GAP BRIDGING (Fig. 5-17)

Condition

Gap between electrodes is bridged with combustion particles

Analysis	*Maintenance*
A. Excessive carbon in cylinder	A. Clean cylinder (Sections 2-21, 2-24)
B. Use of nonrecommended oil and/or fuels	B. Use proper oil and fuel
C. Improper ratio of oil to fuel (two-cycle only)	C. Mix ratio properly
D. Clogged exhaust ports (two-cycle only)	D. Clean exhaust ports (Section 3-8)
E. Sudden high-speed operation after excessive idling	E. Clean plug and regap, or replace and regap with same heat range plug

FIG. 5-17. A spark plug suffering from gap bridging has a combustion particle between the electrodes.

Spark plugs are gapped by bending the side (ground) electrode. Do not bend the center electrode as this may break the ceramic insulator. Use a round wire spark gap gauge to check for the proper gap recommended by the manufacturer (Section 3-17). The gap is correctly set when the wire gauge can be pulled through the gap with only a slight amount of friction. Ensure that the gap is properly set as specified by the manufacturer.

5-8. FLYWHEEL REMOVAL/REPLACEMENT

The flywheel must be removed on most engines to gain access to the ignition system components. Prior to removing the flywheel, you must remove any manual starting mechanism and shroud. If the engine has an electric starter generator, the flywheel will have a stub shaft on which the drive pulley mounts. The stub shaft also has to be removed. Some engines have an external breaker box and thus it is not necessary to remove the flywheel for access to the ignition components.

Flywheels are removed by different methods on different engines; therefore the following procedures are general. If a *flywheel puller* is available for the particular engine being worked on, obtain and use it. Remove the flywheel nut and the flywheel as follows:

1. To remove the flywheel nut, some means must be used to prevent the flywheel from rotating. There are three methods:
   A. A *piston stop* can be screwed into the spark plug hole in place of the plug. The crankshaft is rotated until the piston rises in the cylinder and meets the piston stop. The stop prevents further movement of the piston, connecting rod, and crankshaft as the flywheel nut is removed.
   B. A *chain wrench* (Appendix A) or a strap wrench can be used to hold the flywheel as the flywheel nut is removed. The chain wrench is a special tool available from some manufacturers for their engines.
   C. The engine can be positioned on the bench with a block of soft wood under one of the flywheel air vanes such that when the flywheel nut is loosened, the flywheel cannot turn. Be careful not to bend the flywheel air vanes.
2. Using a wrench of the proper size, remove the flywheel nut.
3. Use a flywheel puller (Fig. 5-18) to remove the flywheel. Some pullers have two or three bolts that are screwed into holes in the flywheel. A

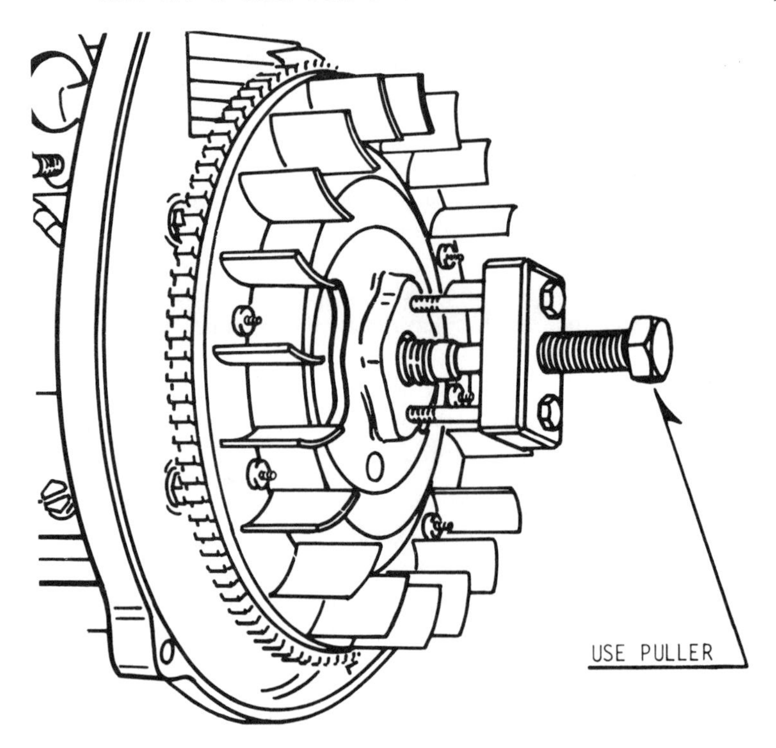

Fig. 5-18. A flywheel puller is used to pull the flywheel from the tapered end of the crankshaft.

center bolt or sometimes additional nuts on the threaded studs are tightened down to pry the flywheel off. If a flywheel puller is not available, place your fingers *under* the flywheel and apply an *upward* pressure. Place a nut (don't use the flywheel nut) on the top of the crankshaft and screw it down until it is *nearly* flat with the end of the crankshaft. While applying an upward pressure with the fingers on the flywheel, strike the top of the crankshaft on the nut with a *soft*-headed rubber or plastic hammer to break the flywheel loose from the tapered position of the crankshaft (Fig. 5-19).

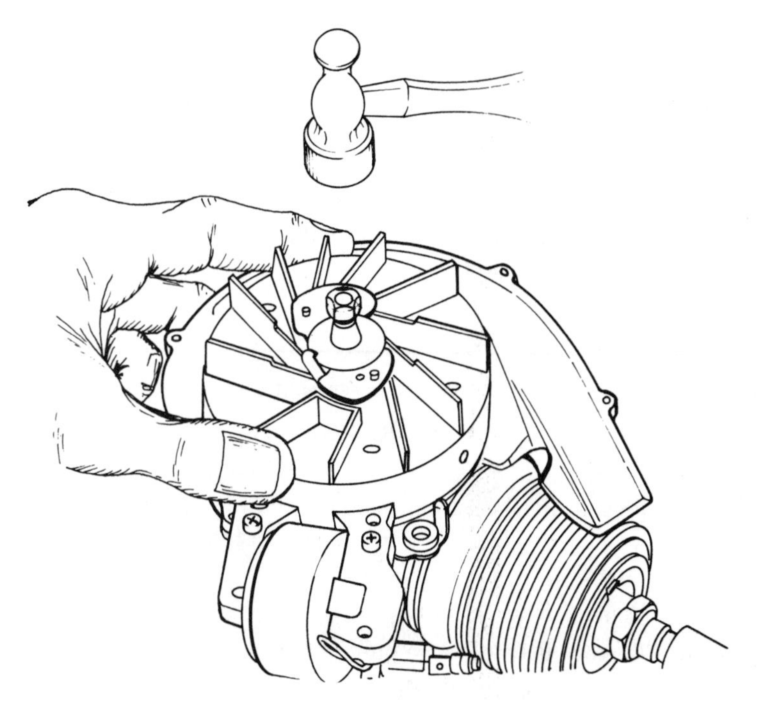

FIG. 5-19. If a flywheel puller is not available, thread a nut on top of the shaft. Lift up on the flywheel while tapping the nut with a soft-headed hammer.

4. After removing the flywheel, note the position of the flywheel *key* (Fig. 5-20). The key and flywheel must be reinstalled in the same position as before removed. The key prevents the flywheel from changing position on the crankshaft and properly aligns the magnets to pass by the ignition coil at the proper time in the engine cycle. This insures the proper positioning of the magnets on the flywheel. If the key is damaged, it should be replaced with a new one.
5. During reinstallation, make sure that the flywheel hub and the crankshaft taper are free of grease and oil; wipe them with a clean cloth dampened with solvent. Insert the flywheel key properly and replace the

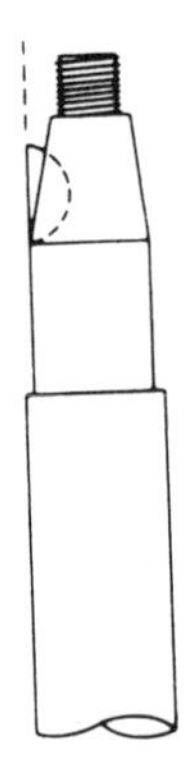

FIG. 5-20. During disassembly, note the position of the flywheel key. Reassemble the key and flywheel in the same manner. This illustration shows only one of the many types of keys and keyways.

flywheel and the nut. Torque the flywheel nut to the value specified by the manufacturer.

QUICK REFERENCE CHART 5-1

FLYWHEEL REMOVAL/REPLACEMENT

1. Prevent flywheel from rotating by using piston stop, chain or strap wrench, or a block of soft wood under flywheel vane. Do not damage vanes.
2. Remove flywheel nut.
3. Remove flywheel. Use puller, if available.
4. Note position of flywheel key.
5. Just prior to reinstallation, clean flywheel hub and crankshaft taper with a rag dampened with solvent.
6. Insert flywheel key and replace flywheel and nut.
7. Torque flywheel nut to specified value.

5-9. BREAKER POINTS

Breaker point maintenance includes inspection of the contact points, readjustment of the gap between the points, and replacement of the points. Since the points time the engine, the opening setting is critical and must be adjusted for the proper gap and timing. Proceed as follows:

1. Visually check the points for pitting, correct alignment, and good contact surfaces. If the points are pitted, corroded, or show unusual wear, replace them.
2. If the points only show traces of dirt or oil, insert a piece of clean bond paper dipped in alcohol or trichloretheylene and pull the paper through the points. Do not touch the points with your fingers because the oil from your skin can act as an insulator and cause the points to break down. Do not let any oil get on the points. Never file the points because they are tungsten-tipped, and cannot be filed. Instead, replace the points.
3. Check the points for good electrical contact. If the points are not mating correctly, rebend the *stationary* arm until the point surfaces correctly meet (Fig. 5-21). If an ignition analyzer is available, check the breaker points by removing them from the engine and placing them on the bench as follows:

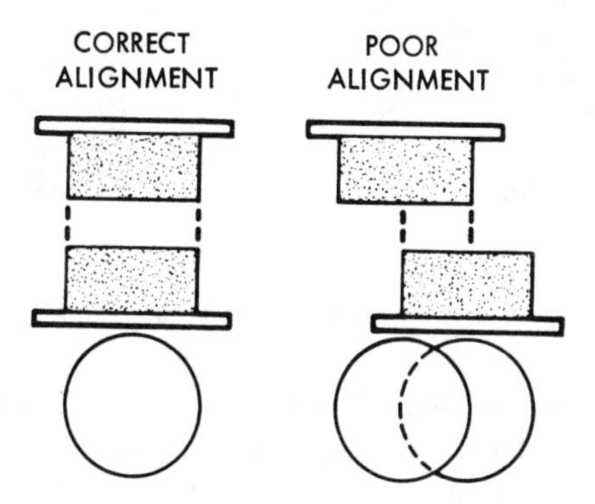

FIG. 5-21. If the breaker points are not contacting correctly, rebend the stationary arm until the points do meet correctly.

A. Connect one test lead from the ignition analyzer to the breaker arm.
B. Connect the second test lead to the breaker assembly screw terminal.
C. The meter readings on the ignition analyzer will indicate if the points are good. If the meter indicates that the points are bad, clean the points as described in Step 2, dry them, and retry the test. If the analyzer still indicates that the points are bad, replace them.

4. Check and adjust the breaker point gap (Fig. 5-22). Rotate the engine crankshaft until the breaker arm is on the high point of the cam; the points are at their widest opening. Loosen the locking screw and turn an eccentric screw, or move the stationary bracket, or adjust the position of the condenser (use whichever procedure is applicable) until the gap is set to that specified by the manufacturer. Check the gap with a clean feeler gauge. When the points are wide open, and properly adjusted, only

a slight drag will be felt when the feeler gauge is moved between the points. Tighten the adjusting screw, rotate the crankshaft through several rotations, and then recheck the gap. Tightening the adjusting screw sometimes alters the gap setting. Ignition timing is discussed in Section 5-13.

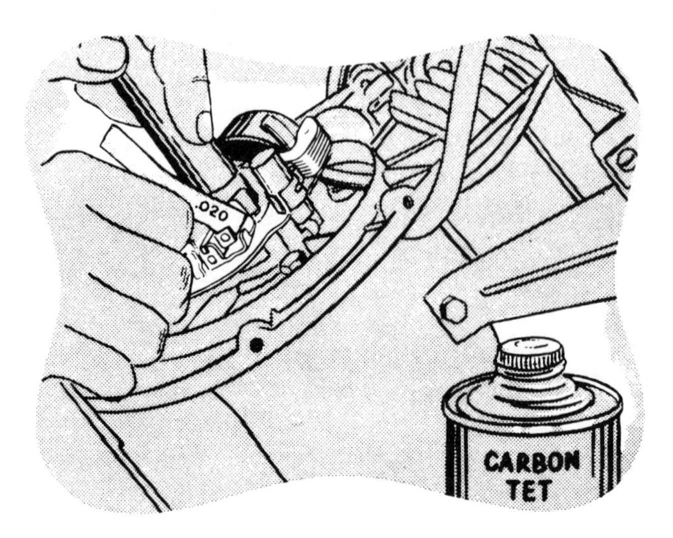

FIG. 5-22. Check the breaker point gap with a flat feeler gauge.

Be sure to set the breaker point gap correctly. If the gap between the points is too wide, the dwell or percentage of time in which the points are closed is decreased. Therefore the points open earlier and close later. Too short a dwell prevents a magnetic field from building to a maximum; the spark is thus of low voltage. If the gap is set too narrowly, the condenser will be unable to prevent the current in the primary ignition circuit from jumping the breaker point gap.

QUICK REFERENCE CHART 5-2

BREAKER POINT MAINTENANCE

1. Check points for pitting, correct alignment, and good surfaces. Replace as required.
2. Clean points.
3. Check that points mate correctly.
4. Set gap.

5-10. CONDENSER

A weak condenser can cause arcing across the breaker points. This will not permit a rapid enough collapse of the magnetic field which surrounds the primary of the ignition coil to induce enough voltage into the secondary coil to produce a good spark. Condenser failures decrease as the length of time of the condenser in the system increases. Most condensers last the lifetime of the engine; therefore it is not necessary to replace the condenser every time the points are replaced. However if the condenser is suspected of being faulty, check it for its capacity, shortage, or leakage and resistance. Before checking a condenser, warm it up to body temperature by holding it in your hand. A leaky condenser shows up better at a higher temperature.

There are three methods that can be used to analyze a condenser to determine if it is good or bad. The first method is to replace the condenser with another one of the same value. The second test requires the use of an ohmmeter, and the third test involves the use of one of the commercial ignition analyzers. The ignition analyzer makes the most complete test.

Check a condenser with an ohmmeter as follows:

1. Disconnect the condenser leads.
2. Set the ohmmeter to a high resistance scale and connect the probes for an ohmmeter (resistance) test.
3. Connect the two leads of the ohmmeter across the two leads of the condenser (Fig. 5-23). A low resistance value should be indicated at first;

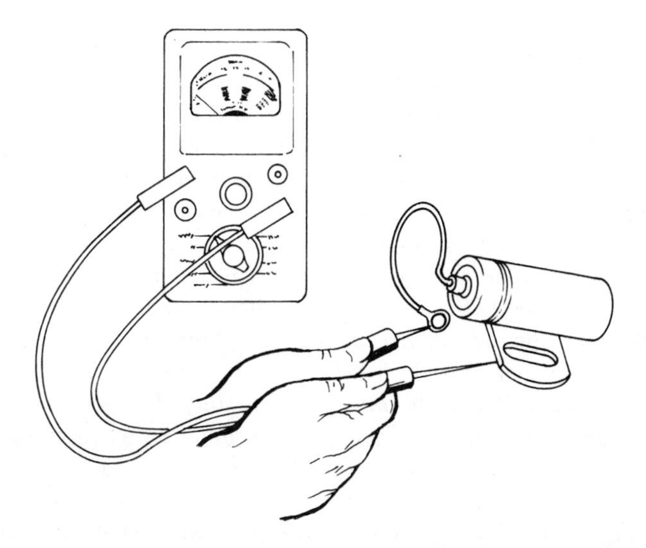

Fig. 5-23. An ohmmeter can be used to check a condenser (capacitor) for a short circuit. At first a low resistance should be indicated, but the reading should quickly rise to a high value. Resistors, diodes, coils, and high-tension leads can also be checked with the ohmmeter.

however, this reading should quickly rise to a high value. The ohmmeter is indicating that the condenser is taking on the charge of the battery. If a low resistance is continually indicated, the condenser is faulty (short-circuited) and must be replaced.

The ignition analyzer checks a condenser for leakage, resistance, and capacity. Use the ignition analyzer to test a condenser as follows (Fig. 5-24 indicates a set-up similar to that used on condensers; an ignition coil is being tested):

1. Remove the condenser from the engine ignition system.
2. Connect one test lead from the ignition analyzer to the condenser mounting bracket or to the outside of the condenser, or to one of two pigtail leads.
3. Connect the second test lead to the condenser pigtail lead or the center pin (post).
4. Replace the condenser if it fails to meet any of the three ignition analyzer tests (leakage, resistance, or capacity).
5. Turn the analyzer switch to the discharge position before disconnecting the leads from the condenser. This discharges the voltage from the condenser.

5-11. IGNITION COIL

The ignition coil does not require regular service. Keep the coil and the terminals clean and tight. If the ignition coil is suspected of being faulty, it can be tested for continuity with an ohmmeter or a multimeter set to the resistance scales, with a flashlight jig continuity checker, or with an ignition analyzer. When an ohmmeter is used to check the primary of the ignition coil, set the meter to the lowest resistance range. (The test setup would appear similar to Figure 5-23 except that the coil is substituted for the condenser.) Connect one probe to the common or ground connector; connect the other probe to the primary coil lead. The meter should indicate that there are almost no ohms of resistance. The secondary is checked by connecting one probe to the ground connection. Set the ohmmeter dial to resistance times 100 or times 1,000. Connect the other probe to the high-tension lead output from the coil. The resistance reading on the ohmmeter may range from several hundred to several thousand ohms because of the large number of turns of thin wire on the secondary. When checking the continuity of either coil, the meter will indicate infinite resistance (no deflection of the ohmmeter when the leads are connected) if the coil is open (no continuity) and must be replaced.

The coil may be tested on the ignition analyzer after it is removed from the engine (Fig. 5-24). The ignition analyzer tests the coil under conditions similar to those of actual operation; it provides an interrupted primary current and measures the induced secondary voltage. The analyzer meter indicates either good or bad. By using the ignition analyzer, the primary and secondary coils are checked for continuity and the coil is checked for leakage and for sufficiently generated voltage to cause ignition when the coil is replaced on the engine. Check an ignition coil on the ignition analyzer as follows:

1. Remove the coil from the engine.
2. Connect the tester ground lead to the ground lead of the coil. Connect the "hot" lead of the analyzer to the coil breaker lead (primary lead) and connect the high-tension lead of the analyzer to the coil secondary high-tension lead output. Set the analyzer coil index adjustment to the specified primary current (see the specific engine's manufacturer's repair manual; for an example, refer to Appendix I, Table I-8).

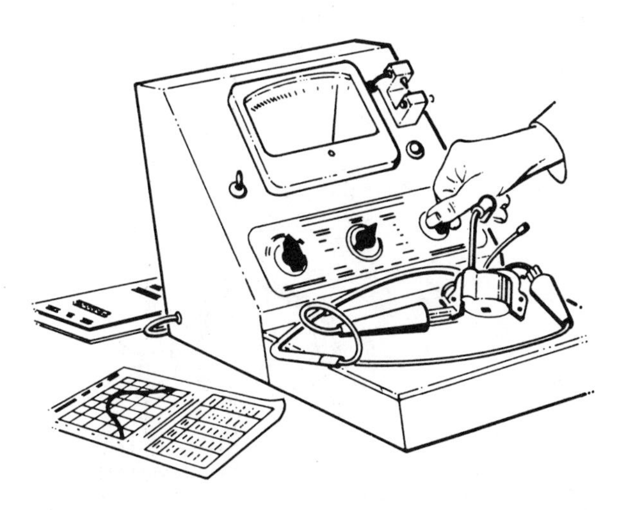

FIG. 5-24. An ignition analyzer can be used to test coils, points, spark plug high-tension leads, and condensers. Use it on a wooden workbench.

3. Note the meter reading. A low reading indicates that the coil must be replaced. No reading indicates a dead coil.
4. Check for voltage leakage from the coil (leakage can be caused by moisture, cracks, or carbon paths). Run the analyzer test probe along the outside of the coil. Replace a coil that shows any voltage leakage.
5. To check the coil high-tension lead to the spark plug for leakage or insulation failures, proceed as follows:

   A. Connect the high-tension lead of the analyzer to the secondary of the coil.
   B. Probe the entire surface of the lead insulation with the grounded probe. Flashover to the probe indicates that the insulation is broken down. Replace the high-tension lead.

When the ignition coil (or solid-state module) is returned to the engine, it is necessary that the correct air gap be set between the coil heels (ends of the lamination) and the flywheel magnets. This air gap is a critical dimension and you must consult the manufacturer's specifications for the proper distance. Use a strip of nonmetallic shim stock to set the coil at the proper gap (for example, a gap of 0.010 inch is used in one manufacturer's lawn mower engine).

The air gap between the flywheel magnets and the ignition coil or between the magnets and a solid-state module are set as follows:

1. Remove the necessary hardware for access to the coil or solid-state module.
2. Loosen the coil or solid-state module hardware.
3. Rotate the flywheel until the flywheel magnets are adjacent to the solid-state module (Fig. 5-25). The magnets will pull the coil or module to the flywheel.

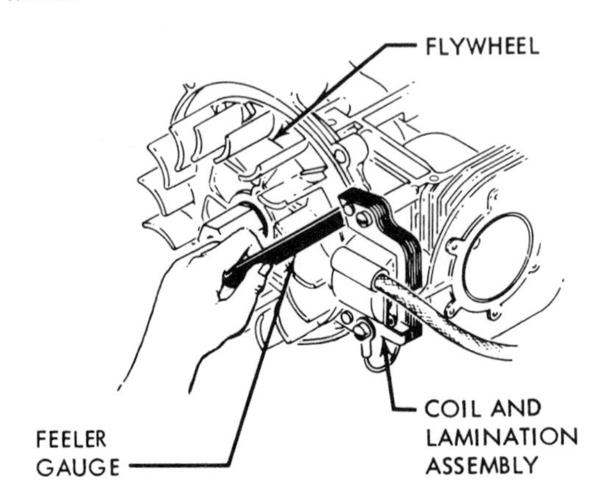

Fig. 5-25. A nonmetallic feeler gauge is used to set the air gap between the flywheel magnets and the ignition coil.

4. Insert a nonmetallic gauge of the proper gap thickness (see manufacturer's specifications—approximately 0.010 inch) between the coil or the solid-state module laminations and the magnets on the flywheel.
5. Tighten the attaching hardware and remove the gauge.

5-12. FLYWHEEL MAGNETS

The magnets on the flywheel can be rather simply checked to determine if they are powerful enough to generate the final ignition spark. Set the flywheel on edge on a flat nonmetallic surface and hold a screwdriver about an inch above the flywheel magnets (Fig. 5-26). The magnets should attract the screwdriver quite strongly.

If they do not attract the screwdriver, the magnets are weak and have to be replaced. Usually, the magnets are of alnico and cannot be recharged.

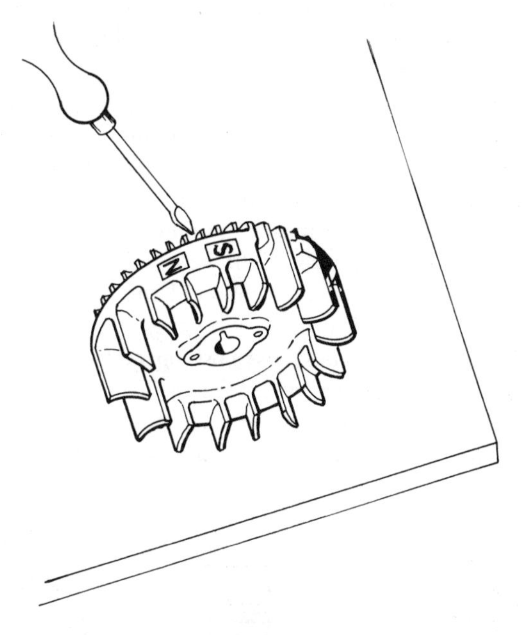

FIG. 5-26. Flywheel magnets are checked by holding a screwdriver about one inch above the magnets. If the screwdriver is attracted, the magnets are sufficiently strong.

5-13. SETTING IGNITION TIMING

An engine's ignition timing is critical. The spark between the electrodes of the spark plug must occur at the proper time to ignite the air-fuel mixture in the combustion chamber; the proper time is just before the piston reaches TDC on the compression stroke. If the spark occurs too early in the cycle, the pressure of combustion will push against the momentum of the flywheel, crankshaft, connecting rod, and piston; hence, the power of the power stroke will be diminished. If combustion occurs after TDC is reached, the compression of the air-fuel mixture will be less. This results in a less combustible mixture and another loss of power. Correct timing involves the positioning of the piston at the exact point where the breaker points begin to open.

Timing is normally set at the factory, but the timing can go wrong and cause engine trouble. If the engine is disassembled, retiming is required. Improper engine timing can be caused by improper installation of the crankshaft and camshaft gears, by a jammed spark adjustment mechanism, by incorrect breaker point settings, by a loose or incorrectly positioned magneto assembly, or by a rotor or flywheel that is incorrectly positioned. The timing is set so that the points open at exactly the moment before TDC is reached on the compression stroke to ensure that the maximum power will be developed during the power stroke.

The ignition timing is set by different methods on different engines. Timing is sometimes set in a *static* condition when the engine is *not* running or in a dynamic condition when the engine *is* running. Different methods of static timing include: the use of a dial indicating timing gauge and a flashlight jig continuity tester; the use of a scale (ruler) and flashlight jig continuity tester; the relocation of the ignition coil armature; and the correct positioning of the timing marks on mating crankshaft and camshaft gears. The dynamic setting of ignition timing makes use of a strobe light. Procedures for timing engines are included here, but the manufacturer's specific procedures should be used if available.

The most accurate static setting of ignition timing is accomplished with a dial indicating timing gauge and a flashlight jig continuity tester. With these devices, it is possible to determine the exact location of the piston at the precise instant that ignition occurs. The dial indicating timing gauge is a depth gauge timing tool (Fig. 5-27) which indicates (in inches) the position of the piston in the cylinder. The manufacturer's timing specifications are given in inches of piston travel before top dead center with timing in the full advance position. Timing can be changed by resetting the breaker gap and also by shifting the position of the stator plate.

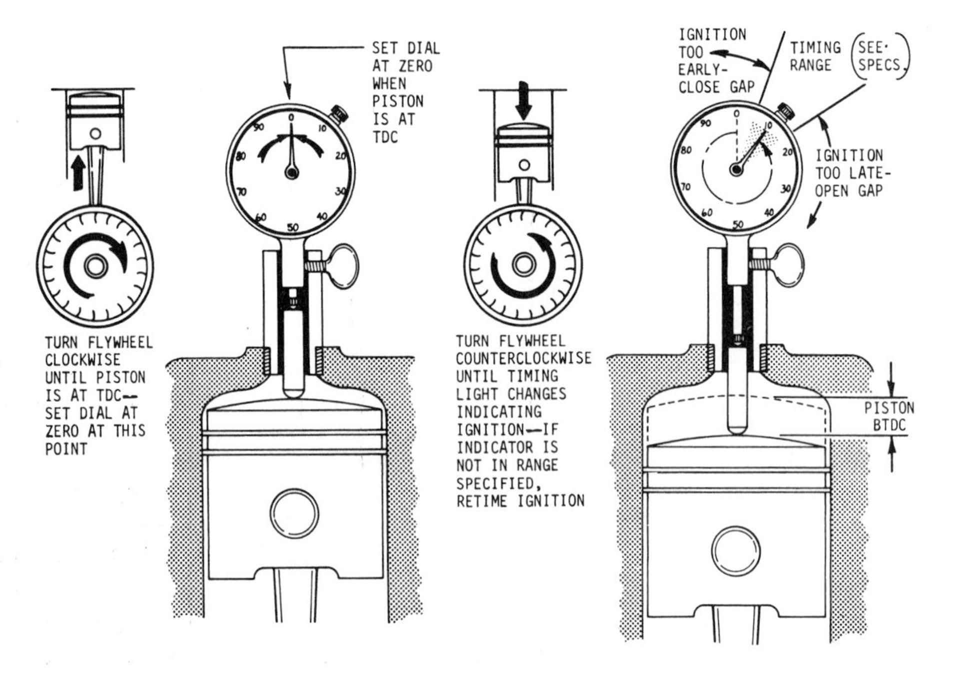

Fig. 5-27. The most accurate setting of ignition timing is accomplished with a dial indicating timing gauge and a flashlight jig.

Use the timing gauge to check ignition timing as follows:

1. Remove the spark plug and install the dial indicating timing gauge in the cylinder head.

2. Connect the flashlight jig timing light between the ignition lead and a good ground. Turn on the light.
3. Adjust the breaker points to the specified gap. Recheck the gap after retightening the adjusting screw. Readjust as required.
4. Turn the flywheel until the piston is at TDC. The pointer of the indicating timing gauge will suddenly reverse direction as the piston passes TDC. Set the pointer to zero when the piston is at TDC.
5. Move the timing advance lever, if applicable, to full advance and hold this position while checking the timing.
6. Turn the flywheel in the opposite direction from Step 4 until a light intensity change is noted. If a change occurs when the pointer of the dial indicating timing gauge is outside the specified timing range, readjust the timing as follows:

   A. First, try to correct the timing by making slight changes to the point gap setting.
   B. If timing cannot be brought into range by changing the point gap, reset the points to the proper gap and then shift the stator until the timing is within the specified range.

If a timing gauge is not available, you can use a setup such as that shown in Figure 5-28. Rotate the crankshaft so that the piston moves in an upward direction and watch on the scale (ruler) until the piston has reached top dead center. The TDC position is reached when the crankshaft can be rotated through a couple of degrees without any movement of the piston. Now rotate the crankshaft so that the piston moves down into the cylinder the correct distance as specified by the engine manufacturer. At this distance, the breaker points should just begin to open.

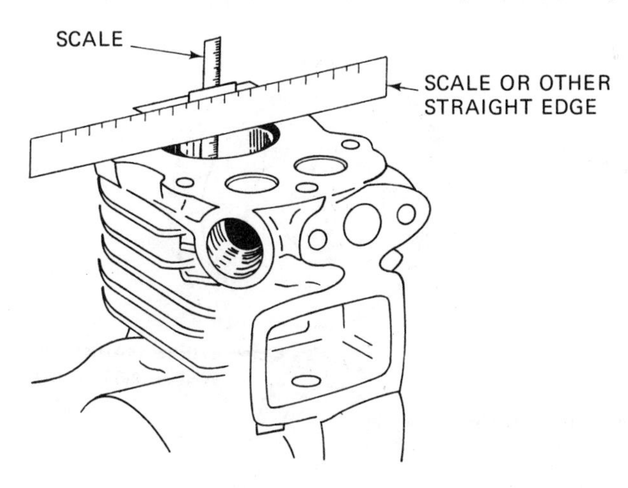

Fig. 5-28. A pocket scale (ruler) and a straight edge can be used to determine piston position if a dial indicating timing gauge is not available.

This opening can be checked by using the flashlight jig continuity check tester or by putting a piece of cellophane or tissue paper between the points; when the points open, the cellophane or tissue will fall out. If the timing is incorrect, loosen the stator plate setscrew and rotate the stator plate until the points just begin to open. Retighten the setscrew.

The timing of some engines is set by relocating the ignition coil armature on its mounting bracket. A mark on the armature lamination is aligned with a mark on the flywheel. To align this type of system, remove the flywheel, set the proper spark gap (Section 5-11), replace the flywheel, and tighten the nut finger tight. Rotate the flywheel in the direction of engine rotation until the points just begin to open. Observe the flywheel mark in relation to the armature mark. If the marks are not in alignment, the armature mounting bolts can be loosened and the armature can be moved until correctly aligned. It may be necessary to remove the flywheel to complete this procedure; if it is necessary, ensure that the crankshaft is not rotated. When the marks are properly aligned so that the spark gap is just beginning to open, tighten the armature mounting screws. When tightening the screws, you must also be sure that the proper air gap remains between the ignition coil laminations and the flywheel. This is done by installing a shim of the correct thickness between the laminations and the flywheel (the shim must be nonmetallic). When all alignments are correct, tighten the flywheel nut (Section 5-8).

In engines where the breaker point cam is driven by the camshaft, the gear on the camshaft and the gear on the crankshaft must align properly. This is done during assembly by aligning the marks that are usually stamped or punched next to one of the teeth or spaces between the gear teeth of each of the gears. The timing marks must coincide exactly (Fig. 5-29) when installed to insure proper timing (also refer to Section 2-25).

Some manufacturers include a dynamic procedure that uses a strobe timing light in conjunction with a timing mark to check the proper timing. The breaker point

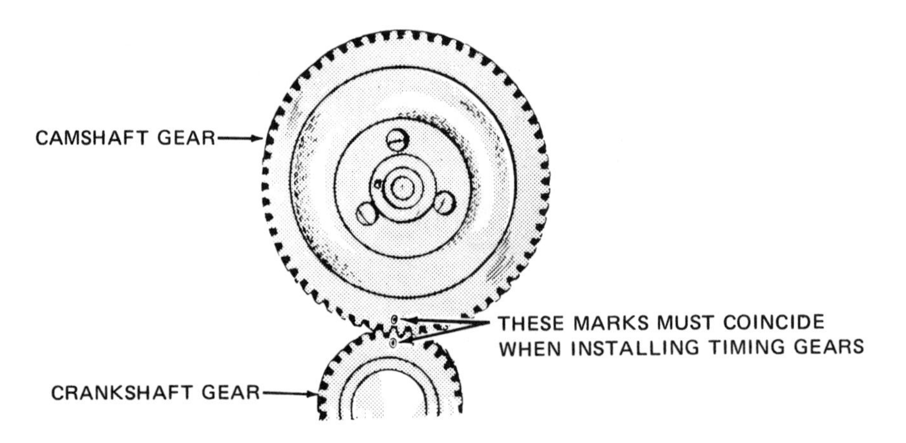

FIG. 5-29. The camshaft gear and the crankshaft gear have timing marks that must be aligned during assembly.

setting must be such that, when checked with a timing (strobe) light connected to the high-tension lead, the flywheel timing mark will line up with another index mark, or will sometimes fall between two index marks (Fig. 5-30). If the timing is not correct, an adjustment is made using a screwdriver to slightly rotate a point opening adjustment screw. When the marks are aligned, the breaker plate screw should be tightened fast. During this test, the engine is operated at a normal operating speed. The strobe light flashes each time that the plug fires. Some engines have the correct timing stamped on the cylinder block (Fig. 5-31).

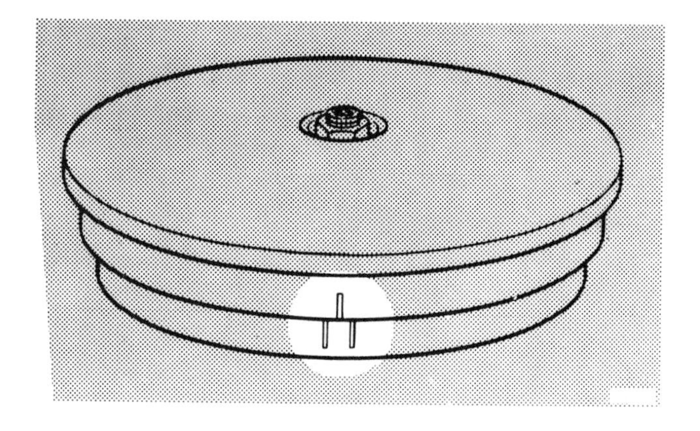

Fig. 5-30. Lineup of one mark in the center of the other two marks indicates correct timing.

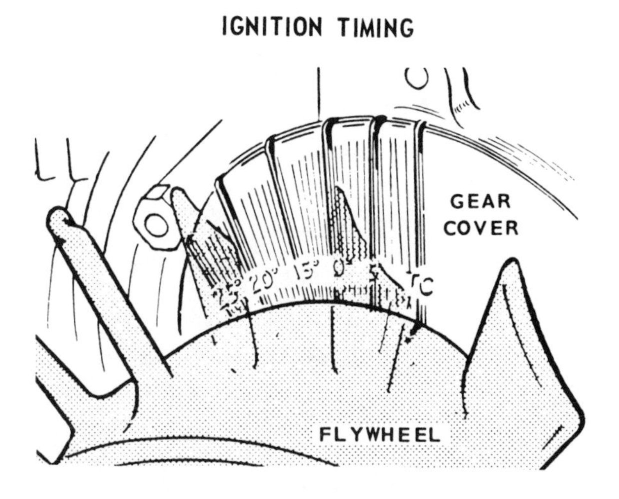

Fig. 5-31. Correct timing is stamped on the cylinder block of some engines. The timing can be checked statically or dynamically.

5-14. FINAL CHECK OF IGNITION SYSTEM

With the ignition system repaired, reassembled, and timed, it should now be given a test which is described in the troubleshooting and tune up procedures of Section 9-3.

6

Starting Systems

Chapter 5 described the ignition system that makes the small gasoline engine run. Once an initial spark is produced, the air-fuel mixture in the combustion chamber ignites and the force of combustion drives the piston down to turn the crankshaft. The flywheel causes the crankshaft to continue to rotate as the gases are exhausted. A new supply of air-fuel mixture is sent to the combustion chamber and is compressed. The ignition system produces another spark and the engine begins another cycle of operation and continues to run until it is out of fuel or until the ignition spark is interrupted.

How does the engine first start? Is the engine cranked? Is a lever kicked? Can a battery be used? Can house current be used? If a battery is used and becomes drained due to difficult starting, how is the battery recharged? Does a battery provide power for other equipment such as lights? Or is there a method of generating electricity? These questions are answered in this chapter on starting systems. Theory of operation and maintenance are included for manual starters, starter motors, AC and DC generators, alternators, magneto-generators, motor-generators, and regulators.

6-1. MANUAL STARTERS

There are four types of manual starters: crank, kick, rope, and quick release spring. The crank starter is the simplest and is used most often to start low-speed multi-cylinder engines. The crank (Fig. 6-1) is placed on a socket on the end of the engine crankshaft and is rotated to crank the engine through one cycle or more until the engine starts. The crank is disengaged when the engine starts.

FIG. 6-1. The crank is placed on a socket on the end of the engine crankshaft and is rotated to crank the engine through one cycle or more until the engine starts.

A kick starter is used on motorcycle and scooter engines. It has an operating lever with an attached gear segment that drives a gear on the crankshaft. The gear segment disengages the crankshaft gear when the operating lever is down.

Figure 6-2 illustrates a kick starter on which the pedal is depressed in a downward stroke. The starter shaft worm moves the starter clutch toward the starter clutch gear. Mating ratchet teeth of the starter clutch gear and the starter clutch engage and transmit the force through the transfer gear, transmission mainshaft, and clutch to the engine.

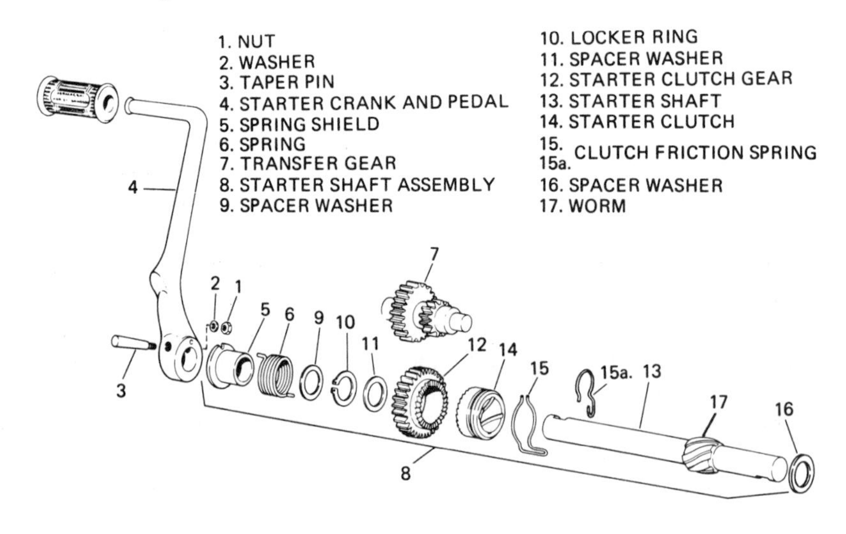

FIG. 6-2. The kick starter is used on motorcycle and scooter engines.

There are two types of rope starters used to start small gasoline engines: the simple rope starter and the retractable (also called rewind or recoil) rope starter. They are used on small engines that operate at relatively high speeds.

The simple starter (Fig. 6-3) uses a rope with a handle on one end and a knot on the other. The knot slips into a groove in a pulley on the crankshaft; the rope is then wound around the pulley. When the rope is pulled to turn the crankshaft and engine through several cycles, the knot at the end of the rope pulls clear of the pulley.

On the retractable rope starter (Fig. 6-4) the rope is permanently attached

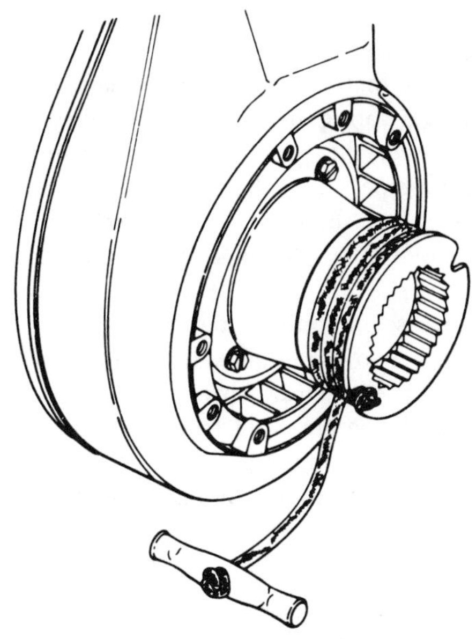

Fig. 6-3. When the rope of this rope starter is pulled, the knot at the end of the rope pulls free of the pulley slot at the end of the starting pull.

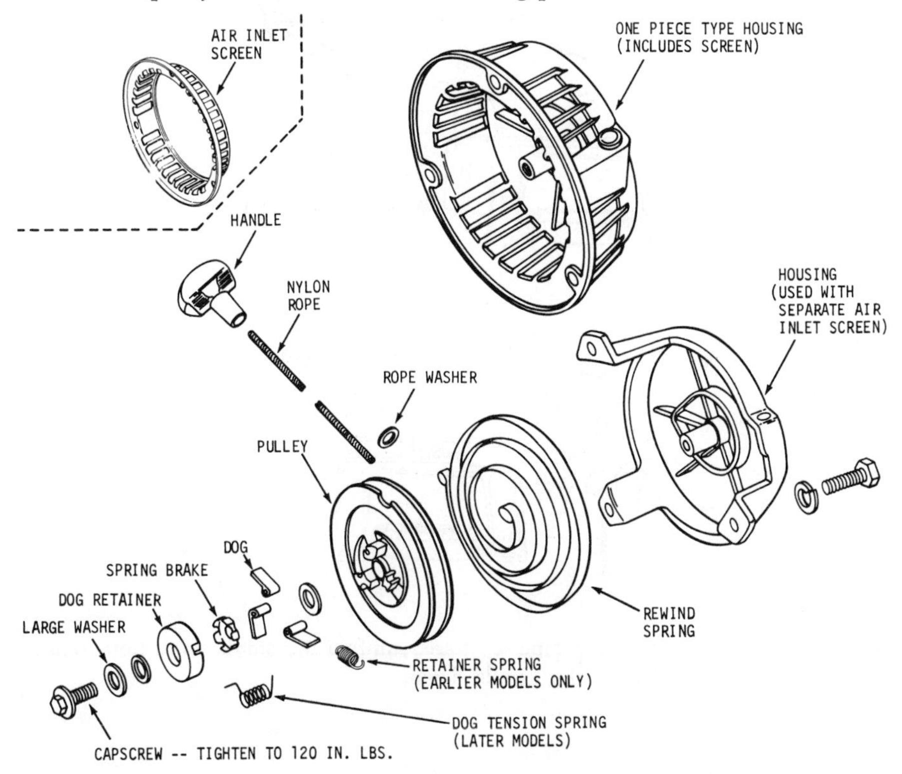

Fig. 6-4. As the rope of the retractable rope starter unwinds, a flat spring attached to the drum is wound. When the force on the rope is decreased, the spring unwinds, thereby rewinding the rope onto the pulley.

to a pulley (drum) that drives the crankshaft. As the rope unwinds from the pulley, a flat rewind spring attached to the pulley is wound. When the force on the rope is decreased, the spring unwinds causing the rope to rewind on the pulley. Pulling the rope causes dogs (pawls) to fly out from the centrifugal force on them. The dogs lock the drum to the crankshaft and cause it to rotate. When the force is removed from the rope, the dogs return inwardly disengaging the crankshaft.

The quick release (impulse) starter (Fig. 6-5) has a folding crank that is permanently attached to the engine. In starting the engine, the operator unfolds the crank and winds the starter; this action winds a heavy starter spring. When the spring is wound, the cranking handle is folded and a release is pressed that allows the spring to unwind rapidly. A ratchet or dog arrangement turns the crankshaft to operate the engine through several cycles (Fig. 6-6).

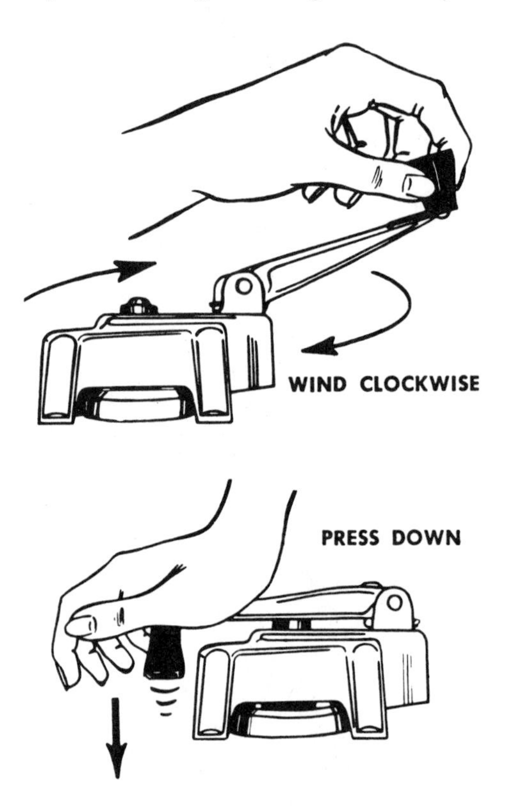

Fig. 6-5. To use the quick-release (impulse) starter, unfold the crank, wind until tight, fold the handle down, and press the handle down.

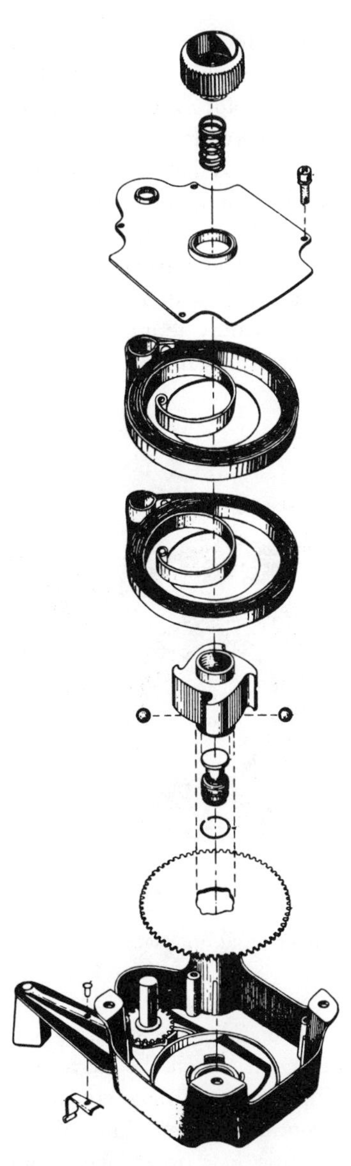

FIG. 6-6. The quick-release starter winds a heavy spring(s). A ratchet or dog turns the crankshaft through several cycles when released.

6-2. STARTER MOTOR

In place of mechanical starters, some small gasoline engines use a motor (Fig. 6-7) to start the engine. The motor is referred to as a *motor, starter, motor-starter,* or *motor-generator* (the motor-generator not only starts the engine, but it also generates electricity to recharge the battery; it is discussed in more detail in Section 6-6). The starter-motor is a machine that transforms electrical energy into mechanical energy. The starter-motor (Fig. 6-8) consists of a magnetic field created by a permanent magnet or, more often, by an electromagnet; the poles are shown as north (N) and south (S). A battery is connected through a starter switch, solenoid (Section 4-5) and commutator to the armature coil; the current causes a magnetic field to be set up in the armature. A commutator is a device for reversing the direction of current in any circuit. The commutator is simply a split flat ring that acts as a reversing switch. Two metal contacts called *brushes* rub against the split rings. The battery terminals are connected through wires to the brushes that are usually made of carbon. The brushes are held by a spring force against the commutator.

Fig. 6-7. A starter motor is sometimes used in place of a mechanical starter to start a small gasoline engine.

The magnetic field force indicated as north and south in Fig. 6-8 is usually produced by an electromagnet. When the starter switch is closed, a current flows through the armature coil causing a magnetic field to be set up; one end of the coil is north and one is south. The south ends and the north ends of the armature field and the electromagnetic field repel one another and the armature rotates. The north and south ends begin to attract each other and this further rotates the armature. Just at the point where the armature would begin to stop because of the magnetic attraction, the rotating split commutator reverses the direction of the current through the armature. The poles of the armature become interchanged and the poles of the armature and field again repel each other causing rotation. This action continues and causes the armature to continue rotating. Rather than a single coil of wire and a commutator that is split in half, the armature actually contains a number of coils and the commutator a number of splits that are in-

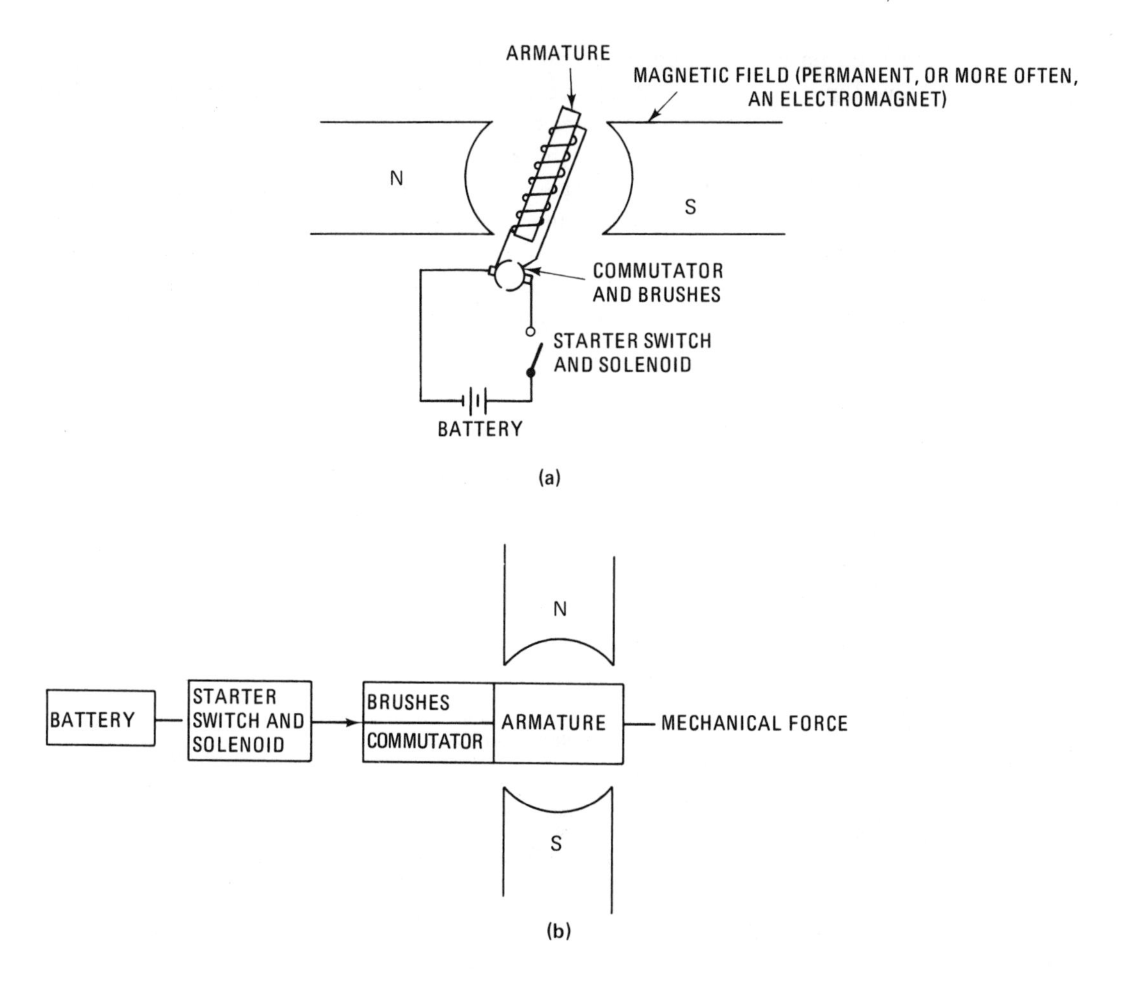

FIG. 6-8. Most starter motors are battery powered although some are powered by 115 volts alternating current.

sulated from each other. This prevents the armature from stopping at a "neutral" magnetic position. The armature coils are wound on a laminated iron rotor to increase the strength of the magnetic field and subsequently the power of the starter.

The starter motor described uses a battery as the source of electrical energy. The electrical energy is transformed to mechanical energy by means of the starter motor to start a gasoline engine. The starter motor draws a large amount of current from the battery and operates at a relatively low speed with high torque. The high current drawn by the motor causes the starter motor to heat up quickly—the starter motor should not be continually operated for a period longer than that recommended by the manufacturer. If the engine does not start within the specified time, discontinue starting for a few minutes and allow the starter motor to cool. Since the starter motor has appreciably discharged the battery by drawing a lot of current, the battery must be recharged by a generator so that the battery will be sufficiently charged to restart the engine when required. Refer to Section 6-3.

When the starter switch is closed, a current from the battery passes through the switch to a solenoid relay causing the relay to energize (Fig. 6-9). The solenoid relay contacts close and the starter motor is connected to the battery; the heavy starting current is drawn from the battery through the solenoid (the solenoid is used because the high current would cause the starter switch to burn up). The

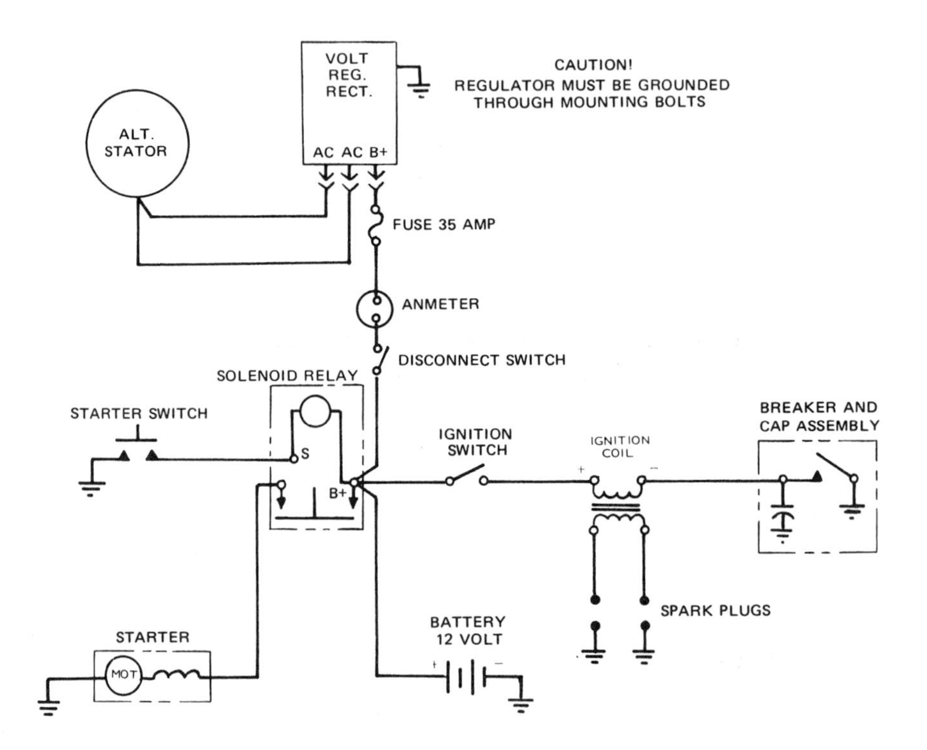

FIG. 6-9. When the starter switch is closed, the solenoid relay is energized. Power from the battery is applied through the solenoid relay contacts to the starter.

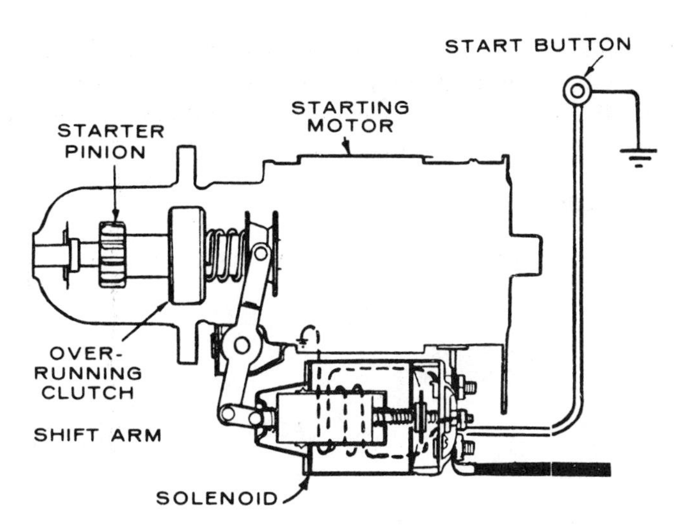

FIG. 6-10. The solenoid may be a part of the starter motor; contacts close to apply power to the motor and a mechanical shift arm engages the starter pinion into a flywheel ring gear.

solenoid may be a part of the starter motor (Fig. 6-10) or it may be separate with another solenoid in the starter motor that causes an arm to push the starter pinion into a flywheel ring gear. The current through the solenoid and the resulting force on the arm overcomes a spring force. The pinion drives the flywheel causing the crankshaft to rotate driving the engine through its cycle of intake, compression, power, and exhaust. When the engine is started and the starting button is released, the spring causes the pinion to disengage from the flywheel. (Additional electrical circuits are shown in Section 6-8 and in Appendix H, illustrations H-8 and H-10.)

Some gasoline engines use a starter that is powered by 115 volts alternating current instead of a battery. This eliminates the need to recharge the battery and thus no electrical generating equipment is needed. But, the engine cannot be used unless there is a source of 115-volt alternating current near by.

6-3. GENERATORS

If a battery is used as the source of electrical energy to power a starter motor to start a gasoline engine operating, a method of recharging the battery must be included in the starting system; a generator is used for this purpose. (On some small gasoline engines such as those used with lawn mowers, the small starting battery is removed and is connected to a small 115 volt AC charger (AC to DC converter) to recharge the battery.) A generator, also called a dynamo, is a machine that transforms mechanical energy into electrical energy. Two outputs are possible from different types of generators: alternating current (AC) and direct current (DC).

The alternating current must be rectified to direct current to charge a battery (the battery current is always direct current). Generators also provide electricity for lights and other engine or machine accessories and may be a part of the electrical system even if a battery is not used to start the engine.

AC GENERATOR

The AC generator (Fig. 6-11) is usually belt driven by the engine. The belt

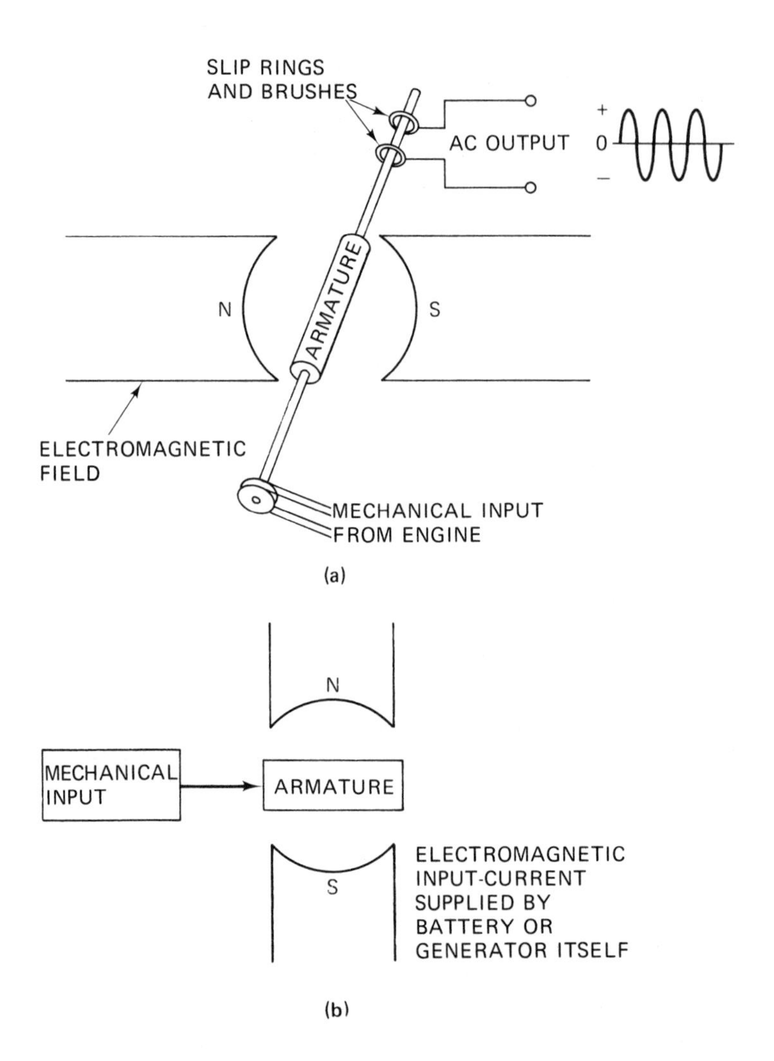

Fig. 6-11. AC and DC generators are used to power electrical accessories and to recharge the battery. An AC generator is shown here; if used to recharge the battery, the AC has to be rectified (converted) to DC.

causes an armature to rotate within a magnetic field generated by an electromagnet. Slip rings are attached to the armature shaft; brushes contact the slip rings and provide an alternating current output. As the armature coil rotates within the magnetic field, the current output builds from zero to a maximum positive value. As the coil rotates, the current induced by the magnetic field starts to decay as the coil is changing position between the poles of the magnetic field. The current decays to zero and then increases in the opposite direction as the coil rotates again toward alignment within the poles of the magnetic field. The current reaches a maximum negative value and then begins to decay as the coil continues to rotate. A number of coils on the armature keep this alternating voltage at a peak value.

The output voltage of the generator increases as the speed of the engine increases. The alternating current output can be used directly for lighting systems, but if the AC is to be used to recharge the battery, it must be rectified (converted) to direct current. The voltage must also be regulated to the recommended battery charging voltage. Refer to Section 6-7.

DC GENERATOR

A DC generator is similar to an AC generator except that the output is *direct current* rather than alternating current; thus, no rectification of the output voltage is necessary to change the AC to DC. Instead of having slip rings and brushes on the armature, the DC generator uses a commutator and brushes. As the armature rotates through the magnetic field of the electromagnet, a current develops in the armature coil. This current is picked off of the commutator by the brushes for the output. The current reaches a maximum and then decays toward zero. Just at the point where the coil output would go to zero and then in a negative direction, a split occurs in the commutator connecting the brushes to the opposite ends of the coil where the current is rising in a positive direction. The output of a single coil is a pulsating voltage between zero and a positive value. In actual operation the armature consists of a number of coils and a number of commutator segments insulated from each other by mica. The brushes pick off the voltage from the commutator (which is connected to the coils) so that the resultant voltage output remains high (rather than alternating from a high voltage to zero and back to a high voltage).

6-4. ALTERNATOR

An alternator (Fig. 6-12) is a generator that produces alternating current. It uses a stator and a rotor to replace the brushes and slip rings. The rotor turns about the stator.

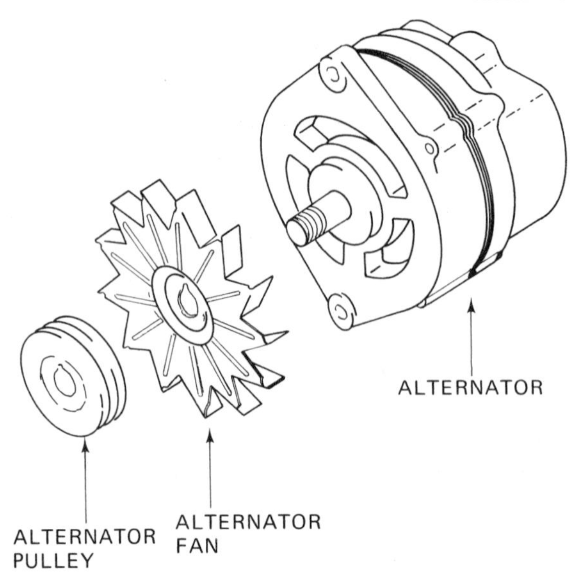

FIG. 6-12. An alternator is a generator that produces AC by means of a stator and rotor.

6-5. MAGNETO-GENERATOR

The magneto-generator (Fig. 6-13) is an alternator that makes use of magnets in the flywheel to generate alternating current electricity to power lights and other accessories. The electricity, after regulation and rectification, can be used to recharge the starting battery. A series of permanent magnets attached to the flywheel pass by two generating (armature) coils connected in series and mounted on the stator plate under the flywheel. The current generated in the coils is alternating. A *bridge rectifier* circuit is used to convert the AC to DC to charge the battery. The DC also powers lights and accessories.

Figure 6-14 also illustrates a magneto-alternator system. Six permanent magnets are mounted on the flywheel providing a rotating magnetic field. A group of coils that makes up the stator is mounted behind the flywheel. As the magnets

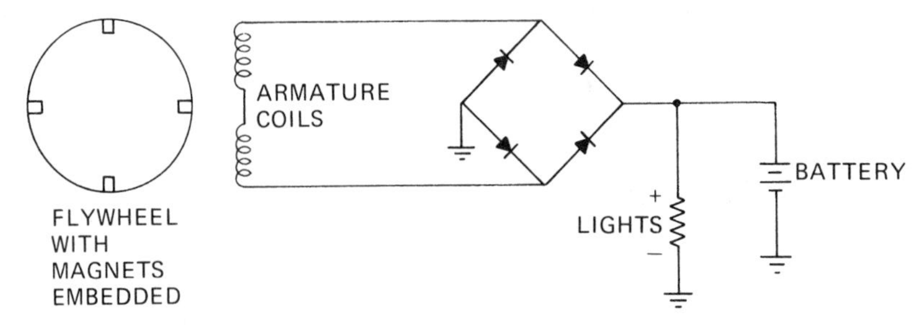

FIG. 6-13. The magneto-generator is an alternator used to power lights and other accessories; the AC electricity is rectified in this bridge circuit to provide power for lights and to recharge the battery.

pass the coils, the coils cut the magnetic field generating AC. A solid-state voltage regulator controls the AC voltage and a full wave rectifier converts the regulated AC to DC for battery charging and for supplying power to the electrical system.

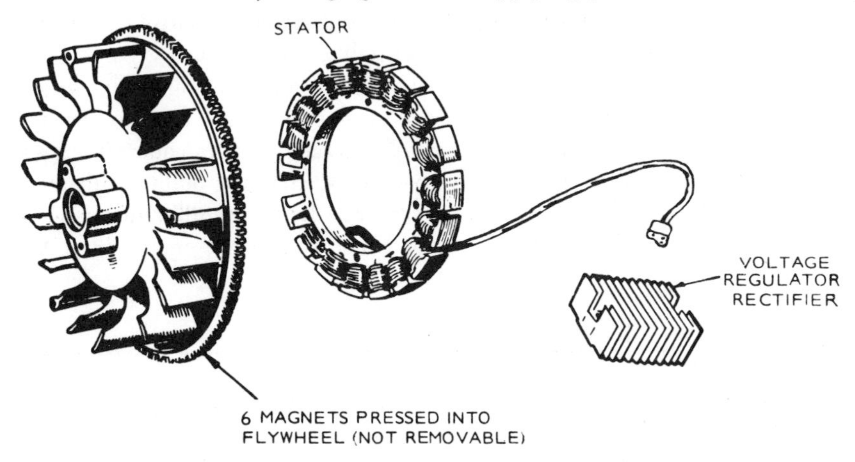

FIG. 6-14. A magneto-alternator system with a DC output consists of magnets on the flywheel, a group of coils, and a solid-state voltage regulator and rectifier.

6-6. MOTOR-GENERATOR

A *motor-generator* (Fig. 6-15) combines the functions of both a starting motor and a generator into one unit. Thus the motor-generator converts electrical energy

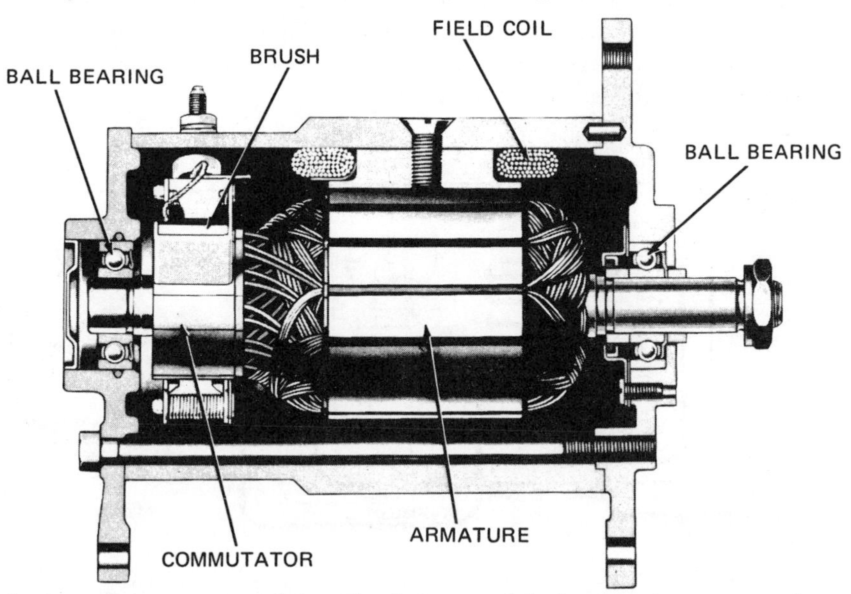

FIG. 6-15. A motor-generator combines the functions of both a starting motor and a generator into one unit.

(battery) to mechanical energy to start the engine and then converts mechanical energy to electrical energy to recharge the battery. A starting switch and voltage regulator are also used with the motor-generator and battery. A motor-generator system offers the convenience of quick and effortless starting as well as available electric power for accessories such as lights, horns, blowers, etc.

When the motor-generator is functioning as a starting motor, series and shunt fields aid each other to produce a strong magnetic field to develop as much torque as possible. Actually a motor-generator can provide either good starting ability with modest charging ability or good charging ability with modest starting ability. Performance variations are obtained by using differently wound armatures and field coils. The motor-generator is coupled to the engine by a V-belt.

6-7. REGULATORS

The DC generator is driven by the engine. When the engine is at normal operating speed, the generator output is equal to or higher than the battery voltage and charging of the battery takes place; however, when the engine is running at slow or idle speeds or is not running at all, the generator output voltage is less than the battery voltage and the battery would discharge through the generator. To prevent the battery from discharging, a *cutout relay* is used.

The output voltage of the generator must be limited to prevent the generator voltage from exceeding a specified maximum voltage. This protects the battery and other voltage sensitive equipment and accessories such as light bulbs, etc. A *voltage regulator* is the device used to prevent an overvoltage condition.

A *current regulator* is a device that limits the generator output so that the current does not exceed its rated maximum output. A cutout relay, voltage regulator, and current regulator are all mounted on a frame and the complete unit is known as a *regulator* (Fig. 6-16).

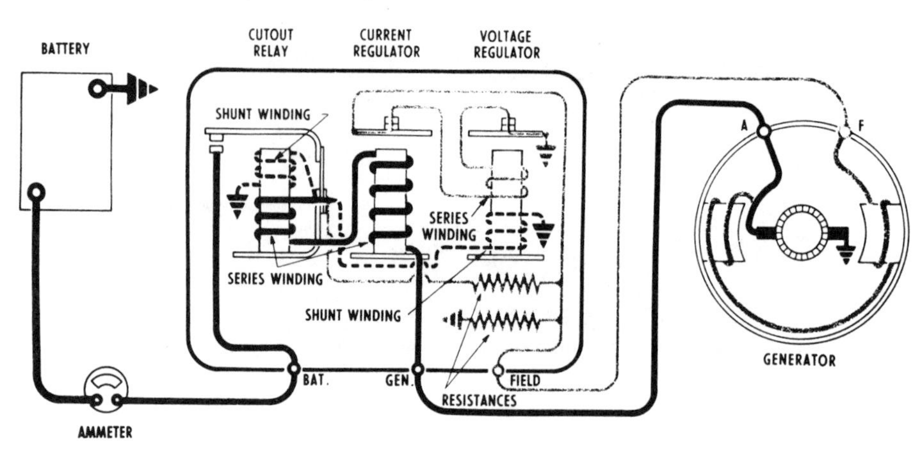

FIG. 6-16. The regulator consists of a cutout relay, a current regulator, and a voltage regulator.

The cutout relay has two coils of wire; when an electric current is passed through either coil, a magnetic field is set up. The generator is connected to one coil and the battery is connected to the other coil (when the cutout relay contacts are closed). When the generator output is greater than the battery voltage, the coil connected to the generator has sufficient current flow through it to cause the relay contacts to close because of the generated magnetic field that pulls a movable contact arm to a closed position. When the generator output decreases below the output of the battery, the magnetic field of the coil connected to the battery cancels the magnetic field of the coil connected to the generator; a mechanical spring force of the movable contact arm opens the generator output circuit to the battery. Thus, the cutout relay prevents the battery from discharging through the generator when the generator output is low because the engine is running at a speed lower than normal.

The voltage regulator has a shunt winding consisting of many turns of fine wire which is connected across the generator. An accelerator or series winding speeds up the action of vibrating contacts. A flat steel armature is attached to the frame by a flexible hinge so that it is just above the end of the core. The armature contains a contact point which is just beneath a stationary contact point. When the voltage regulator is not operating, the tension of a spiral spring holds the armature away from the core so that the points are in contact and the generator field circuit is completed to ground through them.

When the generator voltage reaches the value for which the voltage regulator unit is adjusted, the magnetic field produced by the winding overcomes the armature spring tension, pulls the armature down, and the contact points separate. This inserts resistance into the generator field circuit. The generator field current and voltage are reduced. Reduction of the generator voltage reduces the magnetic field of the regulator shunt winding. The result is that the magnetic field is weakened enough to allow the spiral spring to pull the armature away from the core, and the contact points again close. This directly grounds the generator field circuit and causes the generator voltage and output to increase. This cycle of action again takes place and continues at a rate of many times per second, regulating the voltage to the predetermined value.

The current regulator has a series winding of a few turns of heavy wire which carries all generator output. A flat steel armature is attached to the frame by a flexible hinge so that it is just above the core. The armature has a contact point which is just below a stationary contact point. When the current regulator is not operating, the tension of a spiral spring holds the armature away from the core so that the points are in contact. In this position, the generator field circuit is completed to ground through the current regulator contact points in series with the voltage regulator contact points.

When the generator output reaches the value for which the current regulator is set, the magnetic pull of the winding overcomes the armature spring tension, pulls the armature down, and opens the contact points. This inserts a resistance into the generator field circuit. The generator output and field current are reduced. Reduction of the current output reduces the magnetic field of the current regulator winding. The result is that the magnetic field is weakened enough to allow the spiral spring to pull up the armature and close the contact points again. This directly grounds the generator field circuit, causing the generator output to increase again. This cycle is repeated many times per second, limiting the generator output so that it does not exceed its rated maximum.

6-8. ELECTRICAL SYSTEM—A SUMMARY

Figure 6-17 illustrates an electrical system for a typical small gasoline engine. The system utilizes an electric starter motor, battery, starting solenoid, ignition switch, magneto-ignition system, magneto-generator system, rectifier, light switch, and lights. To start the engine, the ignition switch is placed momentarily to the START position. Current flow causes the starting solenoid to energize; the starter motor is connected through the contacts of the starting solenoid (relay) to the positive terminal of the battery. The starting motor cranks the engine causing one or more cycles of engine operation—intake, compression, power, and exhaust. The magnets in the rotating flywheel induce voltage into the ignition coil which causes a spark to jump across the spark plug electrodes to ignite the air-fuel mixture in the combustion chamber.

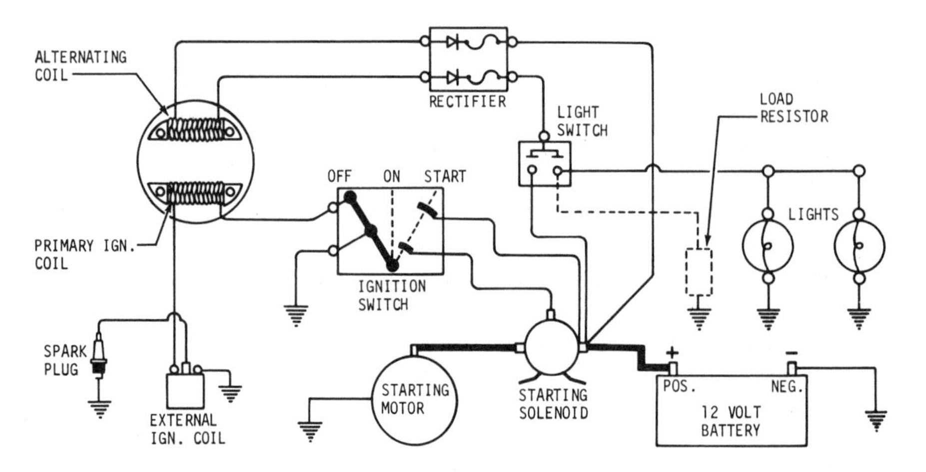

Fig. 6-17. This diagram illustrates an electrical system for a small gasoline engine. Can you correctly trace the currents through the system?

After the engine begins to run, the ignition switch returns (by spring force) to the ON position. This action breaks the current flow to the starting solenoid so that it de-energizes; this in turn breaks current flow to the starter motor so that it disengages from the engine and stops cranking.

The battery charge was somewhat drained by the current necessary to drive the starter motor to crank the engine. The battery needs to be recharged; this is accomplished by the magneto-generator (alternator) and rectifier. As the engine runs, magnets in the rotating flywheel induce AC voltage into the alternating coil. This alternating voltage is converted to direct current voltage by the rectifiers (diodes). The output of one of the diodes is directly to the positive terminal of the battery for recharging. The other diode output is connected to the light switch. When the light switch is closed, the alternator/rectifier circuit provides the power to the lighting system.

Note that the lights can be turned on without engine operation. Power is provided by the battery but without a running engine, the battery would slowly discharge and the lights would become dimmer. During periods when the engine is running at a slow speed, the charged battery keeps the lights bright.

Note that this system does not include a voltage regulator. When there is no voltage regulator, the *load* (the number of lights and accessories drawing current) should remain as the manufacturer recommends. Adding additional accessories would *load* the circuit and cause all accessories to operate below standard—for example, the lights would be dimmer. If less accessories are used, the opposite effect would take place and it could cause light bulbs to burn out.

6-9. STARTER MAINTENANCE

Sections 6-10 through 6-13 cover maintenance of manual starters, starter motors, AC and DC generators, and regulators. Since alternators and magneto-generators are AC generators, their maintenance is covered under AC generators; likewise, since motor-generators are combination starter motors and generators, their maintenance is covered under the applicable individual headings.

6-10. MANUAL STARTER MAINTENANCE

Since manual starters are all mechanical, maintenance basically includes disassembly, cleansing of parts in solvent, inspection of parts for broken gear teeth or other broken or worn parts, replacement, and reassembly. Follow the manufacturer's specific directions; be sure to disassemble slowly and note the relationship of parts. Guidelines for use in maintenance of the crank, kick, rope, and quick release manual starter are presented in the following paragraphs.

CRANK STARTER MAINTENANCE

The crank is placed onto the end of the crankshaft and when rotated, it rotates the crankshaft through its cycle. Inspect the mating surfaces of the crank and crankshaft. File off any burrs. If the crank mating surface becomes worn, replace the crank.

KICK STARTER MAINTENANCE

Kick starter maintenance includes disassembly, inspection for broken or worn parts, replacement as required, and reassembly. Disassembly of a kick starter is shown in Figure 6-2. A nut, washer, and pin are removed first. The starter crank and pedal are removed next with a pulling tool. The spring and spring shield are then removed. The cover is removed exposing the drive sprocket which is removed with a puller. The magneto-generator is then removed. The starter assembly is removed by removing the transfer gear from the transmission. The starter shaft assembly is slid out. The starter shaft is then disassembled.

Clean the parts in a solvent and dry them. Inspect all parts for damaged teeth and wear. Replace as necessary and reassemble.

ROPE STARTER MAINTENANCE

There are two types of rope starters: the simple rope starter and the retractable rope starter. Maintenance of the simple rope starter includes replacement of the rope, handle, and inspection of the pulley groove that holds the knot of the rope. If the groove has burrs, remove them with a metal file. If the rope breaks, replace it with a new rope of the same length. Secure the new rope in the handle and tie a knot into the end.

Replace frayed ropes on retractable rope starters immediately. It is easier to replace the rope when it is in one piece; when the rope breaks, the pulley is free to unwind violently which can result in a broken spring or other damage. To repair the retractable rope starter, disassemble the retractable starter (Fig. 6-4) from the engine. Release the spring tension *slowly*. Remove the rope pulley assembly, the rope, and finally the recoil spring. Clean and inspect each of the parts. Refer to Table 6-1 for problems. Replace parts as required. Coat the recoil

Table 6-1
TROUBLESHOOTING CHART FOR RETRACTABLE ROPE STARTERS

Symptom	*Probable Cause/Suggested Remedy*
1. Rope does not recoil.	A. Bent or broken spring; replace spring. B. Spring disengaged; engage spring. C. Not enough end play; remove shims. D. Starter housing damaged or binding starter pulley; replace housing. E. Broken rope; replace with proper length rope. F. Insufficient tension on spring; rewind spring.
2. Noisy when running.	A. Hub rubbing on cup or too much end play; use additional washer or shim stock to get clearance.
3. Starter frozen up. Will not pull.	A. Starter spring broken and jammed on hub; replace spring. Check hub for damage and replace if necessary. B. Improper lubrication or dirty; disassemble and clean; lubricate with lubriplate.

spring, the inside dome of the housing, and the rope pulley shaft with lubriplate grease. Correctly install the rewind spring so that it will wind in the correct direction. Install the rope pulley making sure that the spring hooks into the pulley. Assemble the pawls (also called dogs) or washers to the recoil housing shaft or rope pulley. Wind the rewind spring by turning the rope pulley. Wind the spring completely tight and then back off one turn. Hold the pulley to keep it from unwinding and install the starter rope and handle. If the rewind spring is wound in the wrong direction or if the spring is not backed off one turn, the spring will fail on the first pull of the rope.

QUICK RELEASE SPRING STARTER MAINTENANCE

When disassembling the quick release starter (Fig. 6-6), make certain that the starter spring is released. Turn the handle a few turns and press the release button. Remove the ratchet. Turning in a counterclockwise direction, remove the screws which hold the bottom cover assembly to the starter housing. Grasp the starter housing top and lightly tap the starter assembly (legs down) on a clean workbench holding the starter at arms length with your fingers on the main housing assembly only. Tapping will remove the bottom cover assembly spring and cup, plunger, and large gear.

NOTE

Do not remove the power spring from the power spring cup. These are serviced as power spring and cup assemblies only. Mark the position of each spring and cup assembly on the double spring starter and replace in same relationship.

The main housing assembly is serviced as a complete assembly and includes the handle assembly, shaft, small gear, pawl and spring, etc.

Remove the bottom cover assembly from the internal assembly. Remove the plunger assembly from the power spring and cup. Clean the parts in solvent and then inspect for cracks, breaks, or excessive wear on all parts. Replace as necessary.

Prior to reassembly, apply lubriplate grease to the mating surfaces of all internal working parts including both sides of the large gear and to the inside of the bottom cover. Reassemble the plunger assembly by sliding the plunger into the bushing and sprocket assembly. Install a ball from each side and slide the retainer into place to complete the assembly. Install the large gear in the housing with the beveled side of the gear teeth facing the housing opening. Hook the pawl with the gear teeth and apply side pressure to allow the gear to move into place.

Hold the large gear in place centered to the housing with the handle shaft gear and large gear teeth meshed. Then install the plunger assembly through the large gear and housing so that the release end of the plunger assembly protrudes through the starter housing top. Install the power spring and cup assembly. On double spring starters, reinstall the springs in the same position as originally assembled.

Use a screwdriver and position the center end of the spring over the plunger assembly. Care should be taken so that the spring is not pushed out of the spring cup. After the spring is hooked over the plunger assembly, push the power spring and spring cup assembly against the large drive gear. If the starter has two springs, the second spring can also be installed as outlined.

The starter cover plate assembly is now installed. Be sure that the bushing in the cover plate assembly has its thrust side toward the inside. Install the hold-down screws and tighten securely. Install the plunger spring into the plunger assembly. This spring pushes up the plunger and failure to install it causes the starter to be inoperative. Install the drive ratchet gear securely by hand. It is not necessary to tighten it with a wrench because it will automatically tighten up with use. Turn in a clockwise direction.

6-11. STARTER MOTOR MAINTENANCE

Many times what appears to be starter motor trouble is really trouble in another item such as the battery, poor, dirty, or loose electrical connections, ignition switch, or ignition solenoid (relay). Table 6-2 provides a troubleshooting guide for the starter motor. Subsequent paragraphs discuss brush replacement, commutator cleaning, resurfacing the commutator, resoldering the armature coils, tests for short circuits in the armature, and lubrication.

The most frequent maintenance required on a starter motor (Fig. 6-18) is inspection of the brushes and commutator. This is accomplished on most starter motors by removing a cover band. Observe the length and condition of the brushes, check

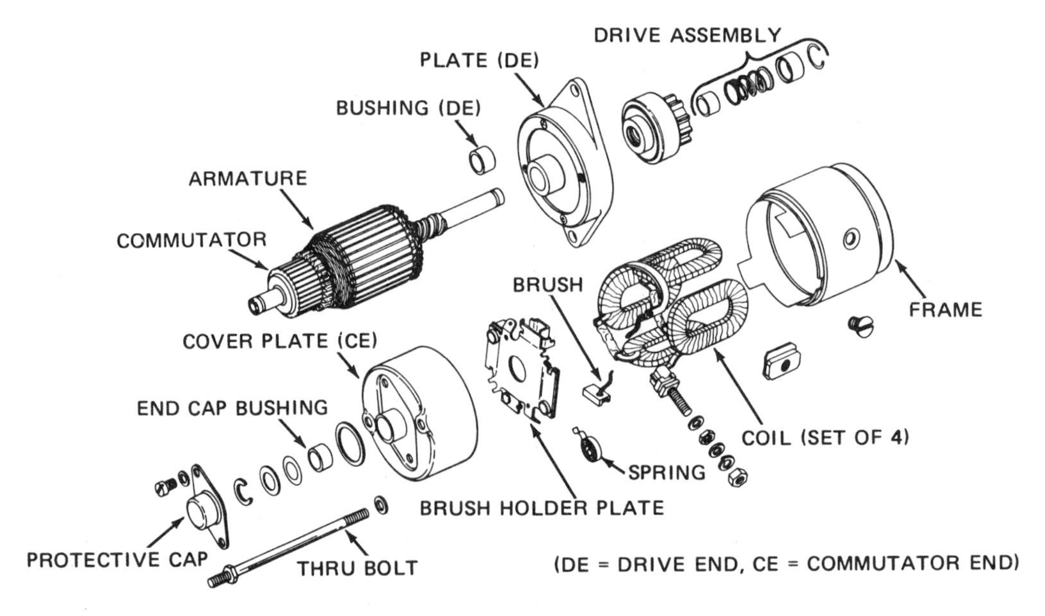

FIG. 6-18. The illustration shows a disassembled starter motor.

Table 6-2
TROUBLESHOOTING GUIDE FOR THE STARTER MOTOR

Symptom	*Probable Cause/Suggested Remedy*
1. Starter does not turn or turns too slowly.	A. Battery low or dead; recharge or replace as needed. B. Electrical connections poor or broken. Clean terminals, tighten connections. C. Faulty starter switch—replace. D. Brushes, commutator dirty or excessively worn. Clean or replace as needed.
2. Starter stops when pinion engages.	A. Battery charge low—recharge. B. Battery cables too long or connections causing excessive voltage drop. Shorten cable. Clean connections. C. Starter solenoid defective—replace solenoid. D. Brush tension too low due to excessive wear or weak springs. Replace. E. Engine seized or locked up. Repair engine.
3. Starter spins, but will not engage.	A. Pinion sticking in retracted position due to dirt or grease on splined shaft—clean. B. Chipped teeth on pinion and/or ring gear. Replace. C. Burrs forming on gear teeth to block engagement—file edges.
4. Starter does not disengage properly after engine starts.	A. Pinion dirty or return spring broken. Clean or replace as needed. B. Gear teeth dirty or damaged. File off burrs or replace.

the brush spring tension, and inspect the commutator for dirt, pit marks, an out-of-round condition, and high mica insulation between the commutator segments. If the brushes are unevenly worn or are less than one-half of their original length (less than approximately 5/16 inch on a typical starter motor), replace them. The brush springs should have sufficient tension to hold the brushes firmly against the commutator.

To clean or do repair work to the commutator, the starter motor must be disassembled. Clean the commutator with extra fine sandpaper (Fig. 6-19). A piece of sandpaper may be attached to a flat piece of wood; revolve the commutator to clean it. Be sure to brush out and vacuum out all of the dirt from the sanding.

Removal of pitted, grooved, and out-of-round conditions of the commutator are accomplished by rotating the armature and commutator in a lathe while the com-

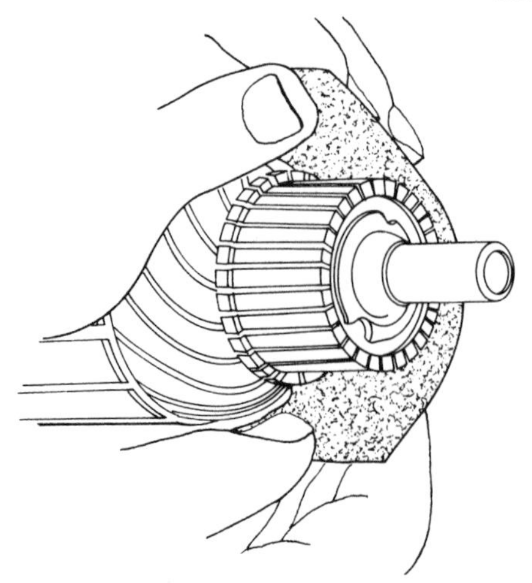

Fig. 6-19. Clean the commutator with extra fine sandpaper. You may also back the sandpaper with a piece of wood and revolve the commutator to clean it.

mutator is resurfaced by a cutting tool. Once this is accomplished, the mica insulation between the commutator sections must be scraped down 1/32 inch below the surface of the sections.

With the starter motor disassembled, inspect the soldered connections from the armature coils to the commutator sections. Overheating of the motor could have caused the solder to melt and the connections to loosen. Resolder any loose connections with a soldering iron and *resin* core solder (do not use acid core solder).

Armature coils may short circuit. The complete armature may be tested at one time for short-circuits in the armature coils in a test instrument known as a growler (Fig. 6-20). With the armature in the growler, a thin strip of iron or steel is held

Fig. 6-20. Armature coils can be checked for short circuits in a test instrument known as a growler.

over the armature. A short circuit will cause the strip to vibrate. If a short circuit is indicated, the armature must be rewound or replaced. If a short circuit is suspected in the field coil, replace the coil.

Grounds in the armature are checked with an ohmmeter (Fig. 6-21). One probe of the ohmmeter is placed on the shaft or core and the other is placed on the commutator sections. If the meter indicates a low reading, the armature is grounded.

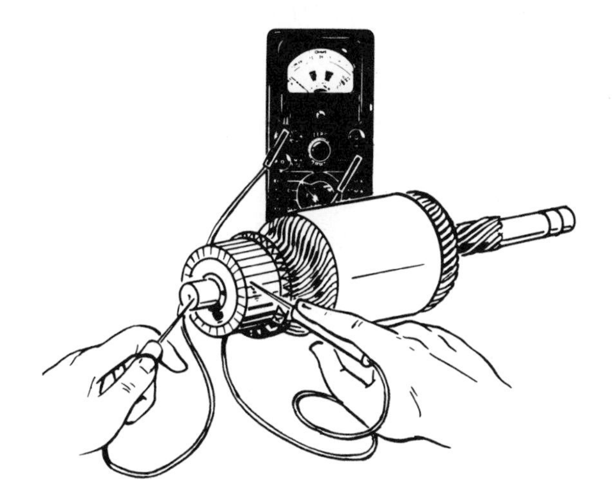

FIG. 6-21. Grounds in the armature are checked with an ohmmeter.

The field coils (Fig. 6-22) and the armature are tested for open circuits (no continuity) with an ohmmeter. Connect the ohmmeter probes to the ends of the coils. If the ohmmeter reading is high, the coil is open and must be replaced.

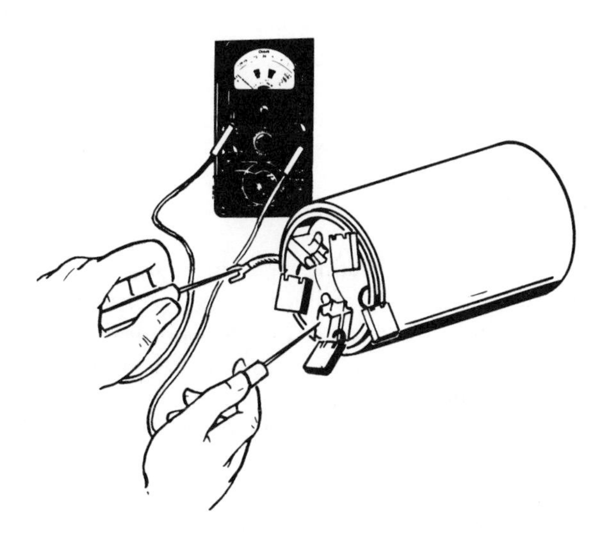

FIG. 6-22. An ohmmeter is used to test for opens (no continuity) in field coils and armatures.

Starter motor lubrication is restricted to periodic lubrication of the drive with no. 30 lubricating oil. Most bushings are impregnated with a lubricant and therefore need replacement only if the starter motor is disassembled. Bronze bearings can be replaced if they can be pressed out; otherwise, the frame must be replaced.

QUICK REFERENCE CHART 6-1

STARTER MOTOR MAINTENANCE

1. Review Table 6-2 for troubles other than with starter. If the trouble is not found, proceed to Step 2.
2. Disassemble starter.
3. Check brush length and brush spring tension. Replace, if necessary.
4. Clean commutator.
5. Remove pits, grooves, or out-of-round condition of commutator. Scrape mica sections to 1/32 inch below commutator sections.
6. Check armature coil soldered connections. Resolder with resin core solder if necessary.
7. Check armature for short circuits.
8. Check armature for grounds.
9. Check field coils.
10. Lubricate as specified by manufacturer.
11. Reassemble.

6-12. GENERATOR MAINTENANCE

Generator maintenance is divided into two areas: AC generators and DC generators. Alternating current generators include alternators and magneto-generators.

ALTERNATING CURRENT GENERATORS

When the engine troubleshooting procedures have indicated trouble in the alternating current generator, the following can be checked or tested: alternator coils for opens and correct resistance, brushes and brush spring tension, and the DC rectifier circuit used to convert the AC output to DC. First check that the brushes are of sufficient length and that the springs are pressing the brushes securely against the slip rings. Replace the brushes if they are less than one-half of their original length.

Before making further tests, ensure that the battery is fully charged (Section 3-19); if it is not, replace the battery with one that is fully charged.

To determine if the regulator and rectifier are operating properly, place a DC voltmeter across the battery terminals with the engine off. Record this voltage which should be approximately 12 volts (or 6 volts for a 6-volt battery). Start the engine and set the speed to about 2000 rev/min or higher. The battery voltage indication should rise within a few minutes to a value of approximately 13.5 volts or higher. If it does not there is probably a problem in the regulator/rectifier unit. However check the AC generation part of the system first as described in the next paragraph.

To isolate trouble to the AC generator section, disconnect the AC output from the regulator and rectifier. Operate the engine (at about 2000 rev/min or higher) and the AC generator; measure the AC output with an AC voltmeter and compare the value with the value specified by the manufacturer (approximately 17–30 volts AC). If no value is specified or no meter is available, quickly strike the two leads together; if a spark is produced, the AC generator is operating correctly and the rectifying system is defective. If there is no spark, stop the engine. Check the coils with an ohmmeter. Compare the readings with those specified by the manufacturer. The ohmmeter reading is usually less than one ohm; no reading (a reading of infinity) indicates that the coil is open and therefore defective. Replace the coil. If the coils are within limits, the magnets could be defective, but this is not likely.

The bridge rectifier system (Section 6-5) that converts alternating current to direct current is troubleshot as follows. First, connect a DC voltmeter across the output of the bridge circuit. If there is no output with the generator running, place an AC voltmeter across the input of the bridge circuit. If there is no AC input, the generator is defective. If there is an input but no output, then the bridge rectifier is defective.

The diodes in the bridge rectifier (or diode rectifiers) are checked when the generator is not running. Each diode must have one end unsoldered from the circuit. Using an ohmmeter, check the resistance of the diode first in one direction and then in the other (the leads of the ohmmeter are placed on each side of the diode; then their position is reversed). The diodes should read somewhat less than 100 ohms in one direction and over 10,000 ohms with the leads reversed. If absolutely no reading is obtained or if a reading of zero ohms is obtained, the diode is defective and must be replaced.

QUICK REFERENCE CHART 6-2

ALTERNATING CURRENT GENERATOR MAINTENANCE

1. Check brushes and brush spring tension. Replace as required.
2. Ensure that battery is fully charged.
3. Check correct operation of regulator and rectifier.
4. Check the AC generation section of generator.
5. Check bridge rectifier.

DIRECT CURRENT GENERATORS

Since direct current generators utilize coils, brushes, and commutators as used in starter motors, the maintenance is similar. Brushes should be checked for length and tension against the commutator; if the brushes are less than one-half of the original size, replace them.

Inspect the commutator for dirt, pits, grooves, and out-of-round conditions. Clean the commutator with extra fine sandpaper attached to a flat piece of wood. Revolve the commutator to clean it. Brush and vacuum all of the dirt produced by the cleaning.

Removal of pitted, grooved, and out-of-round conditions of the commutator are accomplished by rotating the armature and commutator in a lathe while the commutator is resurfaced with a lathe cutting tool. Once this is accomplished, the mica insulation between the commutator sections must be scraped down 1/32 inch below the surface of the sections.

Inspect the soldered connections from the armature coils to the commutator sections. Overheating could cause solder to melt. Resolder with a soldering iron and resin core solder.

The armature coils may short-circuit. Test the armature in a growler. Also check for grounds and opens. Replace or have the armature rewound if necessary.

QUICK REFERENCE CHART 6-3

DIRECT CURRENT GENERATOR MAINTENANCE

1. Check brush length and brush spring tension.
2. Clean commutator.
3. Remove pits, grooves, or out-of-round condition of commutator. Scrape mica sections to 1/32 inch below commutator sections.
4. Check soldered connections from armature coils to commutator sections. Resolder, as required, using resin core solder.
5. Check armature for short circuits and grounds. Replace as required.
6. Check field coil.

6-13. REGULATOR MAINTENANCE

Regulator maintenance consists of mechanical and electrical checks and adjustments. Mechanical checks and adjustments are made with the battery disconnected and preferably with the regulator off of the engine.

CAUTION

The cutout relay contact points must never be closed by hand with the battery connected to the regulator. This would cause a high current to flow through the units which would seriously damage them.

Electrical checks and adjustments are made with the regulator in operating position on the engine. The engine is operated at the operating speed for constant speed engines and at the governed speed for governed engines. After any tests or adjustments are made, the generator must be polarized after the leads are connected, but before the engine is started. The generator is polarized by momentarily connecting a jumper between the generator and battery terminals of the regulator. A momentary surge of current flows through the generator to correctly polarize it. Failure to polarize the generator may result in severe damage to the equipment since reversed polarity causes vibration, arcing, and burning of the relay contact points.

A fully charged battery and a low charging rate indicate normal generator-regulator operation. A fully charged battery and a high charging rate indicates that the voltage regulator is either not limiting the generator voltage as it should or the regulator is set too high. A high charging rate to a fully charged battery will damage the battery and the accompanying high voltage is injurious to all electrical units. A high charging rate may result from an improper voltage regulator setting, a defective voltage regulator unit, a grounded generator field winding, or high temperature which reduces the resistance of the battery to charge so that it will accept a high charging rate even if the voltage regulator setting is normal.

If the trouble is not due to high temperature, the cause of trouble can be determined by disconnecting the lead from the regulator field terminal with the generator operating at medium speed. If the output remains high, the generator field is grounded either in the generator or in the wiring harness. If the output drops off, the regulator is at fault, and it should be checked for a high-voltage setting or grounds.

If the battery voltage is low and there is a high charging rate, the generator and regulator are operating correctly. The regulator setting can be checked as outlined in the next paragraph.

If the battery voltage is low and there is little or no charging, check for: loose connections, frayed or damaged external wiring; defective battery; high circuit resistance; low regulator setting; oxidized regulator contact points; defects within the generator; cutout relay not closing; open series circuit within regulator; or the generator not properly polarized. If the condition is not caused by loose connections, frayed or damaged wires, proceed as follows to locate the cause of the trouble:

To determine whether the generator or regulator is at fault, momentarily ground the field terminal of the regulator and increase the generator speed. If the output does not increase, the generator is probably at fault and it should be checked. Other causes for the output not increasing may be that the relay is not closing or that there is an open series winding in the regulator. If the generator output increases, the trouble is due to either a low voltage (or current) regulator setting, oxidized regulator contact points which insert excessive resistance into the generator field circuit so that output remains low, or the generator field circuit is open within the regulator at the connections.

Burned resistors, windings, or contacts result from open circuit operation, open resistance units, or loose or intermittent connections in the charging circuit. Whenever burned resistors, windings, or contacts are found, the wiring must be checked before installing a new regulator. If not checked, the new regulator may also fail in the same way. Burned relay contact points may be due to reversed generator polarity. Generator polarity must be corrected after any checks of the regulator or generator, or after disconnecting and reconnecting leads.

The contact points of a regulator will not operate indefinitely without some attention. A great majority of all regulator trouble can be eliminated by a simple cleaning of the current and voltage regulator contact points and readjustment. The large flat point that should be cleaned with a spoon or riffler file is located on the armature of the voltage regulator, and is located on the upper contact support of the current regulator for negatively grounded regulator units. This contact point will usually require the most attention. It is not necessary to have a flat surface on this contact point but all oxides should be removed with a riffler file so that pure metal is exposed and should be thoroughly washed with trichlorethylene or some other nontoxic solution. When cleaning the contacts, it is necessary to remove the attaching hardware and the upper contact support.

The small soft-alloy contact point, located on the upper contact support of the voltage regulator and on the armature of the current regulator for negatively grounded regulators, does not oxidize. This contact point may be cleaned with crocus cloth or other fine abrasive material followed by a thorough washing with trichlorethylene to remove any foreign material remaining on the contact surface. Remove all oxides from the contact points but note that it is not necessary to remove any cavity that may have developed.

CAUTION

Never use emery cloth or sandpaper to clean the contact points.

7

Fuel Systems

The fuel system provides the proper mixture of filtered air and fuel to the engine combustion chamber where the mixture is ignited by the ignition system. The power of the burning mixture drives the piston on the power stroke to drive the crankshaft and connected machinery or tool. A fuel system consists of a fuel tank, fuel filter, air cleaner, carburetor and governor. The fuel tank stores the fuel to be burned. The fuel filter and air cleaner remove dirt and other foreign matter before the fuel and air arrive in the carburetor; this prevents clogging of the tiny jets in the carburetor. The carburetor combines the air and fuel in the proper ratio for all speeds and vaporizes the mixture before the mixture is accepted by the crankcase of two-stroke cycle engines or the combustion chamber of four-stroke cycle engines. A governor is used with most fuel systems to keep the engine operating at a constant speed with changes in the operating load and to prevent engine overspeeding that could damage the engine and the operator.

This chapter describes the theory of operation and the maintenance of fuel systems. The major items covered are: carburetors, governors, fuel tanks, fuel filters, air cleaners (filters), a summary, fuel system adjustments, fuel system maintenance, and gasolines.

7-1. CARBURETORS

This section describes the types of carburetors most frequently used on small gasoline engines. The function and parts of a basic carburetor are discussed first. This is followed by a discussion of float valves, diaphragms, fuel pumps, automatic carburetors, touch-and-start primer carburetor, and vapor return lines.

The mixing of air and liquid fuel into a combustible mixture of the proper proportion for burning in a small gasoline engine is called *carburetion*. The action of carburetion takes place in a device known as a carburetor. The basic carburetor consists of four parts (Fig. 7-1): an air passage, sometimes referred to as the air horn, through which air passes on its way to the crankcase or combustion chamber; a supply of fuel held at a constant level or pressure; a fuel hole opening connecting a supply of fuel to the *venturi* area of the air passage; and a restriction in the fuel line to proportion the fuel to air ratio. The venturi is a narrow passage that causes a low pressure in the area of the air passage just beyond the narrow passage. The venturi is also called the carburetor throat. The low pressure causes the fuel to come out of the fuel hole. This is due to a difference in pressure between the low pressure at the restriction (venturi) and the atmospheric pressure in the fuel tank.

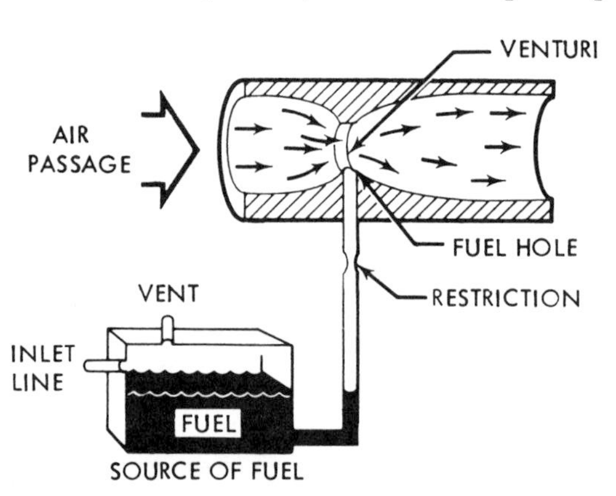

Fig. 7-1. A basic carburetor consists of an air passage, a supply of fuel, a fuel hole opening connecting the supply of fuel to the venturi, and a restriction in the fuel line to proportion the fuel to air ratio.

When the engine is running, air passes through the air horn (passage) because of the difference in pressure between the outside atmospheric pressure and the decrease in pressure caused by the movement of the piston causing a slight vacuum. This pressure difference is, in effect, a suction of air through the air passage. If the air passage remained a constant diameter, the flow of air would also remain constant, but the venturi, or restriction at the carburetor throat, causes the air to speed up. The increased speed and resulting suction cause fuel to be pulled from the fuel hole into the air passage in the form of a fine spray. (There is a decrease in pressure at the venturi and the atmospheric pressure through the vent hole in the fuel supply causes the fuel to flow into the venturi.) As the engine speeds up, the difference in pressure increases and the flow of air and fuel increases.

It is desirable to vary the proportion of fuel to air mixture under certain conditions. To start a cold engine, it is necessary to have a richer mixture—a greater ratio of fuel to air than normal; at idle speeds, it is desirable to have a slightly richer mixture; at full load, a slightly richer mixture provides greater power; and at normal speeds, a less rich or normal mixture gives the best and most economical operation. A slightly richer mixture is necessary at idle speeds because the difference in pressure (suction) is not very great and the fuel does not completely break down into a fine spray. This results in poor vaporization of the fuel or a lean mixture (a lean mixture is a lower than normal ratio of fuel to air). To take care of the range of mixes required, several jets or fuel holes, a choke, and a throttle are used in the air horn. Adjustment screws—main and idle—are used to set up the initial operating and idle speeds, respectively.

The main, or high speed, fuel discharge hole is located in the air horn venturi where the air pressure difference is greatest at high speeds (Fig. 7-2). A main adjustment needle valve replaces the restriction (Fig. 7-1) in the high speed fuel jet and sets the engine for best operation at full power speed. Primary and secondary idle discharge holes are located downstream from the venturi; the secondary discharge hole is located just ahead of the throttle valve. The idle (fuel) adjustment needle is set to restrict the fuel flow for the best air-fuel ratio at idle speed (this adjustment is not available on some carburetors—the idle mixture port is fixed). An air bleed hole is designed and built into the carburetor to make it easier to adjust the idle needle valve and to assist in atomizing the fuel. The air bleed reduces the rate at which the fuel flow increases under a difference in pressure by permitting air bubbles to pass through the idle mixing chamber (Fig. 7-3) and into the passage with the fuel. Some carburetors do not have an air bleed since it is not a necessary part of the carburetor.

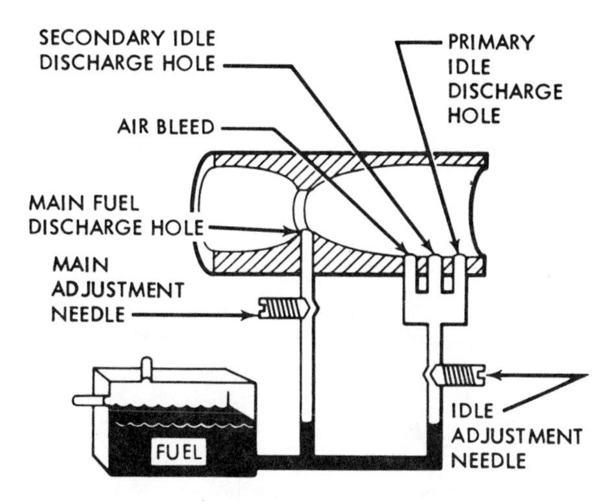

FIG. 7-2. The main fuel discharge hole is located in the air horn venturi. Primary and secondary idle and an air bleed hole are located at the throttle.

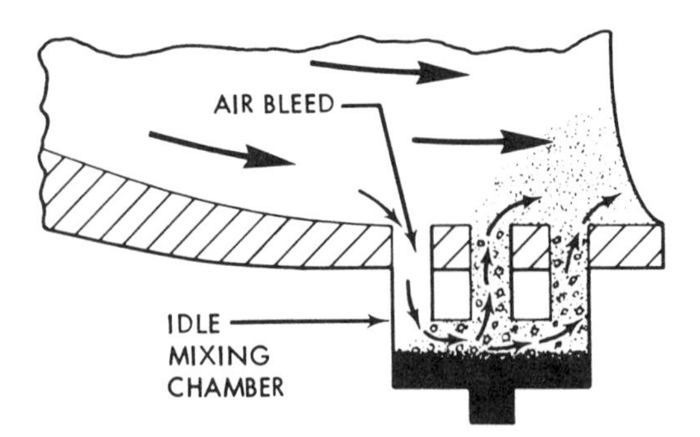

FIG. 7-3. An air bleed hole is designed into the carburetor to make it easier to adjust the idle needle valve and to assist in atomizing the fuel.

The throttle (Fig. 7-4) controls the engine speed by controlling the flow of the air-fuel mixture through the air horn; the faster the air flow, the greater the suction at the venturi resulting in more fuel in the mixture. The throttle is very carefully located in relation to the secondary fuel discharge hole because the control of the air-fuel mixture is so very critical. When the engine is idling, the throttle is almost completely closed (Fig. 7-4); very little air is sucked past it. This results in suction being applied at the primary idle discharge hole. The suction draws fuel from the fuel storage chamber of the carburetor and air is drawn through the secondary idle discharge hole and the air bleed hole. The air and fuel mix in the idle mixing chamber and exit from the primary idle discharge hole to the engine. The mixture is rich. At this idle or low speed no fuel is being drawn from the main fuel discharge hole because the difference in pressure (suction force) is not strong enough.

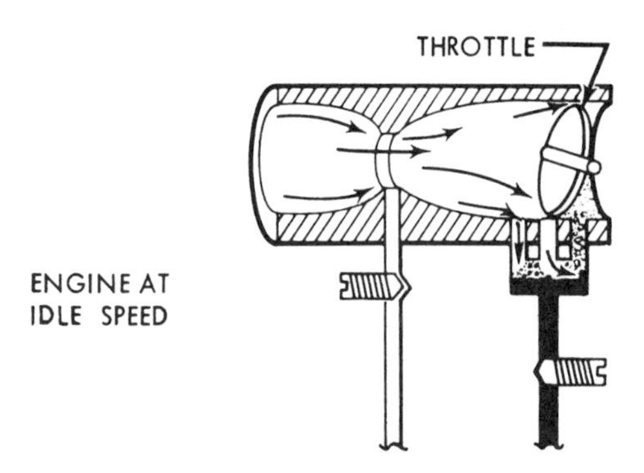

FIG. 7-4. The throttle controls the engine speed by controlling the flow of the air-fuel mixture through the air horn.

As the throttle is opened to increase engine speed, more and more air passes by the throttle plate (Fig. 7-5). The bottom of the throttle plate has changed position and has changed the secondary idle hole from an *air inlet port* to an *air-fuel discharge port*. This increase in air-fuel flow from the idle mixing chamber balances the increased air flow past the throttle keeping the air-fuel mixture to the engine in the correct proportion.

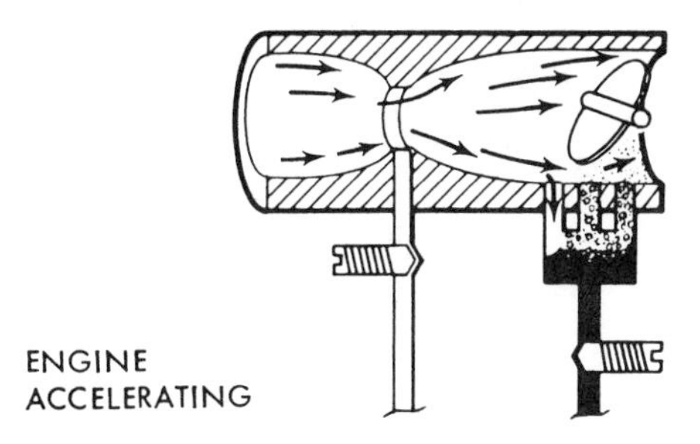

Fig. 7-5. As the throttle is opened further, the secondary idle discharge hole changes from an air inlet port to an air-fuel discharge port.

When the throttle is opened further, the pressure at the venturi increases and more fuel is pulled from the main fuel discharge hole. When the throttle is wide open (Fig. 7-6), almost all of the fuel comes through the main fuel discharge hole. The throttle causes additional turbulence of the air and fuel mixture giving a more thorough mixture.

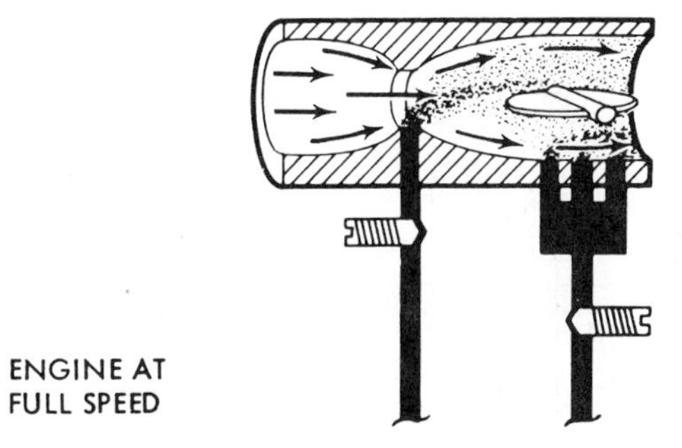

Fig. 7-6. When the throttle is wide open, almost all of the fuel comes through the main fuel discharge hole.

When a cold engine is to be started, the choke is closed. It blocks off the incoming air so that the mixture is rich enough to enable starting of the cold engine. The choke (Fig. 7-7) is an elliptical plate with a small metering hole in it so that when the choke is closed, a small amount of air still passes through it. Large quantities of fuel are drawn through the discharge holes—main, air bleed, secondary, and primary—producing a very rich air-fuel mixture.

A carburetor which has a float valve is shown in Figure 7-8. When no fuel is in the float bowl, the varnished cork or metal float rests on the bottom of it. The needle valve is unseated. As fuel flows into the float bowl, the float rises on top

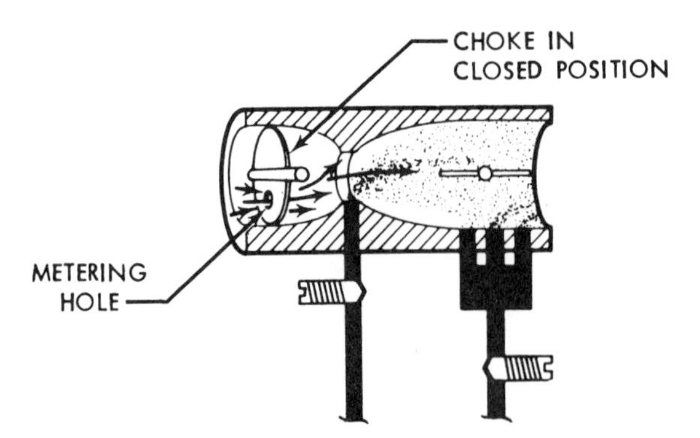

FIG. 7-7. When a cold engine is started, the choke is closed blocking off the incoming air so that the mixture is rich.

of the fuel and lifts the float arm around the pivot pin. The steel needle, which is attached to the float arm with a wire clip, raises. The rubber tipped needle seats in the brass seat, eventually cutting off the flow of fuel into the float bowl.

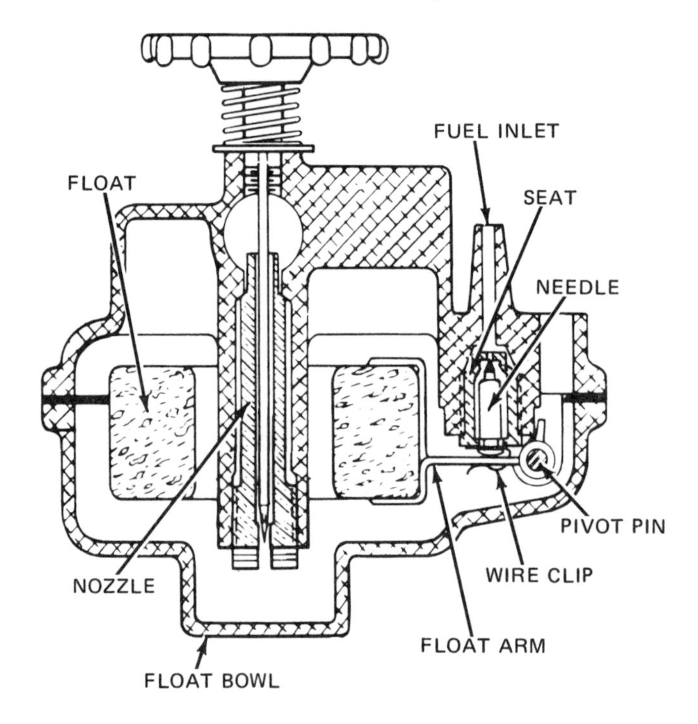

FIG. 7-8. This carburetor uses a float valve to maintain a constant level of fuel in the float bowl.

As the engine uses fuel, the level of the fuel in the float bowl subsides. The float moves downward and moves the needle valve from its seat allowing additional fuel to flow into the float bowl to maintain a constant fuel level in the bowl. Operation is automatic; the amount of fuel flowing into the float bowl is equal to the amount flowing out. The needle and seat are a "matched set" and the need to replace one necessitates replacement of the other.

A diaphragm type carburetor is used with an engine such as used on a chain saw or a recreational vehicle because it can be operated in any position. The float bowl carburetor, which was previously discussed, must remain reasonably upright to prevent starving or flooding of the engine because of the position of the float.

The diaphragm carburetor (Fig. 7-9) consists of valves, an inlet and an outlet, and fuel and air chambers separated by a diaphragm. The air chamber is vented to the atmosphere. As fuel is sucked from the fuel chamber through the fuel outlet to the venturi, the atmospheric pressure presses the diaphragm toward the fuel chamber. The diaphragm presses against the inlet valve control lever and diaphragm spring. Movement of the inlet valve control lever around the pivot pin causes the inlet valve to open admitting fuel into the fuel chamber. When enough fuel has entered the carburetor, the diaphragm moves toward the air chamber closing the inlet valve. In some carburetors, the inlet valve control lever is hooked onto the diaphragm; other carburetors use the diaphragm spring to hold the lever against the diaphragm. The inlet valve can either be a ball or a needle type, but the needle is most often used.

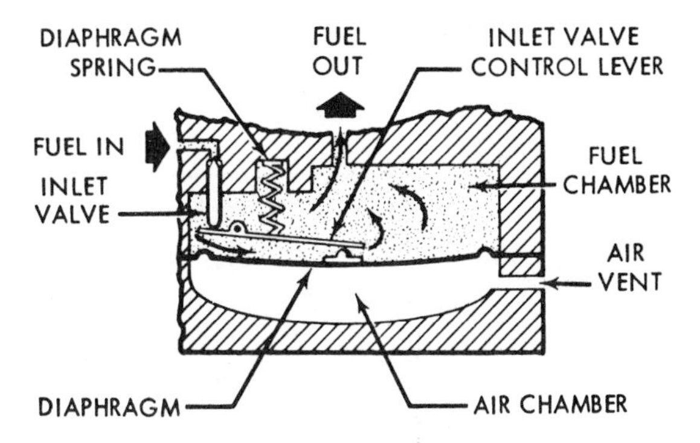

Fig. 7-9. An engine having a diaphragm carburetor can be operated in any position.

Chain saws use a diaphragm pump (Fig. 7-10), which is actually part of the carburetor, to pump fuel from the fuel tank to the fuel supply chamber. Crankcase pulsations provide the pressure to pump the fuel into the carburetor. When the piston moves up in the cylinder, crankcase suction is felt on the dry chamber of the pump. This causes the diaphragm to move toward the dry chamber admitting fuel past the inlet flapper valve into the fuel chamber. When the piston starts downward, a pressure (rather than a suction) is felt on the fuel pump diaphragm. This pressure closes the inlet flapper valve and opens the outlet flapper valve such that fuel is pumped into the carburetor diaphragm section. Repeated up and down strokes of the piston cause the diaphragm to pump fuel to the carburetor diaphragm section.

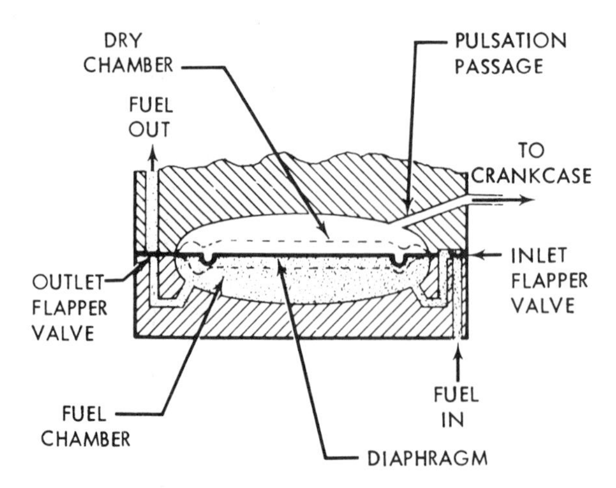

FIG. 7-10. Chain saws use a diaphragm pump that is actually part of the carburetor to pump fuel from the fuel tank to the fuel supply chamber.

Figure 7-11 illustrates the complete carburetor used in one manufacturer's chain saw. The fuel pump, carburetor diaphragm section, and the venturi and air passage are all shown with interconnections. Can you discuss the complete theory of operation of this carburetor?

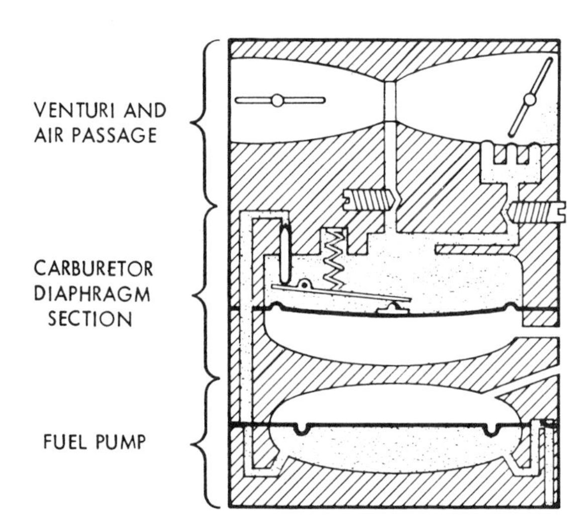

FIG. 7-11. This drawing illustrates a complete carburetor used in a chain saw. Can you discuss the complete theory of operation of this carburetor?

The carburetor shown in Figure 7-12 is completely automatic. There are no normal carburetion adjustments to be made to regulate the amount of fuel entering the carburetor venturi (in essence industry is trying to rid carburetors of adjustments so homeowners do not foul them up). The only adjustment to this automatic carburetor is an atmospheric pressure adjustment that is made if the engine is operated at very high or very low altitudes. The normal setting of this adjustment is three-quarters of one turn from the closed setting (opened slightly more at low altitudes and slightly less at high altitudes). With this setting, the correct amount of air enters the carburetor bowl vent allowing the right amount of fuel to mix with the air entering the carburetor venturi. Keep in mind that it is the governor that controls the amount of fuel entering the engine; the atmospheric pressure adjusting needle is to mix the right amount of fuel with the correct amount of incoming air.

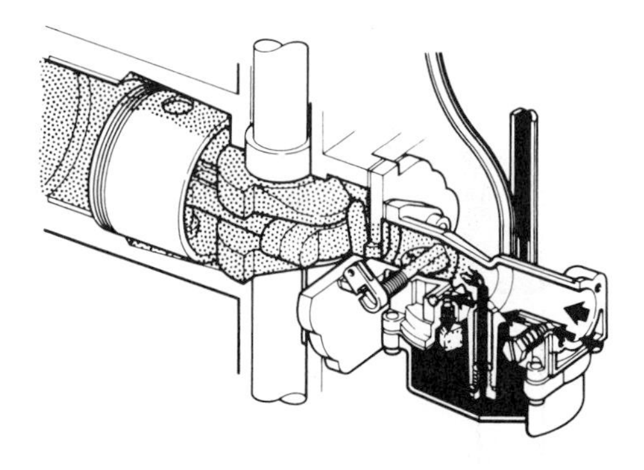

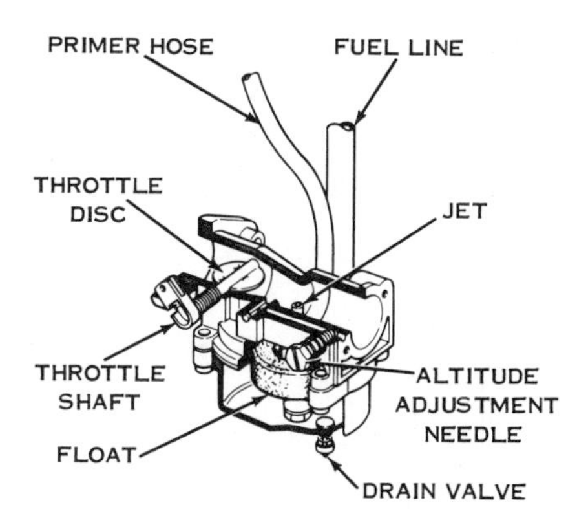

Fig. 7-12. This carburetor is completely automatic. There are no adjustments except for operating altitude.

The touch-and-start primer carburetor (Fig. 7-13) incorporates a primer. When the primer is pressed, air pressure forces fuel from the carburetor bowl into the carburetor throat. When the engine is cranked, the intake valve is opened letting fuel into the combustion chamber for one pull starting. The fuel that was forced from the carburetor bowl during priming is replaced by fuel from the fuel tank.

The touch-and-start carburetor is primed by sealing the bowl vent with a finger. Depress the bulb to pressurize the carburetor bowl. Crank the engine.

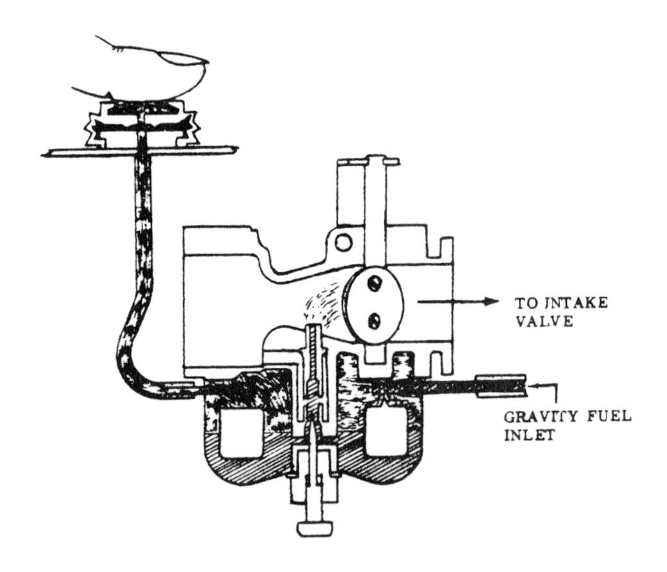

Fig. 7-13. When the primer of the touch-and-start carburetor is pressed, air pressure forces fuel from the bowl into the carburetor throat.

Some engines have a carburetor having a *vapor return line* from the carburetor back to the fuel tank (Fig. 7-14). A vapor return line is used if the area around

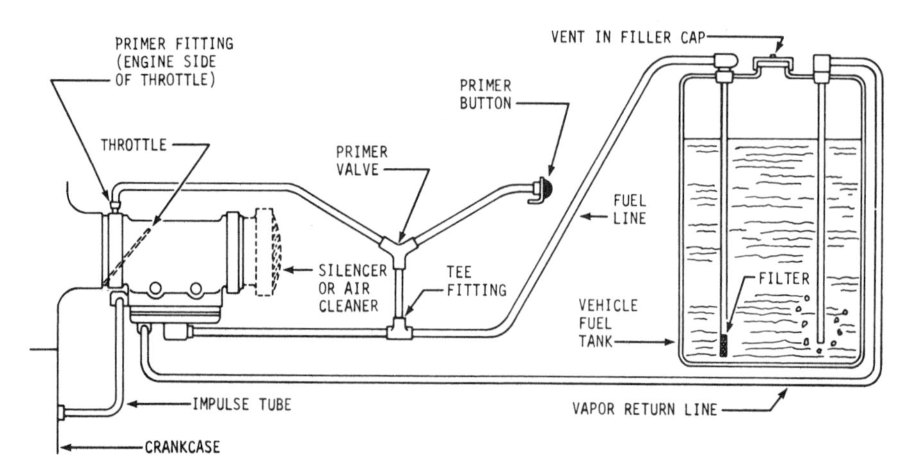

Fig. 7-14. A vapor return line prevents vapor lock; any vapor formed is directed back through a separate return line to the fuel tank where the pressure is relieved.

the carburetor or the air into the carburetor becomes hot enough to vaporize the fuel inside the carburetor. The pockets of vapor formed stop all flow of fuel—the engine is *vapor locked* and will remain vapor locked until the temperature drops low enough for the vapor to return to a liquid state. The vapor return line prevents vapor lock; any vapor formed is directed back through a separate return line to the fuel tank where the pressure is relieved.

There are many designs of carburetors. In addition to the carburetors already illustrated, Figure 7-15 shows some of the many configurations. Exploded views of several carburetors are shown in the maintenance section of this chapter in Figure 7-27.

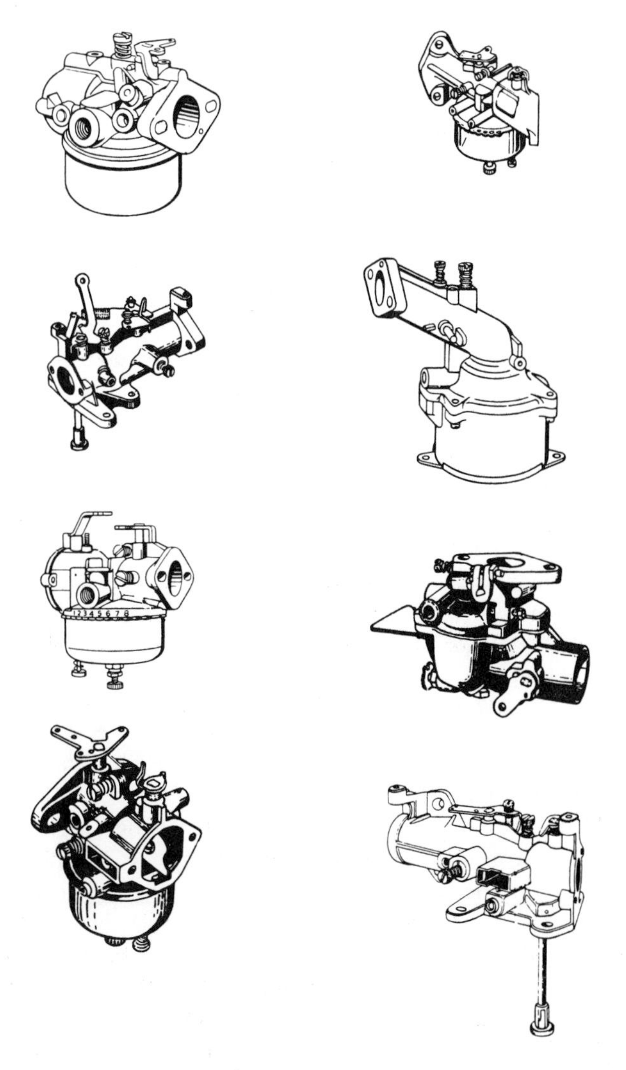

FIG. 7-15. Carburetors come in many designs.

7-2. GOVERNORS

A governor for a small gasoline engine serves two purposes: it keeps the engine operating at a constant speed when there are changes to the load on the engine, and the governor prevents the engine from running at a speed above a predetermined speed set by the manufacturer's engineering design staff. Keeping the engine at a constant speed is desirable, particularly in such applications as when the engine is used to drive an electrical generator. Speed control is maintained from a no load condition to a full load condition. Preventing overspeeding protects the engine and the operator from engine speeds that could cause the engine to tear apart.

There are two types of governors: a pneumatic, or air vane governor, and a mechanical, or centrifugal force governor. Both types are linked to the carburetor throttle and are usually adjustable by one means or another. Initially the throttle presets the engine at operating speed. When the load increases (such as when a lawn mower runs into extra thick and high grass), the engine speed tends to decrease. This causes the governor to operate opening the throttle a little wider to allow more air-fuel mixture from the carburetor into the combustion chamber. This increases the engine power bringing the engine to the desired speed set by the throttle. When the engine speed tends to increase beyond the throttle preset speed because of a light engine load, the governor operates to close the throttle slightly resulting in a slightly lower engine speed. This back-and-forth action of the governor keeps the engine speed relatively constant under all conditions from no load through full load.

Some engines are designed to accelerate freely from idle to governed speed. On these engines, tension on the governor spring does not take place until the governed speed is reached.

Not all engines have a governor—only those operated under varying loads. Engines such as outboards are operated under constant load in the water and do not need a governor. The load is constant at a given throttle speed.

PNEUMATIC, OR AIR VANE, GOVERNORS

The air vane, or pneumatic, governor operates with air pressure. A vane is located near the flywheel (Fig. 7-16) such that the air flow generated by the flywheel blades hits the vane and causes it to change position; the vane is connected through a mechanical linkage consisting of a rod, plate, lever, and spring to the engine throttle. A delicate balance is set up between a spring force and the air (pneumatic) pressure on the vane. A decrease in air pressure caused by a decrease in flywheel speed (an increase in the load on the crankshaft) is felt on the air vane. The spring force overcomes the force on the air vane causing the linkage to move and increase the throttle valve opening in the carburetor. The engine speed increases to compensate for the additional load on the crankshaft. When the load decreases, the engine speeds up; the increased air pressure from the flywheel blades onto the vane causes the governor linkage to close the throttle. Thus the engine speed is maintained relatively constant over the varying load. When the engine is not running, the throttle is open; when the engine is running, the air on the vane causes the throttle to partially close.

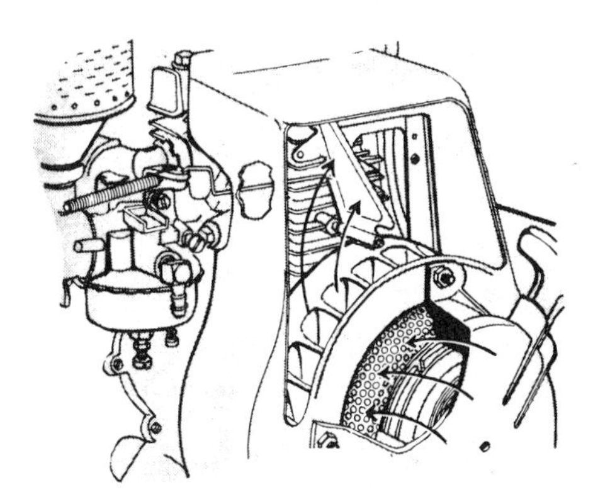

FIG. 7-16. The air vane, or pneumatic governor operates with air pressure.

The vane is made of sheet metal or plastic and pivots within the air-cooling shroud. The engine is not to be operated when the shroud is removed because the air would not be directed to the vane; hence there would be no governor control and the engine could overspeed. The engine speed is set by setting the tension on the governor spring.

MECHANICAL, OR CENTRIFUGAL FORCE, GOVERNOR

Instead of an air vane, mechanical governors use centrifugal weights, flyweights, or flyball weights on a geared shaft, associated linkages, and a governor spring to keep the engine operating at a constant speed. The engine speed depends upon the initial tension applied to the spring either by a speed adjusting lever or by the remote control throttle lever. The weights move in and out by centrifugal force as the engine speed increases and decreases, respectively. An arm and rod linkage from the flyweight assembly causes the throttle to move.

For example, if the speed of an engine is reduced due to an increased load, the flyweights move inwardly because of a decrease in centrifugal force due to the decreased engine speed. Movement through the arm and rod linkage causes the governor spring to overcome the flyweight force. The throttle is opened wider and the engine speeds up to increase the horsepower and keep the speed constant even with the increased load. When the engine speeds up, the centrifugal force causes the flyweights to be thrown outwardly; the throttle is closed slightly and the engine is maintained at a constant speed.

When the engine is not running, the mechanical governor pulls the throttle open. As the engine speed increases, the flyweights move out by centrifugal force to close the throttle. At the governed speed flyweights overcome the spring tension and the throttle will not open any further. The governed speed can be changed by varying the tension on the governor spring. The greater the tension on the governor spring, the higher the governed speed.

The governor shown in Figure 7-17 uses centrifugal weights that are mounted on pivot pins on the side of the camshaft or governor gear. As the engine begins to slow down from an increased load, the sets of weights allow the governor yoke to move toward the cam or governor gear. This lateral movement of the yoke activates the governor shaft assembly which transmits the action through connecting linkage to open the throttle. The movement of the governor shaft and the amount that the throttle is opened will be proportional to the loss of engine speed. The throttle will open slightly to restore the lost speed.

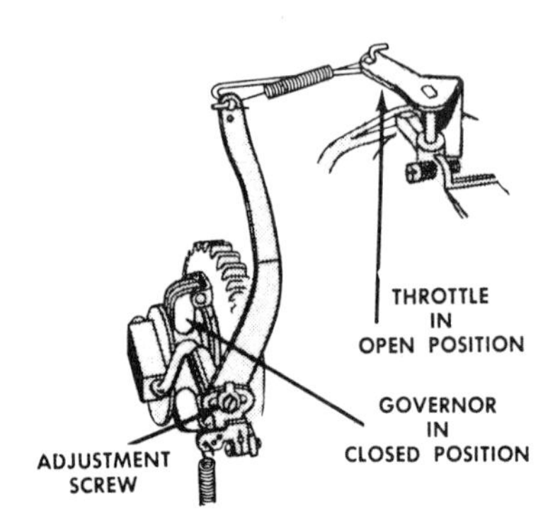

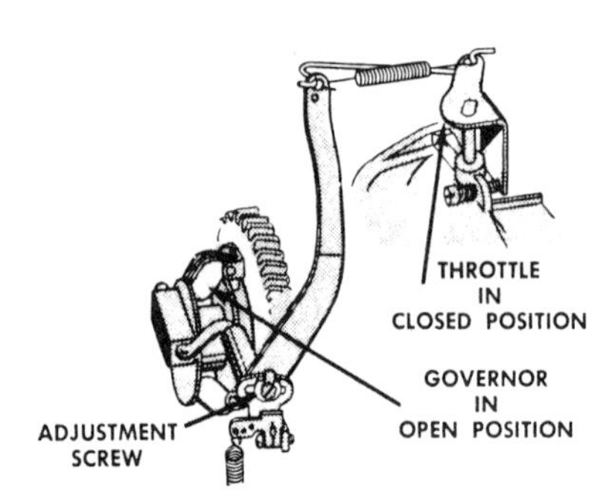

FIG. 7-17. Mechanical governors use centrifugal weights, flyweights or flywheel weights on a geared shaft. Centrifugal weights are shown in these illustrations.

When the load is removed, the governor will reverse the operation to prevent the engine from overspeeding. When the engine is stopped, the governor weights fall dead toward the center of the camshaft or governor gear allowing the governor yoke to move all the way over against the cam or governor gear. This causes the governor shaft to open the throttle wide by means of the governor spring and connecting linkage.

The governor throttle spring is the control or balance acting against the centrifugal force produced by the governor weights. The speed of the engine depends upon the initial tension applied to this spring either by the speed adjusting lever or by the remote control throttle lever.

A governor using flyballs inside a governor cup and spacer is shown in Figure 7-18. The flyballs tend to fly outward because of the centrifugal force caused by

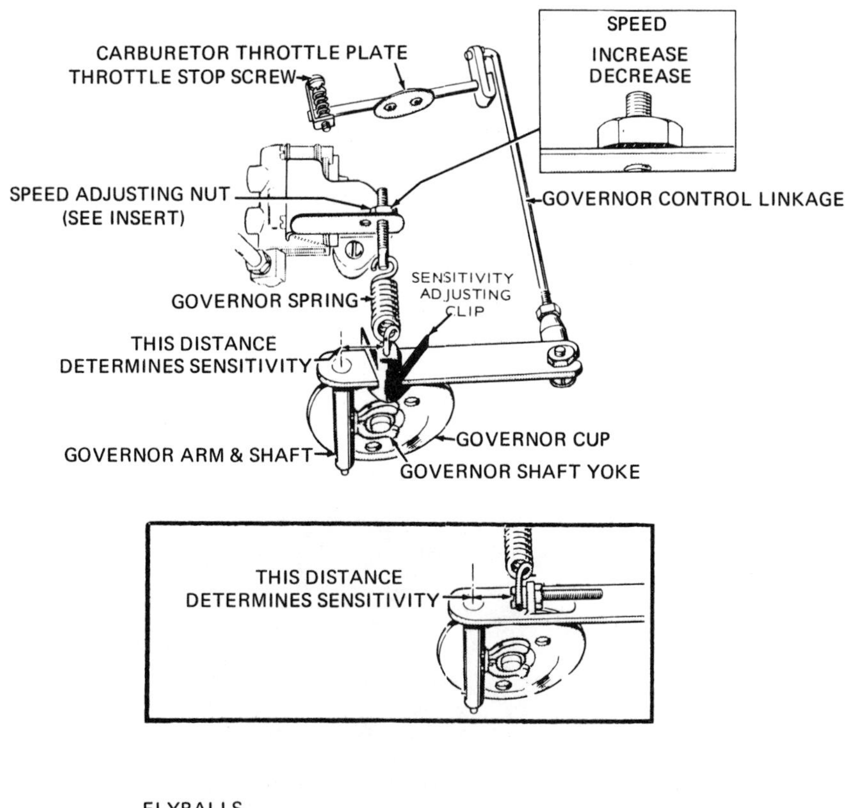

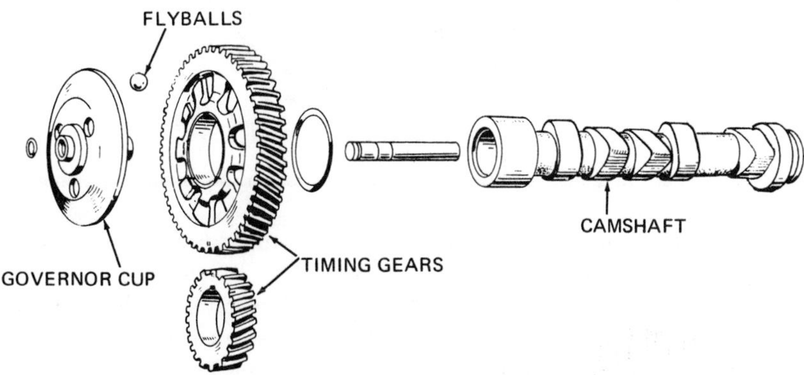

Fig. 7-18. This mechanical governor uses flyballs inside of a governor cup.

the rotating camshaft. The flyballs press against the cup which is against the governor shaft yoke and governor arm and shaft. When the engine speeds up from a decreased load, the camshaft speed increases. This causes the flyballs to move outwardly from centrifugal force; this moves the governor arm and shaft and finally the governor control linkage that closes the throttle valve slightly to decrease engine speed. A decrease in engine speed causes the governor to operate in the opposite order to that of an increased engine speed.

7-3. FUEL TANKS

A fuel tank is a vessel which stores gasoline for four-stroke cycle engines and stores a gasoline-oil mixture for two-stroke cycle engines. Fuel tanks are usually made of metal although there are some plastic tanks (Fig. 7-19). In addition to the storage vessel itself, a tank also has a vented cap, a fuel pickup or valve, and a filter. The vented cap allows atmospheric pressure to act on the fuel and forward it to the carburetor fuel bowl or diaphragm pump when a decrease in pressure is caused by piston movement or a pumping action or when fuel flows by gravitational force from the bottom of the tank to the carburetor. The admission of air also prevents a vacuum from forming in the tank with subsequent stopping of fuel flow.

Fig. 7-19. Fuel tanks take on many shapes. Some are plastic, although most are metal.

Fuel exits from either a pickup tube or a tube through the bottom of the fuel tank. A fuel pickup tube is used when the fuel is pumped from the fuel tank. The fuel is exited from the top of the fuel tank in this case. A flexible weighted pickup tube is used which goes to the deepest part of the tank when the engine tilts. A filter is attached to the end of the pickup tube to prevent foreign matter from exiting from the tank along with the fuel. Refer to Section 3-10 for procedures on cleaning the filter and to Section 7-15 for cleaning the fuel tank.

If the fuel is fed to the carburetor by gravity feed, the fuel tank is located above the carburetor (Fig. 7-19). An output tube is located in the lowest part of the tank for the fuel to run out of. Sometimes a shutoff valve (Fig. 7-20) is attached to the bottom of the tank at the output tube. A filter element is located just inside the tank. Refer to Section 3-10 for procedures for cleaning the filter and to Section 7-15 for cleaning the tank.

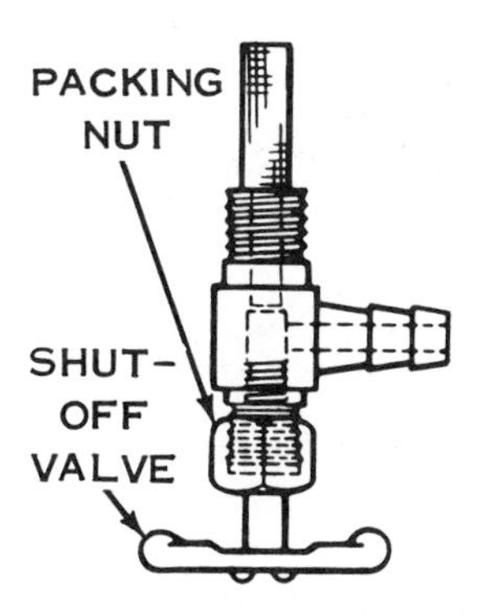

FIG. 7-20. A fuel shutoff valve is sometimes located at the bottom of the tank.

Trouble that is often difficult to detect is caused by a clogged vent cap. The clogging causes a partial vacuum to occur in the tank as fuel flows from the tank, but atmospheric pressure cannot enter the tank. This causes the engine to run a few minutes and then to stop as the vacuum builds up. With the engine stopped, the vacuum decreases and the engine will start again, only to stop running again. To quickly determine if an engine with a condition thus described does have a clogged vent, momentarily remove the tank cap and determine if the problem still exists; if it does not, cover the tank and then clean the vent cap (Section 3-11).

The fuel tank used with some outboard engines is often remote from the engine itself. A fuel pump in the engine is used to lift fuel from the tank to the carburetor. Priming is achieved by squeezing a primer bulb in the fuel line (Fig. 7-21) several times. The tank air inlet and the fuel outlet are seated until the supply line is attached with two valve plungers that depress the valves off their seats. A drain screw is provided in the fuel tank to facilitate draining and cleaning.

Drain and flush the fuel tank at least once a year and at every tune-up or major repair. Drain the tank through the drain screw. Flush with clean gasoline. Remember, do not store gasoline in the tank over a long period of time.

The upper housing and fuel line assembly of the fuel tank of an outboard

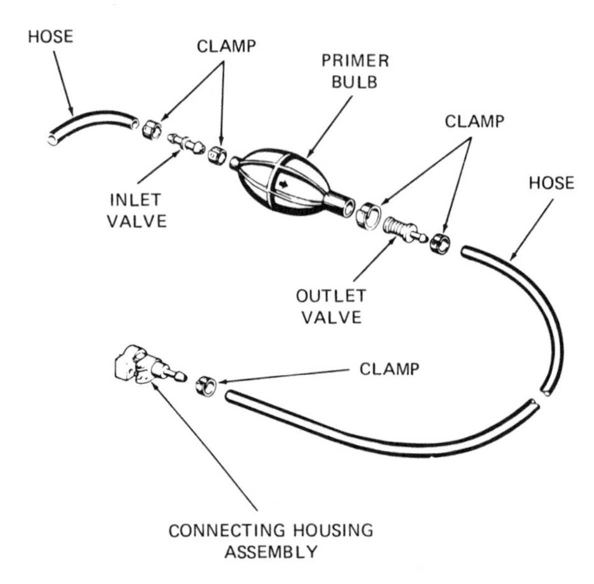

FIG. 7-21. A primer bulb is used to lift fuel from the tank to an outboard engine fuel pump to prime the pump.

engine (Fig. 7-22) connects to the engine via a plastic hose. The hose has a connector that attaches to release valves that seat tightly to prevent gasoline or fumes from leaking out, but opens to provide a clear passage for air to enter

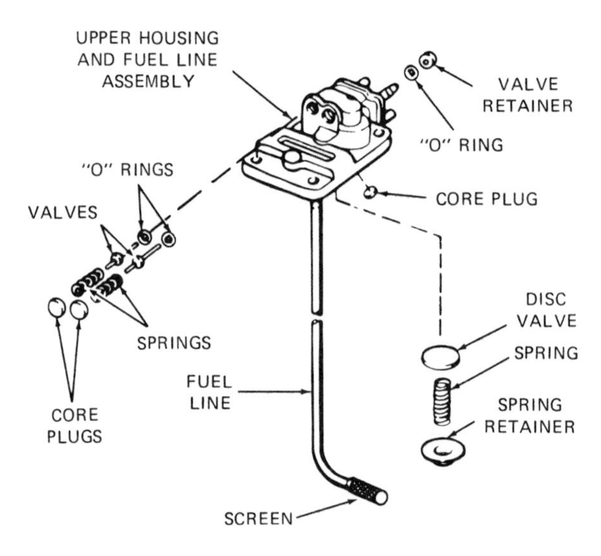

FIG. 7-22. The upper housing and fuel line assembly of the fuel tank of an outboard engine connect to the engine via a plastic hose.

the tank and for fuel to be drawn out by the fuel pump (outboard engines have a diaphragm displacement fuel pump operated by changes in crankcase pressure). Air must enter the tank as fuel exits so that a vacuum does not form. A screen filters the fuel as it is drawn into the engine. The air inlet disk valve prevents fumes from escaping from the tank when the fuel hose is connected, but allows air to enter the tank.

Fuel tank level indicators are incorporated into some fuel tanks; in Figure 7-23, the fuel level indicator is part of the upper housing and fuel line assembly of a portable remote outboard engine fuel tank. The fuel tank level indicator consists of a float, indicating arm, support, indicator lens, and hardware. The float moves up and down with the level of the fuel and moves the indicating arm such that the pointer indicates the fuel level on the indicator lens.

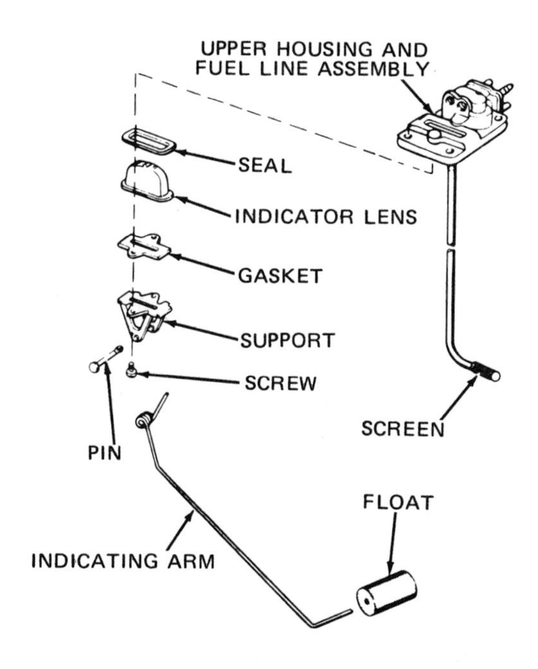

FIG. 7-23. Fuel tank level indicators are incorporated into some fuel tanks.

7-4. FUEL FILTERS (STRAINERS)

Fuel filters, or strainers, are used to filter out dirt and other foreign matter before the fuel enters the carburetor. There are four types of fuel filters: a filter that is fixed in the fuel tank, a weighted filter at the end of a hose in a fuel tank, a bowl type of filter mounted externally to the fuel tank, and a filter at the end of the fuel line at the fuel pump inlet. Refer to Section 3-10 for a further description and cleaning procedures for fuel filters.

7-5. AIR CLEANERS (FILTERS)

Air filters are used to keep dirt from entering the carburetor with the air which is mixed with fuel to develop a highly volatile mixture for combustion in the engine combustion chamber. The dirt must be blocked by the filter so that tiny dirt particles cannot block the carburetor jets and thereby alter the air to fuel ratio causing inefficient operation or loss of operation altogether. Three types of air cleaners are used to clean the air drawn into the carburetor: dry filter; oiled filter; and oil bath air cleaner. Refer to Section 3-9 for descriptions and cleaning procedures.

It is extremely important to service the air cleaner frequently to clean it and reprepare it to trap additional dirt and other foreign matter; failure to do so will decrease the performance level and the life of the engine. Perform maintenance on the air cleaner at least every 10 operating hours—every 3 to 5 hours in dusty environments. Refer to Section 3-9 for service procedures.

7-6. SUMMARY—FUEL SYSTEMS

The main elements of a fuel system are: fuel tank, fuel filter, fuel shutoff valve, air filter, and carburetor. The fuel system provides the correct air-fuel mixture at all engine speeds to the combustion chamber for ignition at the precise time for combustion with resulting power to drive the piston down imparting motion to the crankshaft. A summary of the functions of the main elements of a fuel system and of the carburetor for a typical single cylinder two-stroke cycle engine are shown in Figure 7-24. Study the diagram and refer back to applicable theory sections if you need review.

7-7. FUEL SYSTEM ADJUSTMENTS

Minor engine tune-ups that should be performed on the majority of small gasoline engines brought into the repair shop and can also be performed by the do-it-yourself repairman include carburetor and governor adjustments. These adjustments are also made after fuel system repairs.

7-8. CARBURETOR ADJUSTMENTS

The initial main fuel and idle fuel carburetor adjustments should be made as recommended by the manufacturer (Refer to Fig. 7-2). Remember with two-cycle

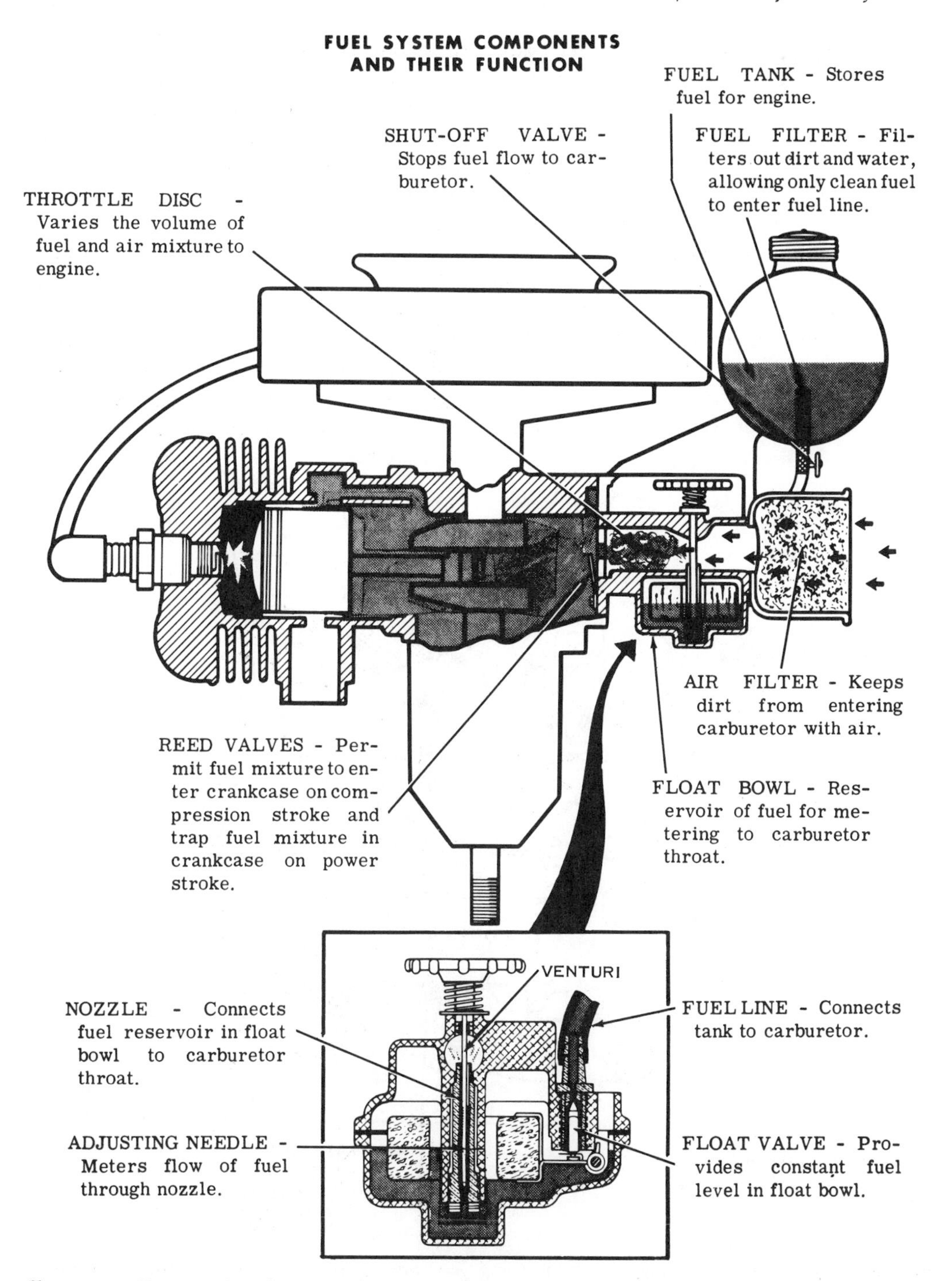

FIG. 7-24. Review this diagram of a two-cycle engine. If necessary, refer back to the text for detailed theory.

engines, a change in carburetor adjustment also changes the amount of lubrication; don't change the carburetor settings unless absolutely necessary. To readjust, stop the engine and turn both the idle fuel and main fuel (high-speed) adjusting needles all the way in. Turn the needles with your fingers—do not force the needles closed; you could damage the needle and seat. Then open (counterclockwise rotation) the needles the number of turns recommended by the manufacturer (usually about 1 to 1½ turns). Turning the adjustments clockwise causes the mixture to become more lean (less fuel in proportion to air); counterclockwise rotation causes the mixture to become richer. Fill the fuel tank to one-half capacity and restart the engine. After the engine is restarted, and operating at normal temperature (and load, if possible), slight readjustment may be made for best performance. Make sure that lean settings are avoided on two-cycle engines because sufficient amounts of lubrication may not be available to prevent the scoring of cylinder walls and other damage. Figure 7-25 illustrates the initial settings for

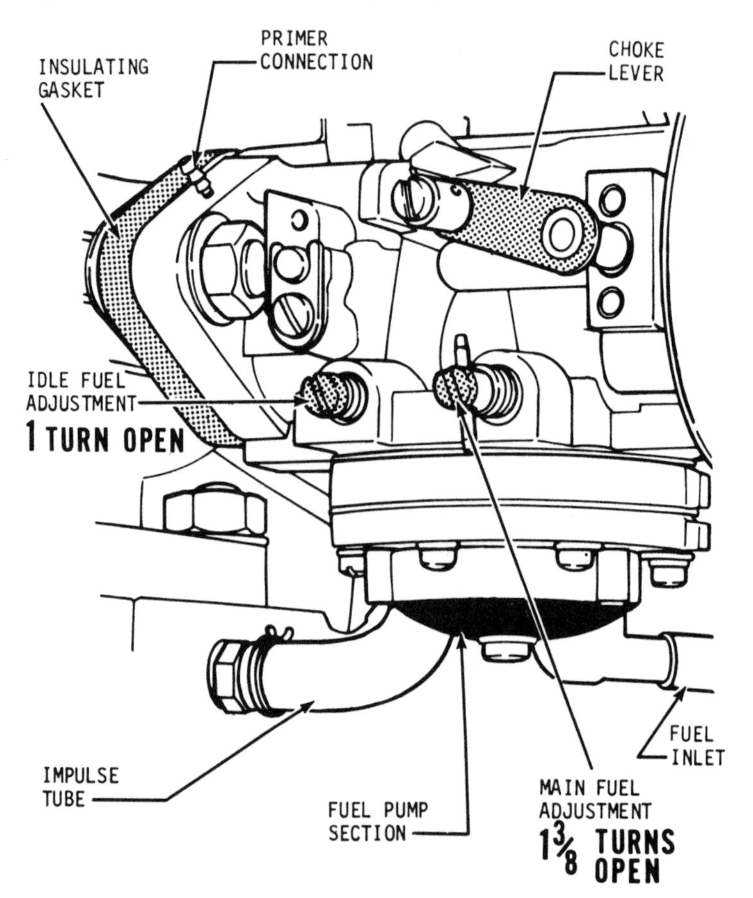

FIG. 7-25. The carburetor is initially set by using your fingers to rotate the main fuel adjustment and the idle fuel adjustment closed; and then by rotating the adjustments 1 to 1½ turns counterclockwise.

the idle fuel and main fuel needle adjustments for one manufacturer's two-cycle single cylinder diaphragm type carburetor engine. General carburetor adjustments are given in the following steps for use when manufacturer's directions for specific engines are not available:

1. Fully open the choke (Fig. 7-26).

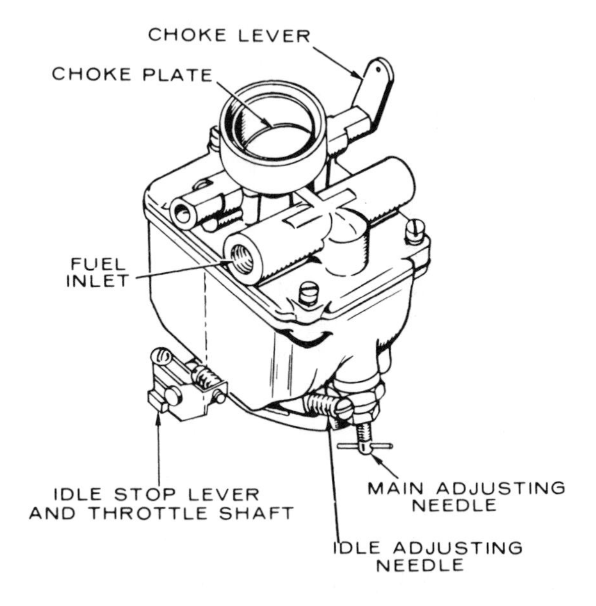

FIG. 7-26. The main fuel adjusting needle is adjusted for smooth engine operation when the throttle is ⅔ to ¾ open; the idle fuel adjusting needle (if available) is adjusted for smooth engine operation with the throttle at idle. The idle speed adjusting screw sets the idle speed.

2. Set the throttle to a position corresponding to two-thirds to three-fourths of the maximum speed.

NOTE

The following adjustments are made in one-eighth turn increments. Wait approximately five seconds between increments to give the engine time to respond.

3. Adjust the main fuel high-speed needle (usually the larger physical size of the two adjustments—the main fuel high-speed and the idle fuel adjustments) to a position where the engine runs its fastest and smoothest. If this adjustment is made with the engine under load, this is the correct setting. If the engine is not under load, open the needle an additional one-eighth turn.
4. Close the throttle control to the idle position.

5. Adjust the idle fuel needle varying the engine speed until the smoothest operation is attained. This adjustment is not available on some carburetors because the idle mixture port is fixed. (Refer to Fig. 7-2.)
6. Adjust the idle *speed* adjusting screw for proper idle speed. Use a vibrating tachometer and the manufacturer's specification, if available. The idle speed is usually about one-half of the maximum operating speed. If the engine has a governor, manually hold the throttle closed while making this adjustment. This idle *speed* adjustment is a mechanical stop adjustment (rather than a fuel mixture adjustment) on the throttle stop lever made to prevent the throttle valve from closing completely causing the engine to stall.
7. Move the throttle quickly from idle to three-fourths of the maximum power position. The engine should accelerate smoothly; if it does not, the fuel mixture is too lean. Readjust the high-speed needle for a richer mixture and accelerate again. If the engine smokes excessively or appears sluggish, the fuel mixture is too rich. Readjust the high-speed needle and repeat the acceleration test.

QUICK REFERENCE CHART 7-1

CARBURETOR ADJUSTMENTS

1. Turn idle fuel and main fuel (high-speed) needle adjustments all the way in by hand.
2. Open needle adjustments 1 to 1½ turns.
3. Start engine and let it reach operating temperature.
4. Apply normal engine load, if possible.
5. Fully open choke.
6. Set throttle to ⅔ to ¾ maximum speed.
7. Adjust main fuel high-speed needle until engine runs its fastest and smoothest. (If engine is *not* under load, open needle additional ⅛ turn.)
8. Close throttle to idle position.
9. Adjust idle fuel needle until engine runs smoothest (this adjustment is not available on some carburetors because the idle mixture port is fixed).
10. Adjust idle *speed* adjusting screw for proper idle speed.
11. Check that engine accelerates quickly from idle to ¾ of maximum speed. If it does not, refer to Section 7-8.

7-9. GOVERNOR ADJUSTMENTS

If the governor is not governing the engine at the proper speed, it should be adjusted. A tachometer is required to set the engine to the desired speed. Some governors are adjusted by rotating a screw; on others, it is necessary to bend the linkage or lever slightly or to change the governor spring to one having the correct tension; do not stretch the existing spring; replace it (Figs. 7-16 to 7-18). Do *not* increase the speed of the engine beyond the design limitations; increased speed can cause damage to the engine and harm to the operator.

7-10. FUEL SYSTEM MAINTENANCE

If the small gasoline engine troubleshooting procedures in Chapter 9 indicate trouble in the fuel system, then it is necessary to perform maintenance on the fuel system. This maintenance may include inspection, cleaning, adjustment, and replacement of faulty parts. Many times, adjustment of the carburetor will repair the fuel system without the need for disassembly, internal cleaning, and repair of internal parts. If adequate preventive maintenance is performed on the engine, there is usually little need for corrective maintenance.

The following paragraphs cover maintenance of carburetors, fuel pumps, governors, and fuel tanks.

7-11. CARBURETOR MAINTENANCE

Carburetors can be completely disassembled for inspection and repairs. Refer to the manufacturer's specific recommendations for a specific carburetor. An abbreviated typical manufacturer's repair procedure is included in this section as a guide. Repair kits containing replacement parts are available for most carburetors.

For carburetor disassembly, select a clean work area; dirt and carelessness are the causes of most carburetor trouble. Be aware that some solvents and cleaners have a damaging effect on the synthetic rubber parts used in carburetors. It is best to use a petroleum product for cleaning. Do *not* use alcohol, acetone, lacquer thinner, benzol, or any solvent with a blend of these ingredients unless the rubber parts and gaskets are removed. If you are in doubt about a solvent, test an old used part in it and observe the reaction. Figure 7-27 illustrates exploded views of a number of carburetors used in small gasoline engines. Study the construction of each.

A typical carburetor is cleaned, disassembled, inspected, repaired, and reassembled as follows (install new gaskets at all times):

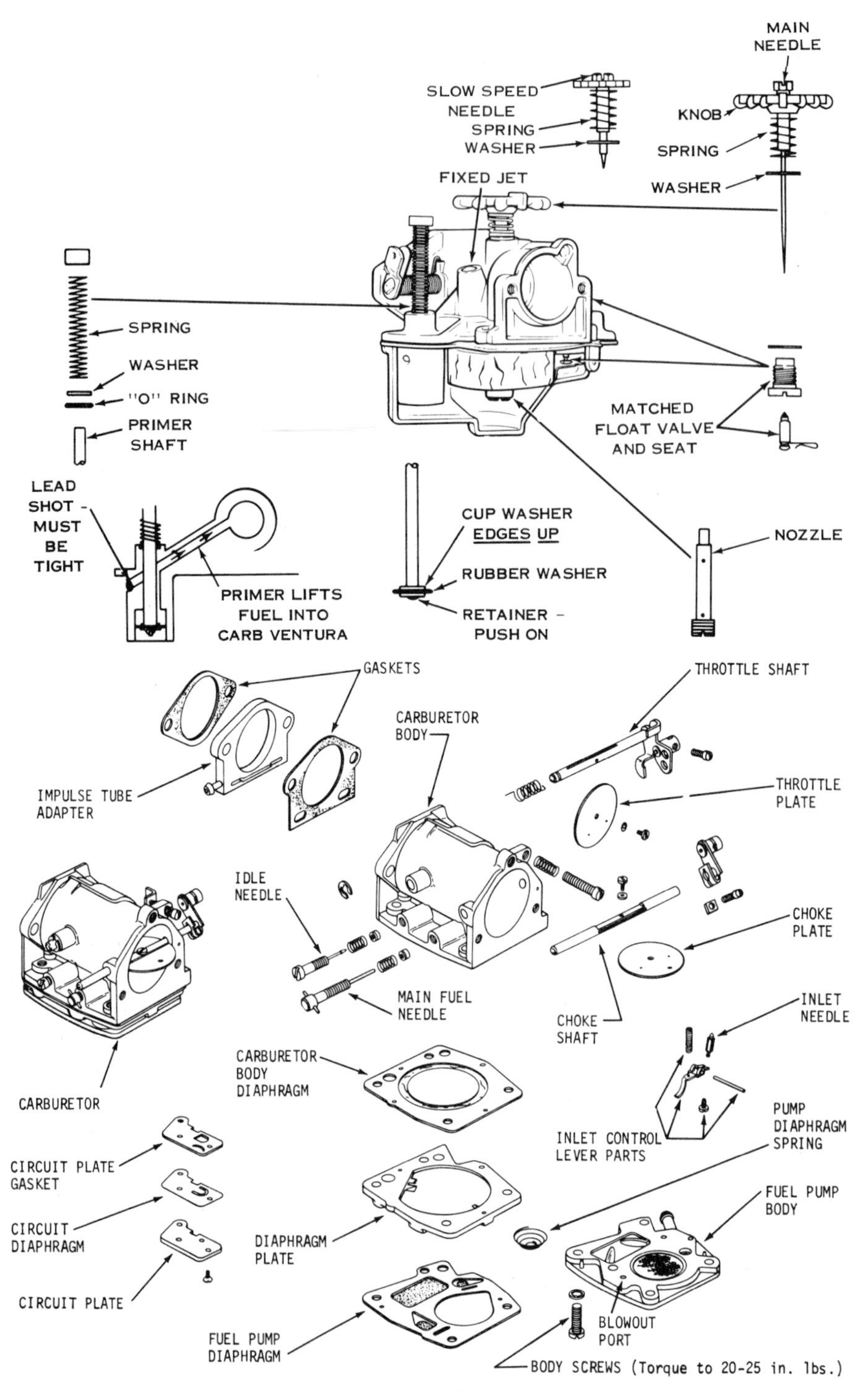

Fig. 7-27. Carburetor designs are many and varied. Study the construction of each of these designs.

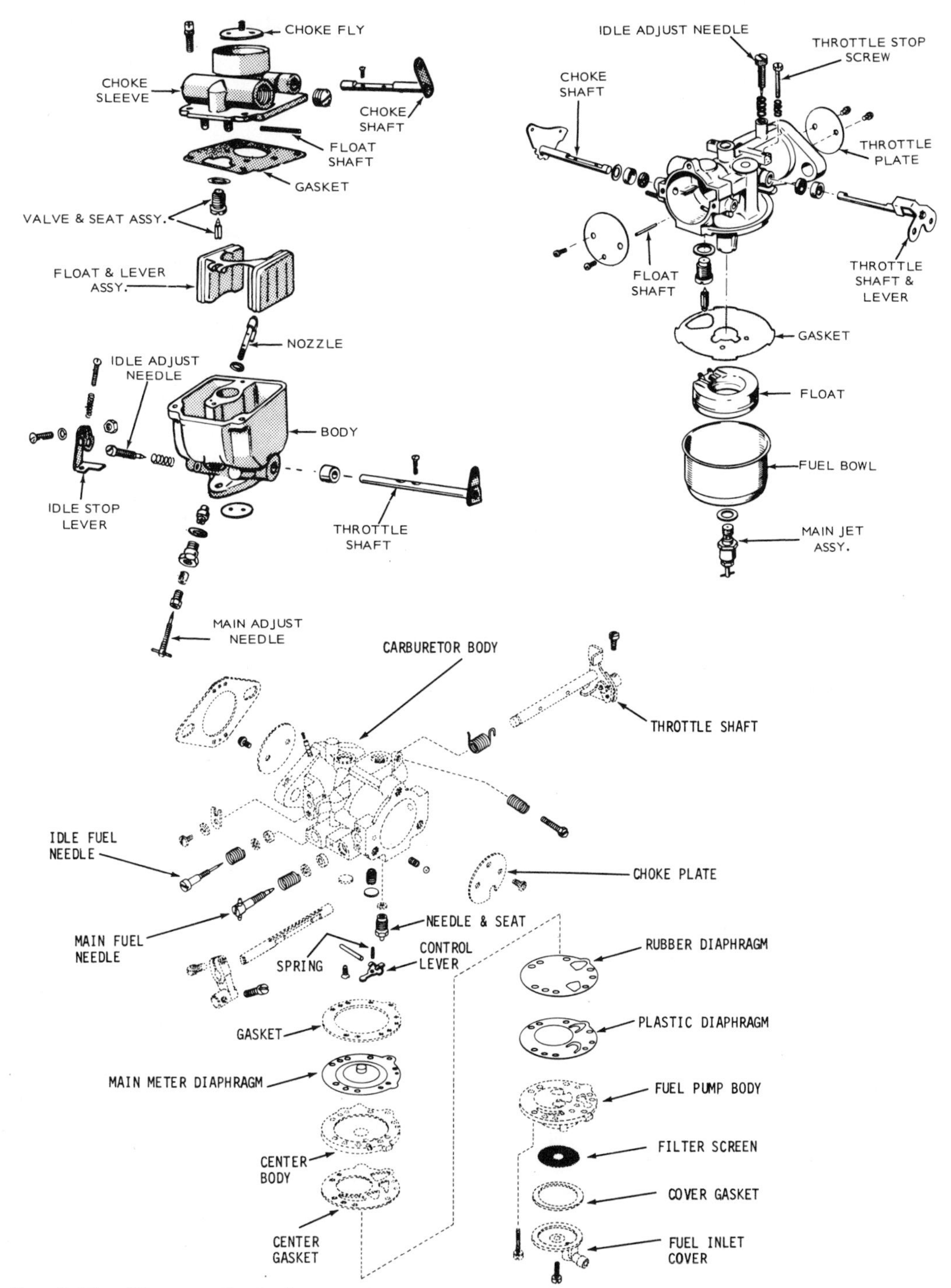

FIG. 7-27. (Continued) Carburetor designs.

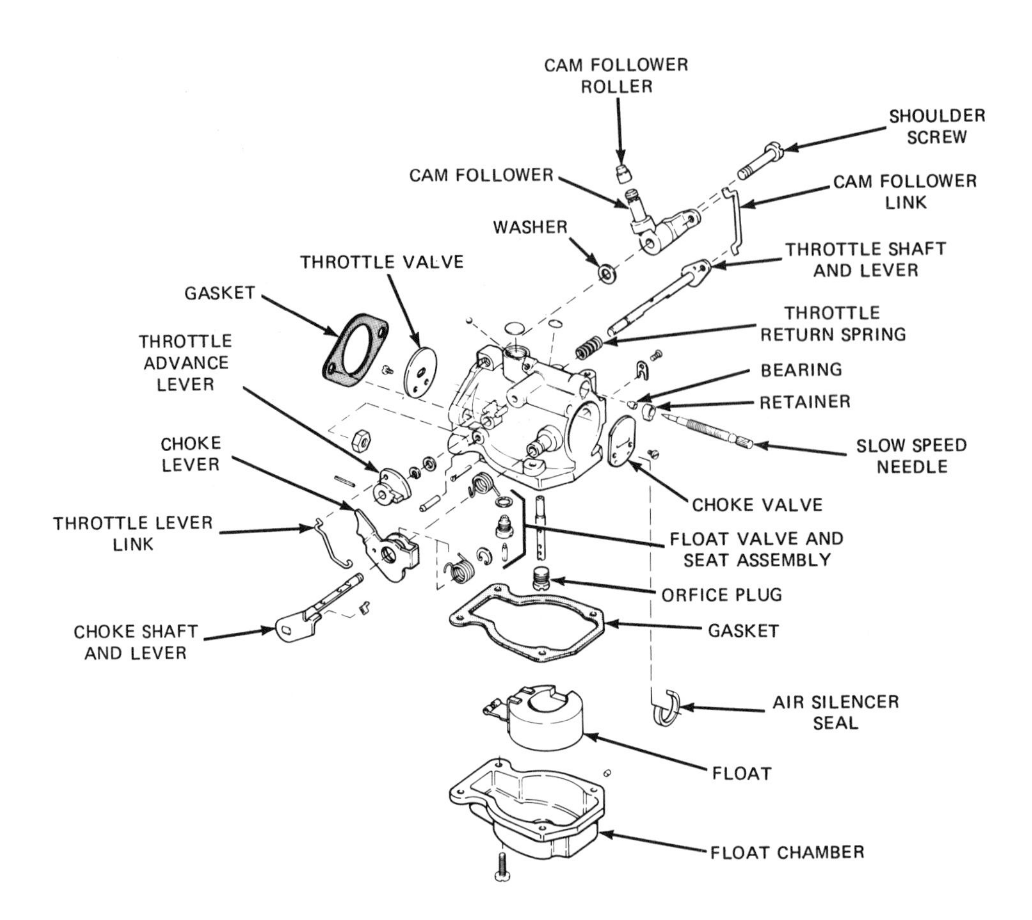

Fig. 7-27. (Continued) Carburetor designs.

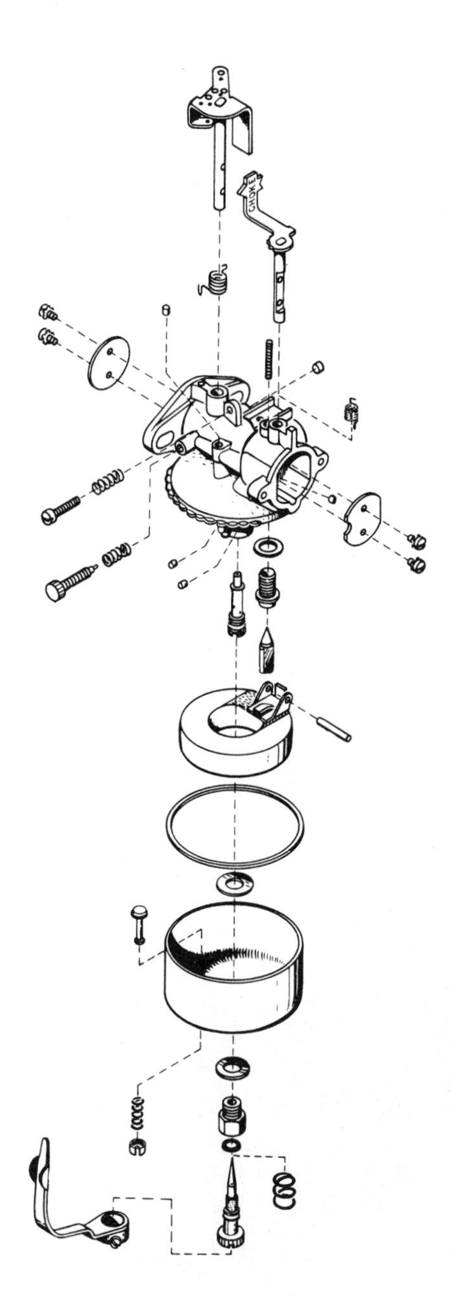

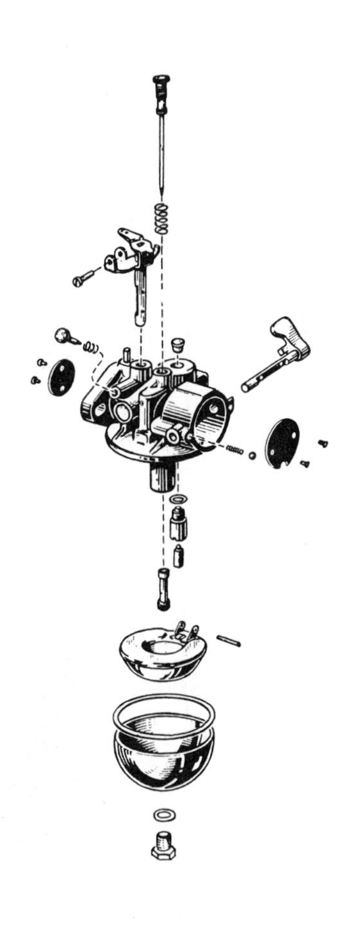

Fig. 7-27. (Continued) Carburetor designs.

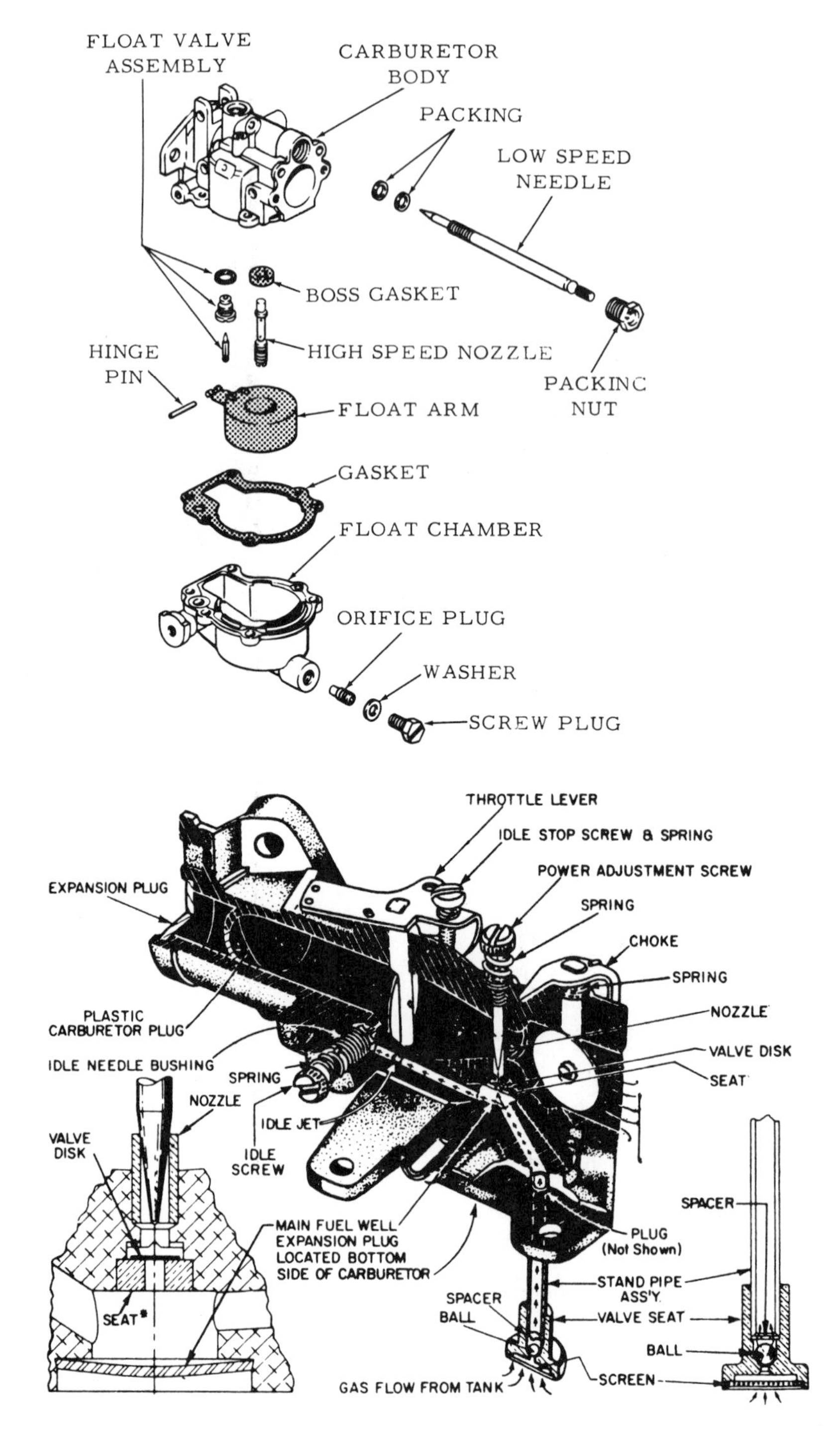

Fig. 7-27. (Continued) Carburetor designs.

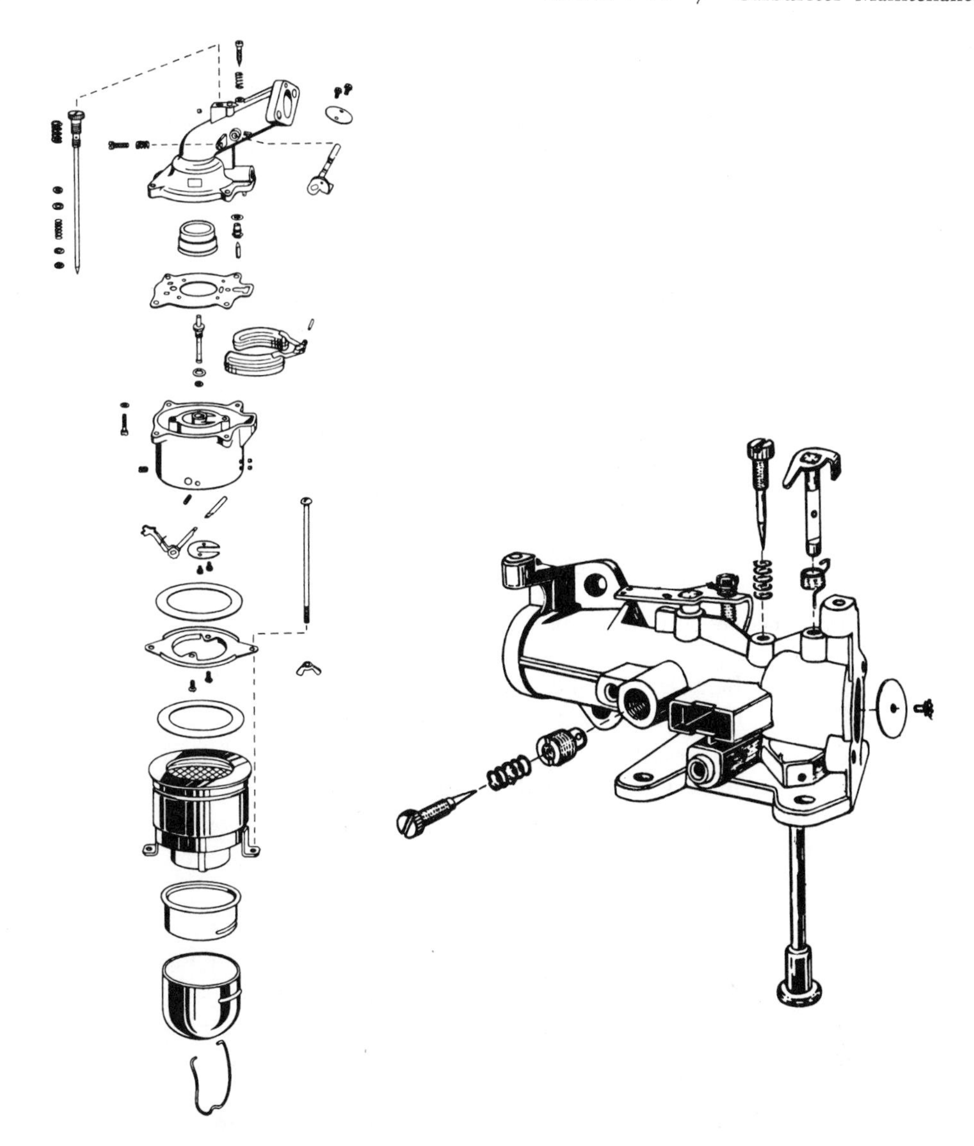

Fig. 7-27. (Continued) Carburetor designs.

1. Clean the carburetor by flushing it with fuel. Blow it dry with compressed air. Do not dry the carburetor or other fuel system components with a cloth because lint may stick to the parts and cause trouble in the reassembled carburetor.
2. Inspect the carburetor for cracks in the casting, bent or broken shafts, loose levers or swivels, and stripped threads. Repair or replace parts or the complete carburetor.

3. Inspect all valve needles for grooves or other defects (Fig. 7-28). If grooved or otherwise defective, replace. Use a magnifying glass for inspection. Inspect the valve seats and associated hardware. Most needle valves and seats are matched pairs; replacement of one necessitates replacement of the other.

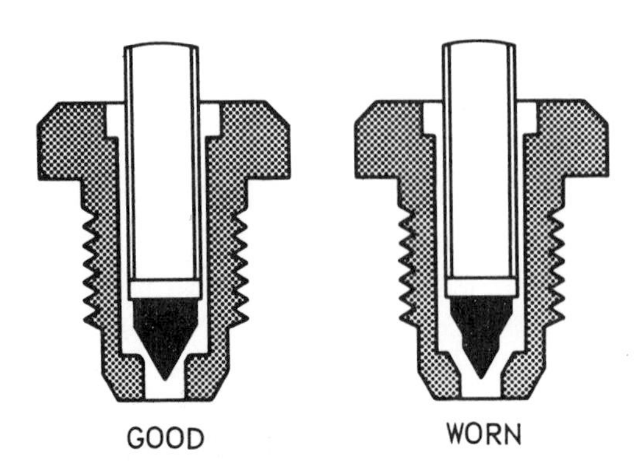

FIG. 7-28. Inspect needle valves and seats for grooves and other defects.

4. Remove the filter cover, cover gasket, and filter screen. Clean the filter screen by flushing it with solvent.
5. Remove the fuel pump cover casting, fuel pump diaphragm, and gasket. Inspect the pump diaphragm—it must be flat and free of holes.
6. Remove the diaphragm cover casting, the metering diaphragm and the diaphragm gasket. Inspect the diaphragm for holes, tears, and imperfections.
7. Remove the fulcrum pin retaining screw, fulcrum pin, inlet control lever, and inlet tension spring. Use caution in removing these parts because the spring pressure may cause the inlet lever to fly out of the casting. Inspect the parts for wear or damage and replace as required. The inlet control lever must rotate freely on the fulcrum pin. Do not stretch the spring.
8. Inspect the idle bypass holes to insure they are not plugged. Do *not* push drills or wires into the metering holes because they could alter carburetor performance. Blow plugged holes clean with compressed air. Remove the main nozzle check valve and inspect it. Replace if defective.
9. Remove the throttle and choke plates. The edges are tapered for exact fit into the carburetor bores. Remove screws and pull the shafts out of the casting. Examine for wear and replace as necessary.
10. Clean all parts, inspect, and reassemble.
11. The reed plate (two-cycle engines) can be cleaned with solvent when the carburetor is cleaned. Refer to Section 2-30.

QUICK REFERENCE CHART 7-2

CARBURETOR MAINTENANCE

1. Select clean work area.
2. Clean carburetor in solvent and blow dry.
3. Inspect for cracks, broken shafts, loose levers or swivels, and stripped threads. Repair or replace.
4. Inspect needle valves and seats. Replace, if required.
5. Clean filter.
6. Inspect fuel pump.
7. Inspect diaphragm.
8. Remove and inspect internal parts.
9. Inspect idle bypass holes. Blow with compressed air to clean.
10. Remove throttle and choke plates. Examine for wear. Replace, if required.
11. Clean all parts and inspect.
12. Reassemble.

7-12. CARBURETOR FLOAT VALVE MAINTENANCE

Some of the problems that can occur with the float valve are listed in Table 7-1. Float valve adjustments are discussed in the next paragraph.

There is a possibility that the float may require adjustment from time to time. This is accomplished by removing the float bowl and gasket. Invert the carburetor and adjust the float by bending the float arm with a pair of needle nose pliers. Do not apply pressure to the cork float. The top of the float cork is set a specified distance from the edge of the carburetor body; refer to the specific manufacturer's recommendations (Fig. 7-29).

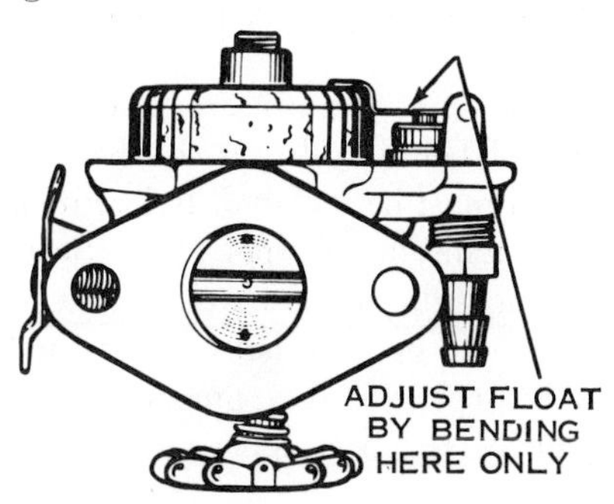

FIG. 7-29. Adjust the carburetor float by bending the float arm. Refer to the manufacturer's specifications for the correct distance.

Table 7-1
FLOAT VALVE PROBLEMS

Cause	*Effect*	*Remedy*
Gum in fuel.	Stops up openings.	* Clean out carburetor with solvent.
Spring wire clip comes off.	Needle may stick shut.	Replace clip.
Needle and seat not matched.	Fuel supply can't be shut off from float bowl.	Replace needle and seat as an assembly.
Float arm not set right.	Set too high—carburetor floods.	Set correctly.
	Set to low—carburetor starves.	Set correctly.
Pivot pin corroded or bent.	Float sticks.	Replace pin.
Float striking nozzle.	Float sticks.	Replace float.
Varnish off float.	Float soaks up fuel, changing floating characteristics.	* Replace float.

* Never allow a strong solvent to come in contact with the float. Sometimes the float is varnished cork. If the solvent removes the varnish, the float will absorb gasoline and its floating characteristics will change.

7-13. DIAPHRAGM FUEL PUMP MAINTENANCE

If fuel is not reaching the carburetor, check the fuel pump. Diaphragm pump failure is usually due to a leaking diaphragm, valve, valve gasket, a weak or broken spring, or wear in the drive linkage. Disassemble, inspect the parts, make necessary replacements, and reassemble. A repair kit is available for some carburetors (Fig. 7-30).

Outboard engine fuel pumps are diaphragm-displacement pumps. If the pump is suspected of being faulty, it can be removed and tested; a faulty pump is not repairable and must be replaced. Before removing and testing the fuel pump, remove, clean, and reinstall the fuel filter (Section 3-10). Also remove the fuel line and blow through all passages and lines with compressed air to be sure they are not clogged. If these steps do not cure the problem, the fuel pump is probably bad and should be replaced. If a pressure gauge and tachometer are available, the fuel pump can be tested as follows (Fig. 7-31):

1. Connect a fuel pressure gauge between the carburetor and the fuel pump.
2. Loosen the fuel tank cap to release any pressure.

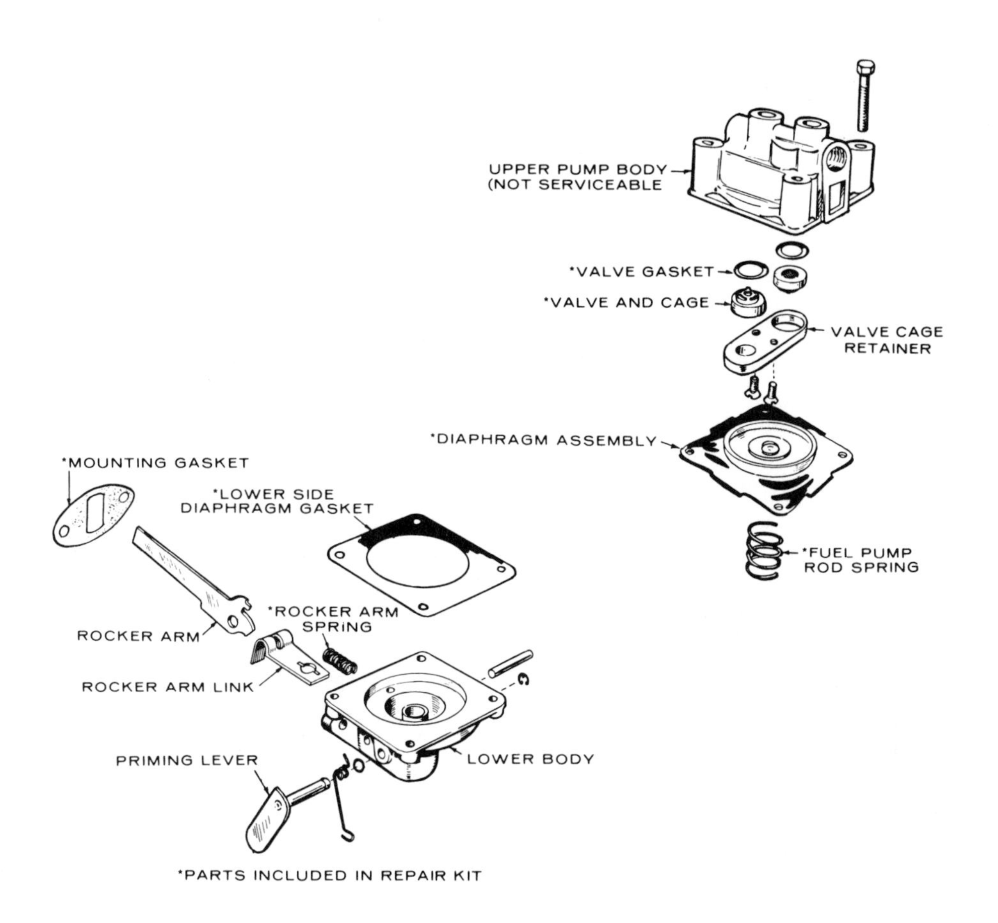

Fig. 7-30. Diaphragm fuel pump failures are usually due to a leaking diaphragm, valve, valve gasket, a weak or broken spring, or wear in the drive linkage.

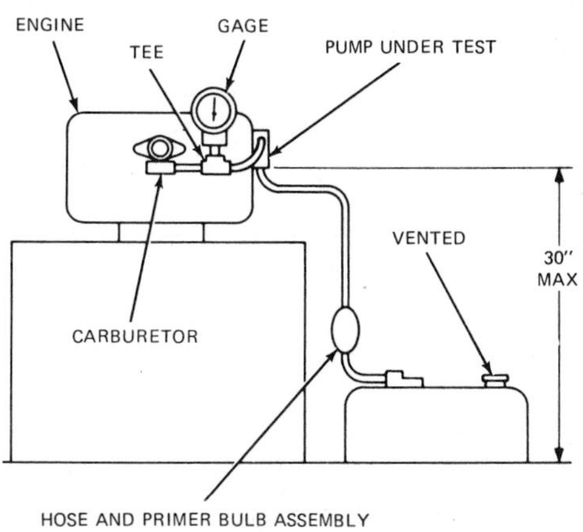

Fig. 7-31. This test setup is used to test an outboard engine fuel pump.

3. Start the outboard engine and check for the following pressures at the indicated speeds:

RPM	PSI
600	1
2500–3000	1.5
4500	2.5

7-14. GOVERNOR MAINTENANCE

Governors should always be repaired to prevent the engine from overspeeding. Replace broken plastic air vanes. Straighten or replace bent metal air vanes. Remove bends or dents in the shroud so that the air blast generated within the shroud by the flywheel blades is properly directed to the air vane.

Inspect the parts of mechanical governors for wear, possible damage, or bent flyweights or flyweight supports. Check that the collar or thimble operates freely on the camshaft or governor gear and that the governor shaft moves freely in the bushing. The flywheel needs to be removed to repair externally mounted centrifugal governors. The engine has to be disassembled to repair internally mounted centrifugal force governors. Adjust governors in accordance with the procedures in Section 7-9.

QUICK REFERENCE CHART 7-3

GOVERNOR MAINTENANCE

1. Replace broken plastic air vanes. Straighten or replace bent metal air vanes.
2. Remove bends or dents from shroud.
3. Inspect parts for wear or damage. Replace, as required.
4. Adjust as per Section 7-9.

7-15. FUEL TANK MAINTENANCE

Fuel tank maintenance includes cleaning the tank and repairing holes in the tank. Clean the vent cap (Section 3-11), filter (Section 3-10), shutoff valve and the tank itself. Close the fuel shutoff valve and disconnect the hose between the tank and the carburetor. Clean the tank by swishing and pouring out the gasoline; partially fill the tank with fresh, clean, gasoline, swish, and discard again. Fill the tank with fresh, clean, fuel (with the proper ratio of fuel to oil for two-cycle engines). Open the fuel shutoff valve and let about a cup of fuel run out. Turn the shutoff valve off, reconnect the hose, and discard the fuel in the cup.

Holes in metal fuel tanks can be repaired by soldering them. First remove the tank from the engine and then discard all fuel and let the tank dry thoroughly. Then, using correct soldering procedures, solder the holes which cause the leaks (be sure to remove all paint and rust from the area to be soldered).

Small holes in plastic tanks can be repaired with a warm soldering iron. Remove all fuel and let the tank dry thoroughly. Apply heat to the area to soften some plastic which can be moved to the area of the hole to cover it.

FUEL TANK LEVEL INDICATOR MAINTENANCE

Remove the fuel level indicator and upper housing and fuel line assembly (Fig. 7-23). Be careful not to damage the indicator float or the fuel line screen. Check for free movement of the indicator on the indicator pin. Inspect the indicator to insure that the float arm is not bent and that the float is not damaged or oil soaked. Lift the indicator lens out of the upper housing and clean it with solvent or soap and water. Check the seal and gasket for cracks or shrinkage that could cause leakage.

UPPER HOUSING AND FUEL LINE ASSEMBLY MAINTENANCE

Dirt may keep the release valves (Fig. 7-22) of the upper housing and fuel line assembly from seating properly. This could cause fuel and fuel vapor to leak out. The valves are cleaned by removing the core plugs and disassembling. Replace the valve seat O-rings with new rings to insure a tight seal. The disk valve spring retainer is staked to the upper housing and is removed, if necessary, by filing off the burrs. Restake the retainer with a small punch.

FUEL HOSE AND PRIMER BULB MAINTENANCE

Check the fuel hose and primer bulb (Fig. 7-21) for cracks. Replace immediately if hairline cracks are indicated.

REASSEMBLY OF FUEL SYSTEM COMPONENTS

Whenever a part of the fuel system, particularly the carburetor, is disassembled, new gaskets should be installed. This will prevent problems from arising on reassembly or soon after. When installing nozzles in the carburetor, unscrew the control knob a few turns to avoid accidentally tightening the needle onto the seat too tightly which could damage both the needle and the seat.

7-16. GASOLINE

Gasoline is a hydrocarbon—that is, it is chemically made up of atoms (Section 4-1) of hydrogen and carbon. These atoms are split apart when the air-gasoline mixture is burned. The atoms combine with the atoms of oxygen in the air to form carbon dioxide and water.

Gasoline is refined from petroleum, or crude oil. By itself, gasoline is not satisfactory to give maximum engine performance while also protecting the engine. Thus *additives* are added to the basic hydrocarbon to make a gasoline having the necessary desirable characteristics. Some of these desirable characteristics are:

1. vaporization at low temperatures, different altitudes, and different climates for starting ease.
2. prevention of gum formations.
3. antirust and anti-ice (in the carburetor) properties.
4. antiknock properties.
5. must not deteriorate during short storage periods.
6. must burn cleanly to reduce air pollution.

An *octane rating* is a measure of the anti-knock qualities of gasoline. Iso-octane produces the least knock and is rated at 100. Normal heptane produces the most knock and is rated at zero. A fifty-fifty mixture of iso-octane and heptane results in a mixture rated at 50.

Higher octane gasolines are used with engines having higher compression ratios (Section 2-8). The higher the octane rating, the higher the resistance to knock. The following can be used as a guide:

Compression Ratio	*Gasoline*	*Octane Rating Desired*
5:1 to 7:1	Low grade	70–85
7:1 to 8.5:1	Regular grade	88–94
9:1 to 10:1	Premium	100
10:1 to 10.5:1	Super premium	Over 100

A good quality regular gasoline with an octane rating of at least 92 is recommended for some two-cycle engines. Premium grades may also be used and may be beneficial especially in warmer weather to prevent detonation or after-run conditions. Heavier buildup should be expected in the combustion chamber when using premium gasoline because these fuels contain greater amounts of lead additives which leave deposits. New nonleaded gasolines are not yet approved for some two-cycle engines.

Most small gasoline engine manufacturers recommend the use of regular grade automobile engine gasoline. Additives should not be used. Some manufacturers recommend lead free or leaded regular gasoline. The use of highly leaded gasoline should be avoided, however, as it causes deposits on valve seats, spark plug points, and on the cylinder head; this shortens engine life.

Always use the type of gasoline specified by the manufacturer. The gasoline should be clean and fresh. Do not use gasoline that has been stored over a period of time because stored gasoline becomes gummy and can cause the carburetor and ports within the small gasoline engine to become clogged.

The desirable attribute of gasoline to easily vaporize presents a hazard to the user. When combined with air, it presents an extremely volatile atmosphere. Be sure that gasoline is stored in tightly sealed metal containers so that vapors cannot escape into the air. The metal container also protects against breakage if the container is accidentally dropped. Gasoline, as well as other volatile liquids, can evaporate into the air in a room if the container is not tightly closed. The introduction of a spark such as the turning on of a light switch or motor, a gas pilot, or a cigarette can cause an immediate violent explosion. Refer to Section 1-4 for additional information on the safe handling of gasoline.

7-17. PREIGNITION AND DETONATION

Two terms that you should be familiar with are preignition and detonation; both have undesirable effects that cause engine damage. Preignition is combustion caused by any hot spot such as a glowing carbon particle, a rough metal edge, improperly seated valves, or an overheated spark plug, before the spark from the spark plug is timed to fire the air-fuel mixture in the combustion chamber (Fig. 7-32). Preignition causes loss of power because the too early combustion pressures try to drive the piston down while the crankshaft and flywheel are driving the piston up. Preignition succeeds in slowing the piston down.

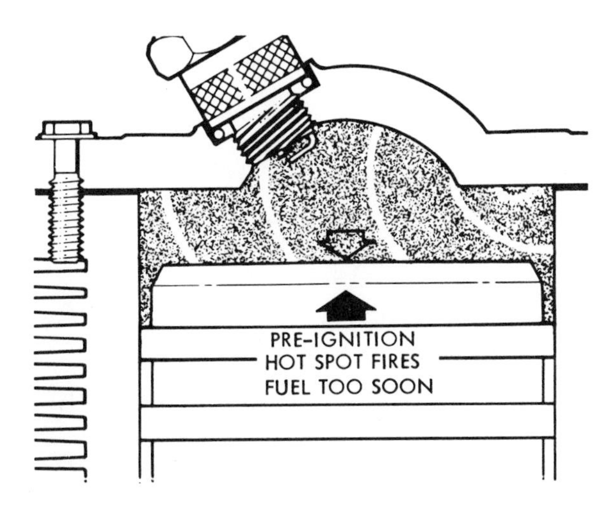

Fig. 7-32. Preignition causes loss of power.

Detonation is combustion caused by a glowing carbon particle, or by a hot spot, at the *same* time as the ignition system spark fires the air-fuel mixture (Fig. 7-33).

Instead of a smooth wave of burning which results in lots of power, detonation causes at least two simultaneous waves of burning and the air-fuel mixture is burned too quickly releasing all the power at once. The excessive heat and pressure try to slap the piston down instead of shoving it down. The slap is heard every time an engine pings.

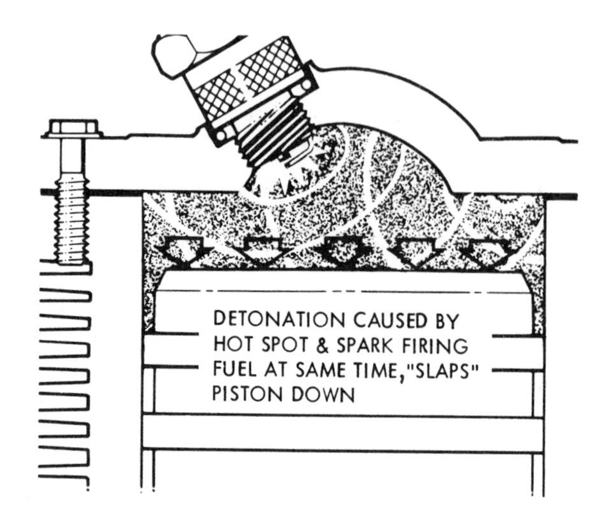

Fig. 7-33. Detonation seriously damages a piston head.

8

Lubricating Systems

Oil is used in small gasoline engines to lubricate, cool, clean, and seal. The oil used must be the correct oil recommended by the engine manufacturer and it must be used in the correct amount and changed periodically (Table 3-1).

This chapter provides information on oils, two-cycle engine lubrication, four-cycle engine lubrication, and additional lubrication points on small gasoline engines and machinery. Lubrication system maintenance consists only of periodic maintenance and of inspection and replacement of parts when the engine is disassembled. Chapter 3 provides periodic maintenance instructions for checking the oil level and changing the oil in the four-cycle engine (Section 3-13); lubricating the two-cycle engine (Section 3-14); and lubrication of nonengine parts (Section 3-15).

8-1. PURPOSE OF LUBRICATING OILS

Oil is essential to engine performance and durability. It provides lubrication, cooling, cleaning, and sealing. Without sufficient oil, engine parts will quickly seize and the engine will burn up.

The main purpose of a lubricating oil is to reduce the friction between moving metal surfaces of engine parts such as between the crankshaft journals and main bearings, connecting rod end cap and crank, or between the connecting rod small end and the piston (wrist) pin. As you know, friction is a resistance to motion between two moving surfaces that contact each other; friction causes heat. The part surfaces are designed with allowances such that there is a small gap between the surfaces for a layer of oil (and for expansion of the metal parts from heat). The layer of oil between the moving parts greatly reduces the friction between the parts. This in turn reduces the heat buildup resulting in less engine wear and increased power. In addition to reducing friction between moving parts, the layer of oil between the parts acts as a shock absorber. For example, when combustion occurs, the powerful force drives the piston down. This force is felt through the connecting rod to the crankshaft. The shock of these forces is partially absorbed by the layer of oil that is sandwiched between the mating parts; the oil is then squeezed out from between the parts.

Lubricating oil also dissipates heat, particularly in four-cycle engines. Oil is splashed or pumped onto all moving mating surfaces. The oil droplets fall off the parts and run down the cylinder and crankcase walls or simply drip to the base of the crankcase. The oil imparts the heat to the metal walls which are air- or water-cooled to dissipate the heat. The splashing or pumping of oil streams onto the engine parts also cleans the parts. Dirt, carbon, microscopic metal pieces, and other foreign matter are rinsed to the bottom of the crankcase. Periodic oil changes remove this matter (called sludge) from the crankcase and replaces the dirty oil with clean oil for continuing rinsing during engine operation. Bathing of the parts with oil also prevents rusting of the parts from moisture caused by condensation and as a byproduct of combustion.

Oil also aids to prevent the power of the combustion force in the combustion chamber from escaping between the piston rings and the cylinder walls into the crankcase. Thus blow-by is reduced resulting in more power transfer to the crankshaft. The oil clings in the piston oil ring(s), the piston grooves, and on the cylinder walls to make an airtight seal against the escaping gases of combustion.

Add quality grade lubricating oils according to manufacturer's recommendations (Sections 3-13 and 3-14); quality oils used as recommended will help ensure that the engine will operate over its expected design life. But even quality oils begin to break down and lose their properties during the first use. This is largely due to accumulation of contaminants—water, sludge, carbon, gum, acids, dirt, lacquer-like substances, and fine metal particles. The oil filter (if there is one) does not remove all of this contamination; therefore the engine oil requires periodic changing.

If the engine oil level drops below the recommended level, engine lubrication is inadequate; there may not be sufficient oil to allow the lubricating system to operate adequately. Friction increases because of lack of lubrication; this causes the parts to heat excessively. Excessive heat causes additional metal expansion that further increases friction—and the cycle continues to spiral. There is no lubricant to be thrown onto the parts; heat dissipation by oil flow to the air- or water-cooled crankcase ceases. More friction also causes more microscopic metallic pieces because of metal moving over metal; these pieces are not cleaned off when the oil is not applied to the surface because of an inadequate supply of oil to the slinger or pressure system.

The lack of lubricating oil is most significant in the four-cycle engine because of the method of operation of the lubricating system (Section 8-5). Two-cycle engines do not have a lubricating *system.* Oil is mixed in proper proportion to the fuel. If the engine has fuel, it has lubricating oil (presuming the oil was properly mixed into the fuel, Section 3-14); if the engine runs out of fuel, it runs out of lubricating oil, but at least the engine cannot run.

8-2. OILS

Oils are made by many different manufacturers and are designated by the Society of Automotive Engineers (SAE) and the American Petroleum Institute (API). Different oils with different *viscosities* and different *additives* are used for specific applications and environments. Engine manufacturers specify the type of oil to be used in their product; the specified oil is determined by designed engine part clearances and expected operating temperatures. After considerable use of the engine, the manufacturer may recommend the use of a heavier weight oil.

Viscosity is the tendency of an oil (or other substance) to resist flowing. An increase in temperature reduces the viscosity—the oil loses body and gains fluidity.

Engine oils are specified by *weight* which is an indication of the thickness of the oil. A *heavier* oil is therefore a *thicker* oil.

Engine oil weights are specified by SAE designations ranging from 5W to 50; the letter W indicates oil that is used in temperatures below freezing. The oils most often used are 10W, 20W, 20, and 30; 10W and 20W oils are used most frequently in the winter and 30 oil most frequently in the summer. A lower number (as 5 or 10) indicates a thinner (less viscous), but a more fluent oil; the lower the number, the lower the temperature at which the oil will flow. *Multiviscosity* oils span several oil viscosities. For example, SAE 10W-30 contains 10W low viscosity (thin) oil, 20W, and 30, a high temperature viscosity oil. A 30 oil is a *heavier* (less viscous) oil than a 20W oil.

Oils are designated for use by the API; the designations range from use under mild conditions (SA) to use under extremely severe (SE) conditions. Most small gasoline engine manufacturers recommend oils with SC or SD designations.

The SD oil (formerly rated MS—severe) protects against operation under most unfavorable or severe conditions such as numerous starts and stops, high loads, high temperature operating, overloading, or operation at extreme or maximum speeds. SD rated oils are adequate for most any application of four-cycle air-cooled engines. The SD oil also protects against engine deposits, rust, corrosion, and wear.

SC oil (formerly suitable for MS) has about the same characteristics as SD, but not as effective. It provides control of high and low temperature engine deposits, corrosion, and part wear.

SB rated oils (formerly MM—moderate) are for operation under mild conditions when minimum protection is required; SB oils should not be used under severe conditions. SA oils (formerly ML—light) have no set performance requirements and are not recommended by small gasoline engine manufacturers. SE oils (for service under extreme conditions) provide more protection against oil oxidation, high temperature engine deposits, rust, and corrosion than do the SD or SC oils. SE oils are used for engines operating in extreme conditions such as short, cold weather operation, and high-speed, hot weather, long operating period uses.

Pure oil could not perform its functions of lubrication and protection under adverse operating conditions; additives must be added. The additives include: pour point depressants; oxidation, rust, and corrosion inhibitors; detergent; foam inhibitors; and extreme pressure agents. Pour point depressants keep the oil a liquid at very low temperatures—they make the oil pour easily in cold weather. Oxidation, rust, and corrosion inhibitors prevent oxidation, rust, and corrosion caused by water, acids, and other by-products of high combustion temperatures. Detergents prevent the formation of sludge, gum, and varnish thus helping to keep the engine clean. Foam inhibitors keep the oil in the crankcase from foaming because of the air whipped into the oil by the revolving crankshaft and other engine parts.

Sludge is a thick solution of water, oil, dirt, and carbon. It is beaten by the rotating crankshaft until it is frothy. The water in the sludge is from condensation of droplets of water from the cold air on the crankcase and as a by-product of combustion. If an engine operates for an hour or more at a time, the water is evaporated. If it is impractical to run the engine for more than an hour, then the engine oil must be changed more frequently to remove the sludge.

In choosing a lubricating oil for a four-cycle engine, always select the oil recommended by the engine manufacturer. In lieu of any manufacturer's instructions, the following can be used as a guide; select any high quality detergent oil having the API classification of SC or SD (previously classified MS) as follows:

---

1. Summer: Above 40°F—use SAE 30. If not available, use SAE 10W-30 or SAE 10W-40.
2. Winter: Below 40°F—use SAE 5W-20 or SAE 5W-30. If not available, use SAE 10W or SAE 10W-30.
   Below 0°F—use SAE 10W or SAE 10W-30 diluted 10 percent with kerosene.
   Below −10°F—use SAE 5W.

Four-cycle engines generally use the same oils as automobiles, but two-cycle engines use special oils for two-cycle engines only. The key words on the oil container are *two-cycle oil* for outboards, snowmobiles, chain saws and lawn mowers. Automobile oils are *not* satisfactory in two-cycle engines. Detergent oils are of no use in two-cycle engines either because oil is not splashed or pumped onto the engine parts. Hence, the contaminating deposits are not washed away.

It is also important that the manufacturer's recommendations be followed in mixing oil and fuel for two-cycle engines (refer to Section 3-14). Improper mixing can cause preignition, carbon deposits to form in the combustion chamber causing plug fouling, ring sticking, scuffing of piston skirts, excessive smoking, and deficient lubrication of engine parts.

8-3. FLOW OF LUBRICATING OIL

Lubricating oils are placed on engine moving parts by separation of the oil from the air-fuel mixture in a two-cycle engine and by a *splash* or *pump* system in the four-cycle engine (two- and four-cycle lubrication are discussed in detail in Sections 8-4 and 8-5, respectively). Once the oil is on the metal part surfaces, it must have a path for flowing to the mating surfaces to lubricate the bearing surfaces where the friction is most prevalent.

Access holes or slots are provided in parts to pass oil in and out of the bearing surfaces. These holes or slots are used for replaceable sleeve type bushings and also when the alloy casting is the actual bearing surface. Where pressure lubricating systems are used, oil is pumped to the main crankshaft bearings; the pressurized oil then flows from the main bearings through passages drilled in the crankshaft to the connecting rod bearings (Fig. 8-1). This oil is discharged through small holes that are aimed at the piston pins and cylinder walls. Oil that is splashed or pumped onto cylinder walls is spread by the oil ring(s).

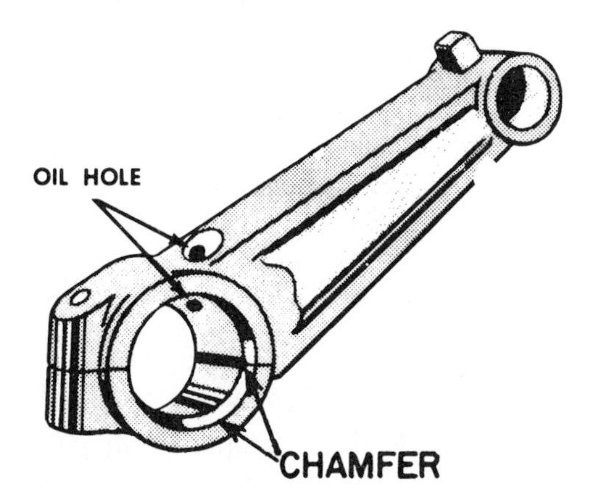

FIG. 8-1. Access holes are provided in parts to pass oil in and out of the bearing surfaces.

Two-cycle engines as well as four-cycle engines have access holes to admit the oil to all bearing surfaces. Some connecting rods (Fig. 8-2) have a wide acceptance slot to allow a larger volume of the oil-gasoline mixture to reach the rod journal area.

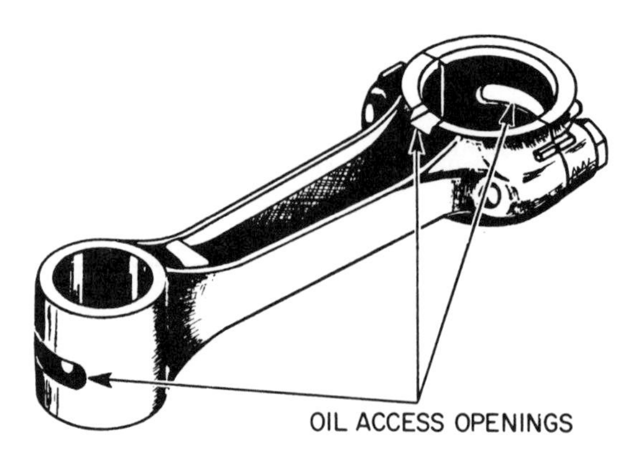

FIG. 8-2. Wide access slots allow a larger volume of the oil-gasoline mixture to lubricate bearing surfaces.

8-4. TWO-CYCLE ENGINE LUBRICATION

The two-cycle engine is lubricated by the oil in the air-oil-gasoline mixture for the engine. On the compression stroke of the piston, a slight vacuum is created in the crankcase. Atmospheric pressure causes the air-fuel mixture from the carburetor to flow into the crankcase. The larger and heavier droplets of oil in the air-fuel mixture vapor drop out of the air-fuel mixture to condense and lubricate the bearing surfaces and other engine parts in the crankcase. When the power stroke takes place, the piston comes down the cylinder opening the exhaust and then the intake port to the combustion chamber. The pressure increase in the crankcase caused by the piston coming down closes the reed valve cutting off the air-fuel mixture to the crankcase. The air-fuel mixture in the crankcase with some oil remaining in it is transferred to the combustion chamber where the tiny droplets of oil in the mixture lubricate the piston, piston rings, and cylinder walls. On combustion, the oil as well as the fuel burns.

Because the oil is always mixed into the fuel, the two-cycle engine is lubricated as long as there is fuel in the fuel tank. There is no concern over oil level. The engine is lubricated regardless of the angle of operation which makes the two-cycle engine ideal for machines such as chain saws. As mentioned the oil is suspended in the fuel vapor and clings to the surfaces of all moving parts keeping them continually coated with a film of oil (Fig. 8-3).

The ratios of two-cycle oil to gasoline are determined by each manufacturer for his specific engine; refer to the manufacturer's instructions for types of oil and mixing ratios and to Section 3-14 for mixing procedures. Most manufacturers recommend SAE 30 SB, SC, or SD (formerly MM and MS) nondetergent oils in the ratios of from 1 part oil to 20 parts gasoline to 1 to 100. One manufacturer's recom-

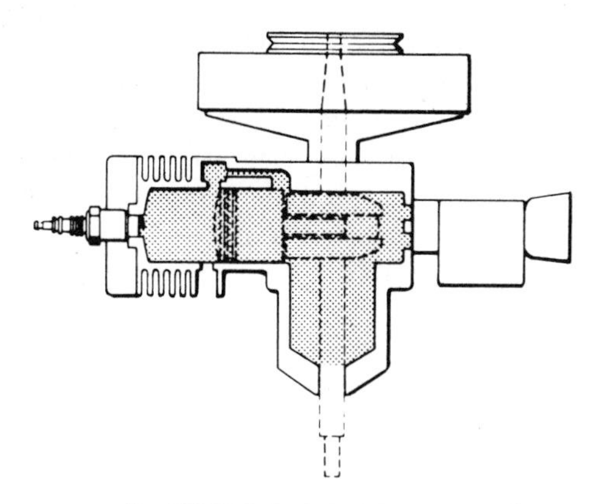

Fig. 8-3. In two-cycle engines, the oil is suspended in the fuel vapor and clings to the surfaces of all moving parts keeping them continually coated with a film of oil.

mendation of oil to gasoline mixture for various numbers of gallons of gasoline is given in Table 3-2.

8-5. FOUR-CYCLE ENGINE LUBRICATION

The four-cycle engine has a separate lubrication system that includes an oil reservoir and a splash or pressure system to distribute the oil. The reservoir *must* be kept above a specified level and must be kept fairly level to prevent a lack of lubrication to moving parts.

The splash system uses a dipper (Fig. 8-4) attached to the connecting rod bearing cap (Section 2-12). As the connecting rod drives the crankshaft in a circular motion, the dipper throws oil from the crankcase (reservoir) onto the crankshaft

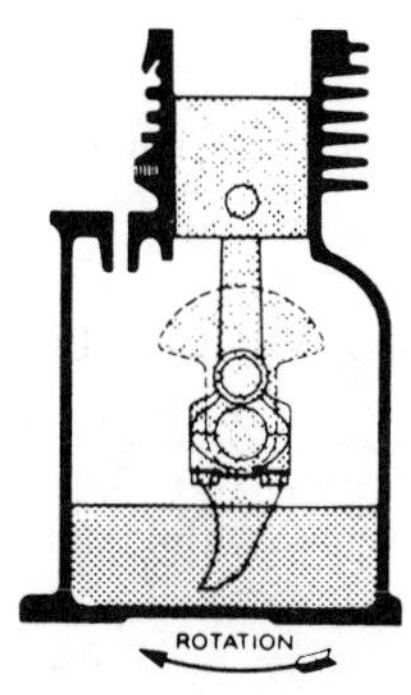

Fig. 8-4. An oil dipper is attached to the connecting rod bearing cap. As the cap rotates, the dipper throws oil from the reservoir onto the four-cycle engine parts.

bearings, camshaft, cylinder wall, etc. Countless droplets of oil are scattered to all parts; the dipper is moving at 2000 to 3000 rev/min. Instead of a dipper, some systems use a revolving slinger (Fig. 8-5) that is driven by a camshaft gear. Still another method is to use a scoop attached to the bottom of the camshaft gear (Fig. 8-6). The scoop sprays oil in a circular path throughout the top of the engine. The dipper, scoop, and slinger protrude approximately 80 percent into the oil; it is therefore important that the proper level of oil is maintained. Some engines have a trough mounted under the dipper. The level of the oil in the trough is maintained constant by a cam operated pump.

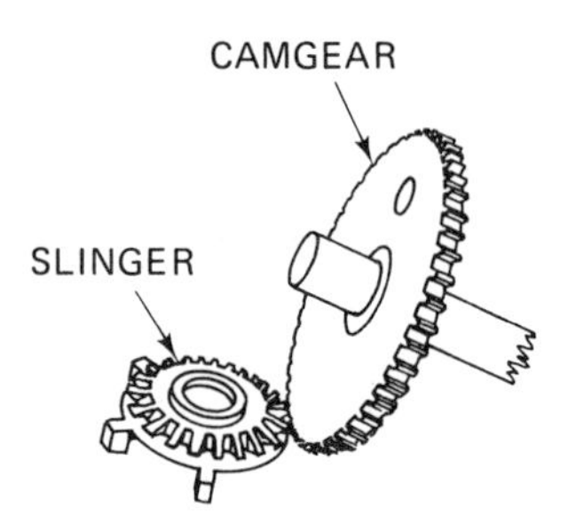

FIG. 8-5. A revolving slinger driven by a camshaft gear is another method of lubricating the four-cycle engine.

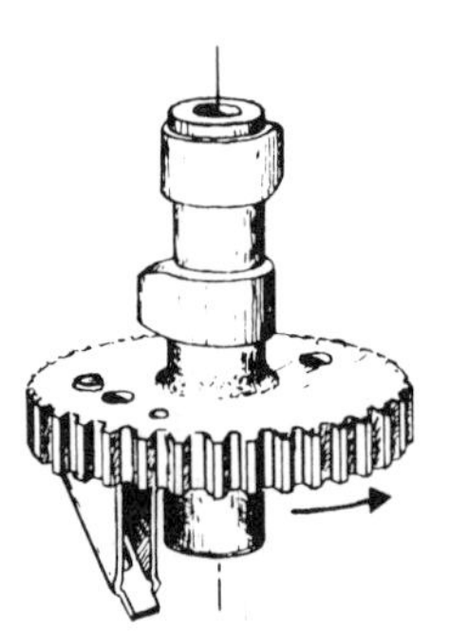

FIG. 8-6. An oil scoop attached to the bottom of the camshaft gear is also used to lubricate four-cycle engines.

Pressurized lubricating systems utilize a pump to distribute the oil through open spaces, drilled passages, and slots to bearings, gears, tappets, etc. Some of the types of pumps used are: ejection, barrel type, and rotary dual-gear. The ejection pump is cam operated; it pumps oil at the connecting rod. Some oil goes through the connecting rod holes to lubricate the bearings at the crankshaft and piston pins. The remainder of the oil is thrown by the moving connecting rod to the other internal parts of the engine.

A barrel type pump has a plunger that draws oil through an intake port, compresses the oil, and then squirts the oil onto the main bearing and connecting rod bearing. Oil is splashed by the movement of the crankshaft and connecting rod to the other engine parts. The barrel type pump is driven by the camshaft.

The rotary dual-gear oil pump (Fig. 8-7) is driven by the camshaft by means of a drive pin located in the hub of the camshaft. Oil is forced through a line from the pump outlet to the top of the engine block to connect with drilled passages toward the top main bearing.

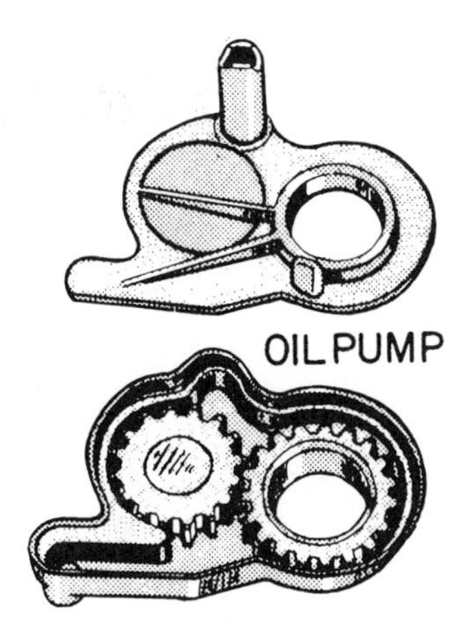

FIG. 8-7. Three types of oil pumps are ejection, barrel, and rotary dual-gear. The illustration shows a rotary dual-gear pump.

8-6. MAINTENANCE

It is obvious that there is no maintenance for the two-cycle engine lubrication system. Maintenance of the four-cycle engine lubrication system is easy and encompasses only periodic maintenance (Section 3-13), inspection, and replacement of parts. Unfortunately, the engine owner probably won't know there is any malfunction in the lubricating system until it is too late and there is extensive damage to other parts of the engine.

Any time that the crankcase is opened (Section 2-20), the lubricating system—the dipper, slinger, scoop, trough, or pump—should be inspected for damage. Replace damaged parts. Oil access holes and slots are to be cleaned to allow proper lubricant flow. After cleaning, apply a light coating of the recommended oil to the mating surfaces before reassembly.

Ensure that the proper lubricating oil in the proper amount is added to the engine (Sections 3-13 and 3-14). Check periodically for proper oil level. Change the oil periodically. These preventive periodic maintenance procedures will prevent problems in the lubricating systems.

9

Troubleshooting, Tuning, and Overhauling

The first eight chapters of this book have presented the theory of operation and maintenance of the small gasoline engine and supporting systems. Each chapter covered a specific part of the engine from the simplest part to the most complex part and from the simplest maintenance to the most complex maintenance. Chapter 3 provided specific operating procedures and specific periodic maintenance procedures. This chapter is designed to tie all of your knowledge together to enable you to locate and repair troubles in a small gasoline engine.

The following list is presented to help you quickly locate the test, troubleshooting table, and tune-up and overhaul procedures contained in this chapter:

Section 9-2 Four Basic Engine Tests
Section 9-3 Test No. 1. — Ignition System
Section 9-4 Test No. 2. — Spark Plug Condition
Section 9-5 Test No. 3. — Fuel Supply
Section 9-6 Test No. 4. — Compression
Section 9-7 Troubleshooting
Table 9-1 No Ignition Spark
Table 9-2 Weak Ignition Spark
Table 9-3 Weak Compression
Table 9-4 Engine Fails to Start or Starts Hard

Table 9-5	Engine Missing Under Load or Lack of Power
Table 9-6	Engine Surges or Runs Unevenly
Table 9-7	Engine Overheating
Table 9-8	Engine Noisy or Knocks
Table 9-9	Engine Vibrates Excessively
Table 9-10	Causes of Engine Failure
Table 9-11	Summary Chart of Engine Troubleshooting
Section 9-8	Engine Tune-up and Overhaul
Table 9-12	Minor Engine Tune-up
Table 9-13	Major Engine Tune-up
Table 9-14	Minor Engine Overhaul
Table 9-15	Major Engine Overhaul

The procedures in the tests and tables are to aid you in troubleshooting, repairing, and performing tune-ups and engine overhauls. The tests and tables are guides —they will not suit the purposes of every repair shop nor of every engine; but they can be used as a guide for the majority of engines.

9-1. THE CUSTOMER AND THE ENGINE

One of the most important questions you can ask a customer when he brings an engine for repair is, what seems to be the problem? Note down all of the problems/symptoms that the customer can give you. If he doesn't have any ideas, probe him with several questions. Does the engine start? Does it stall out? Does it start hard? Does anything appear to be broken? Does the engine lack power under load? Does it overheat? Is the engine noisy? Does it vibrate excessively? By asking these questions, you have a basis for further trouble analysis.

If the customer would like an immediate "ballpark" estimate of the problems and cost of repair, you can quickly perform the four tests of Sections 9-3 to 9-6. If he does not demand an immediate estimate, you should wait until you are ready to perform the repairs on the engine; then, armed with the verbal information given to you by the customer and noted on the repair tag, you are ready to perform the four tests to isolate the trouble to the ignition system, fuel system, or to a lack of compression. All engine troubles fall into these categories: *ignition, fuel,* and *compression.* It really doesn't matter if you don't perform the tests of Sections 9-3 to 9-6 in sequence; the result is that you will find a problem in one or more of the three areas.

Once you know the area(s) of the problem, you refer to Tables 9-1 through 9-9 which list engine symptoms in their titles. Each of the tables then presents the probable cause of the trouble, the maintenance action to be performed, and reference to a section in the text that details the maintenance for that remedy.

Finally, once you have repaired the trouble, you should then perform the applicable tune-up or overhaul procedure to return the engine to "as new" condition for your customer. Tune-up and overhaul procedures are given in Tables 9-12 to 9-15.

9-2. FOUR BASIC ENGINE TESTS

Four basic engine tests are presented in Sections 9-3 to 9-6 to help localize trouble to the ignition system, fuel system, or to a lack of compression. The procedures refer to the troubleshooting tables 9-1 through 9-9 for trouble analysis and maintenance actions.

9-3. TEST NO. 1. — IGNITION SYSTEM

Check the output of the ignition system with a wide gap spark plug. Proceed as follows:

1. Disconnect the high-tension lead from the spark plug.
2. Connect the high-tension lead to a test spark plug with a gap set at between 5/32 and 3/16 inch (0.155 to 0.190) (Fig. 9-1).
3. Ground the side spark plug electrode to the engine bare metal (Fig. 9-2).
4. Crank the engine over and look for a sharp snappy spark to jump the

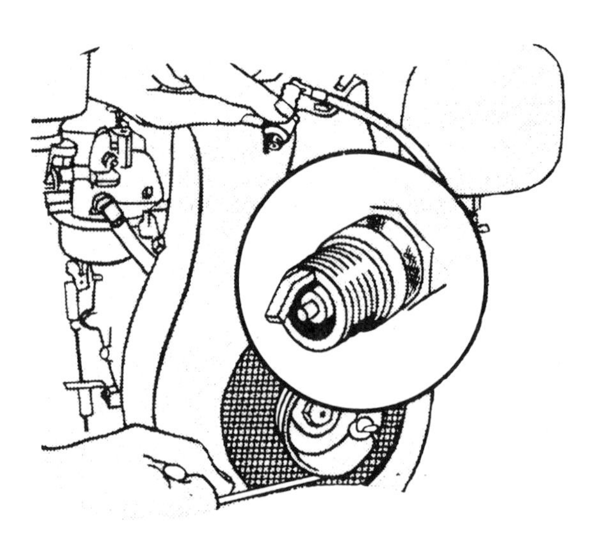

FIG. 9-1. A test spark plug can be made by adjusting the gap to between 5/32 and 3/16 inch. Construction of a test spark plug is discussed in Appendix A.

FIG. 9-2. Ground the spark plug threads to bare metal on the engine. Crank the engine. If a spark jumps the gap, the ignition system is good.

electrode gap. If the spark jumps, the ignition system is operating correctly —however, it is possible that the timing could be incorrect.

5. If a spark does not jump the electrode gap, the ignition system cannot generate an adequate voltage; refer to Table 9-1. If the spark cannot jump the wide gap, but can jump a narrower gap, then refer to Table 9-2.

9-4. TEST NO. 2. — SPARK PLUG CONDITION

Check the condition of the spark plug.

1. Remove the plug.
2. Visually inspect the plug. If the plug is wet with fuel, it indicates that fuel is getting to the combustion chamber (Fig. 9-3). If the plug is dry, check the engine back to the carburetor input. Visually inspect the plug for carbon buildup, burned electrodes, cracked insulation, and carbon between the electrodes.

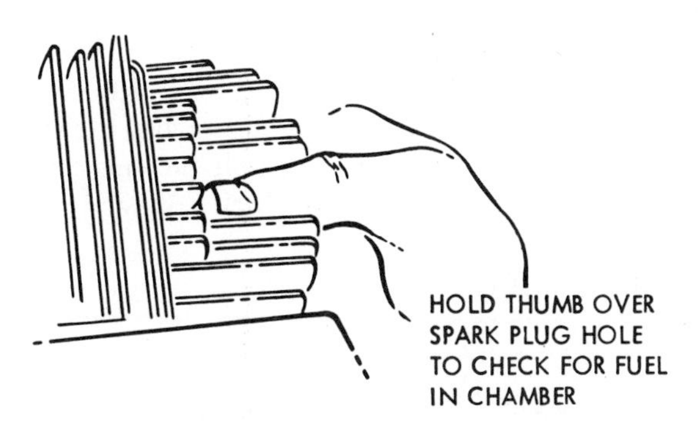

FIG. 9-3. If the plug doesn't look wet with fuel, you can place your thumb over the spark plug hole and crank the engine to see if fuel is getting to the combustion chamber; your thumb should get damp. Be sure the high-tension wire is not near your hand or the hole.

3. Many small gasoline engine troubles are diagnosed by examining the spark plug. Refer to Table 5-1 for spark plug diagnosis.
4. Check the spark plug gap. Regap the plug, as required (Section 3-17). Refer to Table 9-2.

9-5. TEST NO. 3. — FUEL SUPPLY

Check the supply of fuel to the carburetor.

1. If the carburetor is equipped with a bowl drain, press the valve and let a small amount of fuel leak out into a flat container (Fig. 9-4).

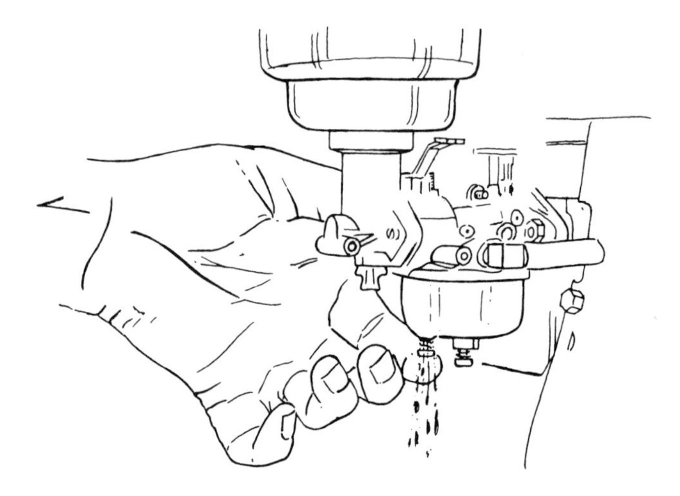

FIG. 9-4. If fuel can be let out of the carburetor bowl drain valve, it indicates that fuel is getting from the tank and lines to the carburetor.

2. If fuel does not leak out of the carburetor bowl, it indicates an obstruction in the fuel supply tank or line. Clean the tank and lines (Section 7-15).
3. If fuel leaks out of the carburetor bowl, check the fuel for puddles of water or other foreign matter. If present, consideration should be given to servicing the carburetor, fuel tank, and fuel line (Sections 7-11 and 7-15).
4. Refer to Table 9-4.

9-6. TEST NO. 4. — COMPRESSION

Compression tests can be made by one of two methods: without the use of a compression gauge, and with a compression gauge. Using a compression gauge is the accurate method and the method to be used if you suspect weak compression.

Check compression without a compression gauge as follows:

1. Remove the high-tension lead from the spark plug.
2. Turn the engine over slowly by hand. As the piston reaches TDC on the compression stroke, considerable resistance against turnover of the engine should be felt. Once the piston passes TDC, it should be pushed down the cylinder rapidly indicating good compression.
3. The lack of resistance just prior to TDC and the lack of compression to rapidly push the piston down the cylinder just after TDC indicates lack of compression. Proceed to make a compression test using a compression gauge.

Perform a compression test using a compression gauge as follows:

1. Remove the spark plug from the cylinder head.
2. Hold or screw (depending upon the type of compression gauge) a compression gauge into the spark plug hole (Fig. 9-5).

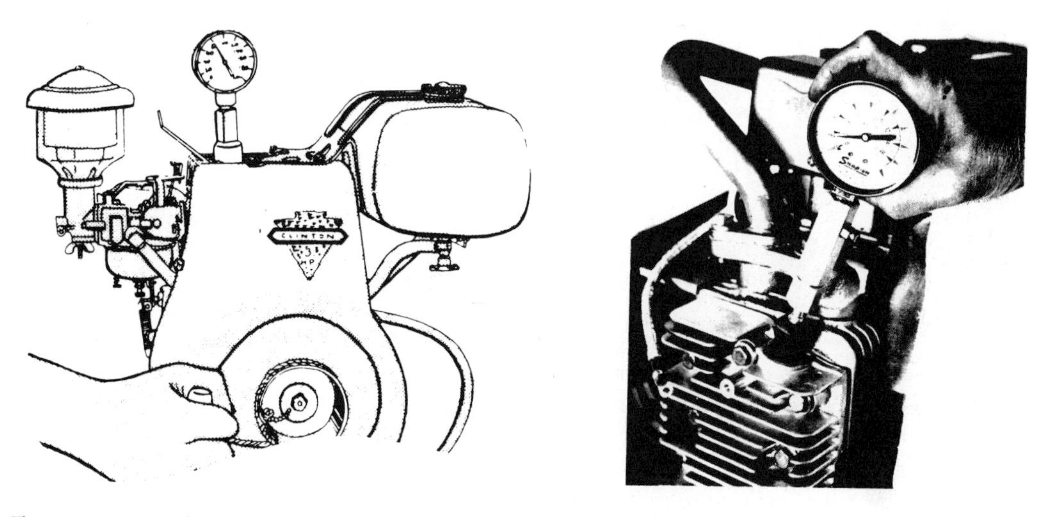

FIG. 9-5. Check the engine compression with a compression gauge.

3. Crank the engine over at normal cranking speed. Crank until the compression gauge needle does not rise any more (at least six cyles of operation).
4. The gauge should read:

   A. over 60 psi (pounds per square inch) for two-cycle engines.
   B. over 65 to 75 psi for four-cycle engines through 4.5 horsepower.
   C. over 70 psi for four-cycle engines above 4.5 horsepower.

5. If the compression reading is below the value specified by the engine manufacturer, refer to Table 9-3. Readings may be as high as 120 psi.

Hot spots, caused by excessive friction—lack of lubrication, bulging metal, or overheating—can cause the piston rings and cylinder wall to weld together. Since the piston keeps moving, it pulls the ring free but some of the metal remains behind attached to the cylinder wall. The continued movement of the rings causes the area around the scuffed spot to become burnished and excessively worn and the damaged ring can be so weakened that compression and combustion pressure blow by (Fig. 9-6).

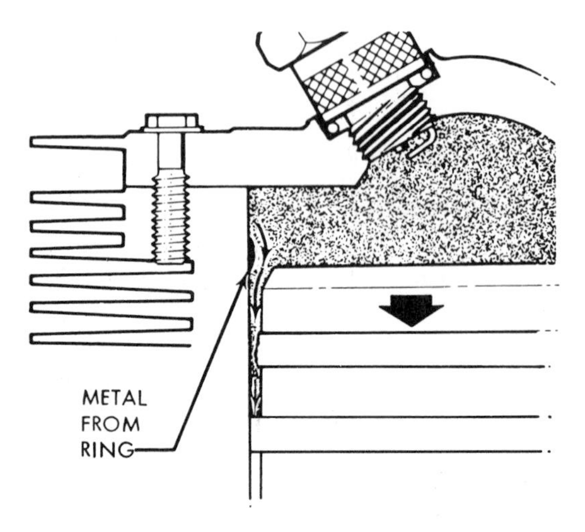

FIG. 9-6. Hot spots can cause a piece of the piston ring to weld to the cylinder wall. The continued movement of the rings over the spot eventually causes wear that allows the compression and combustion pressures to blow-by.

Replacement of the rings only does not always end compression problems. If the ring grooves of the piston are worn, pressure can move around the piston rings. If the grooves are badly worn, new rings may flutter in the grooves and eventually break. The broken ring may then jam between the piston and the cylinder wall and score both badly; a major overhaul will then be required. Therefore when installing new rings, make sure that a new piston isn't also needed.

9-7. TROUBLESHOOTING

Tables 9-1 through 9-9 are titled by engine trouble symptoms. The results of the four tests in Sections 9-3 to 9-6, the customer's description of the trouble, and your analysis of the trouble will lead you into these tables. Each table lists probable causes of the trouble symptom described in the table title, maintenance actions to be performed, and a reference to a section or table in the text that describes that particular maintenance action. Table 9-10 lists engine failures and their causes. Table 9-11 is a summary chart of engine troubleshooting (courtesy of ONAN, Division of Onan Corporation).

Table 9-1
NO IGNITION SPARK

Cause	*Maintenance Action*	*Refer to Section*
1. Switch turned off.	Turn ignition switch on.	—
2. Engine not turning over.	Check manual or electrical starter; check starting battery.	6-9
3. Ignition leads disconnected or broken.	Inspect, reconnect, or replace.	—
4. Bad plug.	Replace plug with plug recommended by manufacturer. Gap plug correctly.	3-17
5. Ignition switch faulty.	Replace switch.	—
6. Breaker points oxidized, pitted, or burned.	Check points. Replace if required. Set gap.	5-9
7. Breaker points stuck; breaker cam or shaft broken.	Check points. Replace if required. Set gap.	5-9
8. Condenser faulty.	Check condenser. Replace if required.	5-10
9. Ignition coil faulty.	Check ignition coil. Replace if required.	5-11
10. Solid state module faulty.	Replace.	—
11. Magnets in flywheel have lost strength.	Check strength. Replace magnets.	5-12

Table 9-2
WEAK IGNITION SPARK

Cause	*Maintenance Action*	*Refer to Section*
1. Plug wet.	Refer to Table 5-1. Perform maintenance actions.	Table 5-1
2. Plug gap incorrect.	Regap plug.	3-17
3. Plug fouled with carbon.	Refer to Table 5-1. Perform maintenance actions.	Table 5-1
4. Wrong plug.	Replace plug with type recommended by manufacturer. Gap plug.	3-17
5. Breaker points dirty or bad.	Check points. Replace and reset gap.	5-9
6. Breaker point gap incorrect.	Check gap. Reset, as required.	5-9
7. Timing incorrect.	Check timing and reset, as required.	5-13
8. Condenser weak.	Check condenser. Replace, as required.	5-10
9. Flywheel magnets have lost strength.	Check strength. Replace magnets or flywheel and magnets.	5-12

Table 9-3
WEAK COMPRESSION

Cause	*Maintenance Action*	*Refer to Section*
1. Spark plug loose.	Check gasket, replace if necessary. Tighten spark plug to torque value specified by manufacturer (Fig. 9-7 and 9-8).	3-17
2. Cylinder head loose.	Tighten cylinder head bolts to torque value and in sequence specified by manufacturer. Repeat compression test.	2-21
3. Head gasket leaking.	Replace gasket. Tighten cylinder head bolts to torque value and in sequence specified by manufacturer. Repeat compression test.	2-21
4. Valves sticking.	Inspect valves, seats, and guides. Recondition or replace as required.	2-26
5. Piston rings sticking.	Inspect cylinder, piston, and rings. Recondition or replace as required.	2-23, 2-24
6. Cylinder badly worn.	Recondition or replace cylinder. Replace piston and piston rings.	2-24
7. Burned piston.	Replace piston and piston rings. Inspect cylinder. Recondition cylinder, if required.	2-23, 2-24

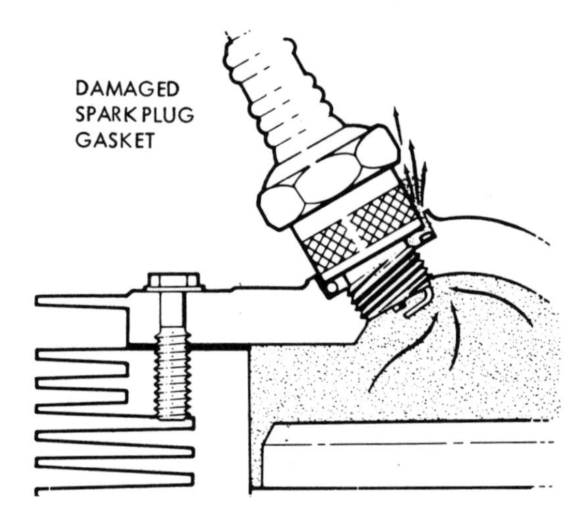

Fig. 9-7. If the spark plug gasket is damaged, the compression and combustion pressures can blow out of the cylinder head. Replace damaged gaskets and torque the plug.

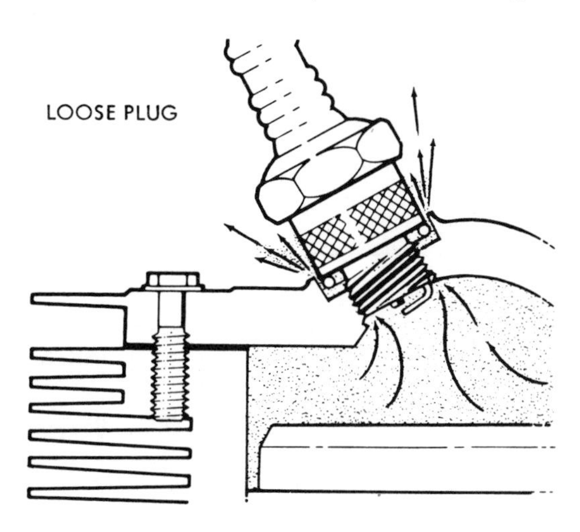

FIG. 9-8. A loose spark plug can likewise allow compression and combustion pressures to blow out of the cylinder head. Torque the plug to the specified value.

Table 9-4
ENGINE FAILS TO START OR STARTS HARD

Cause	*Maintenance Action*	*Refer to Section*
1. No fuel in tank.	Fill tank with clean, fresh fuel.	3-14, 7-16
2. Fuel shutoff valve not open.	Open fuel shutoff valve.	—
3. Primer bulb (if applicable) and line not operating.	Place finger over lower end of primer base and press primer bulb. Resistance should be noted in bulb depression. Remove finger, there should be no resistance present when bulb is depressed.	7-15
4. Engine not cranking over.	Check starting system.	6-9
5. Fuel line to carburetor blocked.	Clean fuel line or remove and replace with new.	—
6. Water or foreign liquid in tank.	Drain tank. Clean carburetor and fuel lines. Dry spark plug points. Fill tank with clean, fresh fuel.	7-11, 3-14, 7-16
7. Stale fuel in tank.	Drain tank. Clean carburetor and fuel lines. Dry spark plug points. Fill tank with clean, fresh fuel.	7-11, 3-14, 7-16
8. No spark or insufficient spark to jump electrode gap.	Check points, condenser, coil, high-tension lead, flywheel keyway, and magnet charge. Rework or replace as necessary.	5-6

Table 9-4 (Continued)

Cause	*Maintenance Action*	*Refer to Section*
9. Spark plug fouled or defective.	Replace spark plug.	3-17
10. Stop device in the off position.	Move stop device to on position.	—
11. Engine flooded.	Open choke. Remove air cleaner, clean and service.	3-9
12. Choke valve not completely closing in carburetor.	Adjust control cable travel, and/or speed control lever.	—
13. Carburetor idle needle or power needle not properly adjusted.	Reset idle and power needles to the recommended preliminary settings.	7-8
14. Carburetor throttle lever not open far enough.	Move speed control lever to fast or run position; check for binding linkage, or unhooked governor spring.	7-9, 7-14
15. Low or no compression.	Check the following:	
	(A) Blown head gasket.	2-21
	(B) Damaged or worn cylinder.	2-24
	(C) Valves stuck open, burned, not properly adjusted, or bad seats. Rework, or replace as necessary.	2-26
16. Not cranking engine over fast enough.	(A) Starter broken or weak.	6-9
	(B) Too much drag on driven equipment. Replace broken or weak spring, and remove belts, chains, and/or release clutch.	—
17. Carbon blocking exhaust ports (2-cycle engine).	Remove muffler and clean carbon from ports.	3-8
18. Reed, rotary, or poppet valve broken or damaged (2-cycle engine).	Replace reed, rotary, or poppet valve or assembly.	2-30
19. Oil seals leaking.	Replace oil seals.	2-29
20. Carburetor dirty.	Remove and clean carburetor in a recommended cleaning solvent.	7-11

Table 9-5
ENGINE MISSING UNDER LOAD OR LACK OF POWER

Cause	*Maintenance Action*	*Refer to Section*
1. Weak or irregular spark to spark plug.	Check points, condenser, coil high-tension lead wire, flywheel, keyways, and flywheel magnet charges.	5-6
2. Defective spark plug.	Remove and replace.	3-17
3. Choke not completely open.	Open lever to full choke position.	—
4. Carburetor idle or power needle not properly adjusted.	Reset idle and power needles to the recommended preliminary settings.	7-8
5. Restricted fuel supply to carburetor.	Clean tank, open gas tank cap vent, or clean and/or replace fuel lines.	7-15, 3-11
6. Valves not functioning properly.	Reseat or reface valves, clean guides and stems of valves and reset valves to tappet clearance.	2-26
7. Stop device not in the positive on position.	Move stop device to the on position and/or adjust.	—
8. High-tension lead wire loose or not connected to spark plug.	Adjust high-tension lead wire terminal, and/or connect to spark plug.	—
9. Air cleaner dirty or plugged.	Clean and/or replace air cleaner element.	3-9
10. Not enough oil in crankcase (4-cycle engine).	Drain and refill with the proper type and quantity.	3-13
11. Improper fuel oil mix (2-cycle engine).	Drain tank and carburetor, and refill with the correct clean, fresh fuel mix.	7-15
12. Engine needs major overhaul.	Overhaul engine.	Table 9-15
13. Too much drag on driven equipment.	Adjust clutches, pulleys and/or sprockets on driven equipment.	—
14. Obstructed exhaust system or muffler not the type designed for engine.	Remove obstruction and/or replace muffler with correct one.	3-8
15. Weak valve springs (4-cycle engine).	Replace weak valve springs with new.	2-26
16. Reed, rotary, or poppet valve assembly not functioning properly (2-cycle engine).	Replace and/or adjust reed, rotary, or poppet valve assembly.	2-30
17. Crankcase gaskets or seals leaking.	Replace gaskets and/or seals in question.	2-29

Table 9-6
ENGINE SURGES OR RUNS UNEVENLY

Cause	*Maintenance Action*	*Refer to Section*
1. Fuel tank cap vent hole obstructed.	Remove obstruction and/or replace with new cap.	3-11
2. Carburetor float level set too low.	Reset float level.	7-12
3. Restricted fuel supply to carburetor.	Clean tank, fuel lines, and/or inlet needle and seat of carburetor.	7-15, 7-11
4. Carburetor power and idle needles not properly adjusted.	Readjust carburetor power, and idle needles.	7-8
5. Governor parts sticking or binding.	Clean, and if necessary, repair or replace governor parts.	7-14
6. Engine vibrates excessively.	Check for bent crankshaft and/or out of balance condition on blades, adaptors, pulleys, sprockets and clutches. Replace or rework as necessary.	2-27
7. Carburetor throttle linkage or throttle shaft and/or butterfly binding or sticking.	Clean, lubricate or adjust linkage and deburr throttle shaft or butterfly.	7-11

Table 9-7
OVERHEATING

Cause	*Maintenance Action*	*Refer to Section*
1. Carburetor settings too lean.	Reset carburetor to proper setting.	7-8
2. Improper fuel.	Drain tank and refill with correct clean, fresh fuel.	7-16
3. Over speeding and/or running engine too slow.	Reset speed control and/or adjust governor to correct speed.	7-9
4. Overloading engine.	Review the possibility of using larger horsepower engine.	—
5. Not enough oil in crankcase (4-cycle).	Drain and refill with the proper type and quantity.	3-13
6. Improper fuel mix (2-cycle).	Drain tank and refill with correct clean, fresh mix.	3-14
7. Air flow to cooling fins, head, and block obstructed.	Clean debris from rotating screen and/or head, and cylinder cooling fins.	3-8

Table 9-7 (Continued)

Cause	*Maintenance Action*	*Refer to Section*
8. Engine dirty.	Clean grease and/or dirt from cylinder block and head exterior.	3-8
9. Too much carbon in combustion chamber.	Remove head and clean carbon deposits from combustion chamber.	2-21
10. Obstructed exhaust system or muffler not the correct type designed for engine.	Remove obstruction and/or replace muffler with correct type.	3-8
11. Engine out of time (4-cycle engine).	Time the engine.	5-13

Table 9-8

ENGINE NOISY OR KNOCKS

Cause	*Maintenance Action*	*Refer to Section*
1. Piston hitting carbon in combustion chamber.	Remove head and clean carbon from head and top of cylinder.	2-21
2. Loose flywheel.	Torque flywheel nut to recommended torque.	5-8
3. Loose or worn connecting rod.	Replace rod and/or crankshaft if tightening rod bolt won't correct.	2-22, 2-27
4. Loose drive, pulley blade, or clutch on power take-off end of crankshaft.	Replace, tighten or rework as necessary.	—
5. Main bearings worn.	Replace worn bearings and/or crankshaft if necessary.	2-28
6. Rivet holding oil distributor to cam gear hitting counterweight of crankshaft.	Replace cam gear and/or grind head of rivet off.	2-25
7. Rotating screen hitting housing flywheel.	Center screen on flywheel.	—

Table 9-9
ENGINE VIBRATES EXCESSIVELY

Cause	*Maintenance Action*	*Refer to Section*
1. Engine not mounted securely.	Tighten mounting bolts.	—
2. Bent crankshaft.	Replace crankshaft.	2-27
3. Blades, adaptors, pulleys, and sprockets out of balance.	Rework or replace parts involved.	—

Table 9-10
CAUSES OF ENGINE FAILURE

Failure	*Cause*	*Refer to Section*
1. Broken or damaged connecting rods, and scored pistons.	Engine run low on oil (4-cycle engine).	3-13
	Engine operated at speeds above the recommended rev/min.	7-14
	Oil pump, line and passage obstructed with debris (4-cycle vertical shaft engines).	8-5
	Oil distributor broken off.	8-5
	Not enough oil in fuel mix (2-cycle engine).	3-14
	Oil in crankcase not changed often enough (4-cycle engine).	3-13
2. Excessive wear on parts. This covers valves, valve guides, cylinders, pistons, rings, rods, crankshafts and main bearings.	Air cleaner not serviced often enough.	3-9
	Oil not changed often enough in crankcase.	3-13
	Air cleaner element improperly installed in air cleaner body, or element needed replacing.	3-9
	Air cleaner body not making good seal to carburetor.	—
3. Main bearing failure.	Engine run low on oil.	3-13
	Excessive side loading of crankshaft.	—
	Oil in crankcase not changed often enough (4-cycle engine).	3-13
	Blades, adaptors, pulleys, and sprockets out of balance.	—

Table 9-11
SUMMARY CHART OF ENGINE TROUBLESHOOTING

GASOLINE ENGINE TROUBLESHOOTING GUIDE — CAUSE	Backfire at Carburetor	Bearing Wear	Black Exhaust	Blue Exhaust	Burned Valves	Connecting Rod Wear	Cranks Slowly	Cylinder Wear	Engine Stops	Failure to Start	Governor Hunting	High Oil Pressure	Low Oil Pressure	Loss of Coolant (Water Cooled)	Mechanical Knocks	Misfiring	Overheating (Air Cooled)	Overheating (Water Cooled)	Piston Wear	Poor Compression	Ring Wear	Sticking Valves	REFER TO SECTION
STARTING SYSTEM																							
Loose or Corroded Battery Connection							●			●													3-19
Low or Discharged Battery							●			●													3-19
Faulty Starter							●			●													6-9
Faulty Start Solenoid										●													4-5
IGNITION SYSTEM																							
Ignition Timing Wrong	●				●					●					●	●	●	●					5-13
Wrong Spark Plug Gap										●					●	●							3-17
Worn Points or Improper Gap Setting									●	●						●							5-9
Bad Ignition Coil or Condenser										●						●							5-11,5-10
Faulty Spark Plug Wires										●						●							5-11
FUEL SYSTEM																							
Out of Fuel – Check									●	●													–
Lean Fuel Mixture – Readjust					●					●	●					●	●	●					7-8
Rich Fuel Mixture or Choke Stuck	●		●							●					●	●							7-8
Engine Flooded	●		●						●	●													3-1
Poor Quality Fuel	●		●		●					●					●	●							7-16
Dirty Carburetor	●								●	●	●					●							7-11
Dirty Air Cleaner	●	●	●					●		●						●			●		●		3-9
Dirty Fuel Filter										●	●					●							3-10
Defective Fuel Pump									●	●	●					●							7-13
INTERNAL ENGINE																							
Wrong Valve Clearance					●					●					●	●				●			2-26
Broken Valve Spring					●					●					●	●				●		●	2-26
Valve or Valve Seal Leaking				●	●						●					●				●			2-26
Piston Rings Worn or Broken				●						●					●					●			2-23
Wrong Bearing Clearance		●				●	●						●		●								2-22
COOLING SYSTEM (AIR COOLED)																							
Poor Air Circulation																●	●						3-8
Dirty or Oily Cooling Fins																●	●						3-8
Blown Head Gasket									●	●						●				●			2-21
COOLING SYSTEM (WATER COOLED)																							
Insufficient Coolant																		●					–
Faulty Thermostat														●				●					2-18
Water Passages Restricted																		●					2-31
Defective Gaskets														●									–
Blown Head Gasket									●	●				●		●		●		●			2-21
LUBRICATION SYSTEM																							
Oil Level Low		●				●		●					●		●		●	●	●		●		3-13
Oil Too Heavy							●					●											8-2
Dirty Crankcase Breather Valve		●		●		●						●											3-12
THROTTLE AND GOVERNOR																							
Linkage Out of Adjustment										●	●												7-9
Linkage Worn or Disconnected											●												7-9
Governor Spring Sensitivity Too Great											●												7-9
Linkage Binding											●												7-9

After the trouble has been repaired, refer to the applicable tune-up or overhaul procedure. Perform the additional maintenance actions necessary to return the engine to an "as new" condition.

9-8. ENGINE TUNE-UP AND OVERHAUL

Service on small gasoline engines can be categorized into four categories: minor engine tune-up, major engine tune-up, minor engine overhaul, and major engine overhaul. The majority of engines brought into the repair shop require a *minor* engine tune-up (Table 9-12). When work is performed on the carburetor or ignition system, then a *major* engine tune-up (Table 9-13) is required. The major tune-up

Table 9-12
MINOR ENGINE TUNE-UP

Task	*Section*
1. Clean and regap, or replace and gap spark plug.	3-17
2. Tighten spark plug and cylinder head bolts to the torque value specified by the manufacturer.	3-17, 2-21
3. Test compression.	9-6
4. Clean air cleaner.	3-9
5. Adjust carburetor.	7-8
6. Clean fuel tank, line, and filter.	7-15
7. Adjust governor speed.	7-9

Table 9-13
MAJOR ENGINE TUNE-UP

Task	*Section*
1. Clean and regap, or replace and gap spark plug.	3-17
2. Tighten spark plug and cylinder head bolts to the torque value specified by the manufacturer.	3-17, 2-21
3. Test compression.	9-6
4. Clean air cleaner.	3-9
5. Remove carburetor and overhaul.	3-11
6. Clean fuel tank, line, and filter.	7-15
7. Adjust governor speed.	7-9
8. Inspect reed, rotary or poppet valve (two-cycle engine).	2-30
9. Test condenser. Replace if necessary.	5-10
10. Test coil. Replace if necessary.	5-11
11. Install new breaker points.	5-9
12. Clean carbon from muffler and exhaust ports (two-cycle engines).	3-8
13. Adjust carburetor.	7-8

includes performance of the tasks in the minor tune-up. If an engine is using oil and the compression is low, then a minor engine overhaul (Table 9-14) should be performed. When an engine requires replacement of the main bearings in addition to the work performed under a minor engine overhaul, it is considered a major overhaul (Table 9-15).

Table 9-14
MINOR ENGINE OVERHAUL

Task	*Section*
1. Check connecting rod and bearing. Replace if necessary.	2-22
2. Check piston pin. Replace if necessary.	2-23
3. Replace piston rings.	2-23
4. Reseat valves.	2-26
5. Check cylinder and hone with fine hone to remove glaze.	2-24
6. Perform work under major engine tune-up.	Table 9-13

Table 9-15
MAJOR ENGINE OVERHAUL

Task	*Section*
1. Check connecting rod and bearing. Replace if necessary.	2-22
2. Check piston pin. Replace if necessary.	2-23
3. Replace piston rings.	2-23
4. Reseat valves.	2-26
5. Check cylinder and hone with fine hone to remove glaze.	2-24
6. Check crankshaft. Replace if necessary.	2-27
7. Replace main bearings.	2-28
8. Perform work under major engine tune-up.	Table 9-13

10

Math and Measurements

The small gasoline engine is designed and built to aid man in doing work quickly, efficiently, and with less fatigue. Engines also enable man to enjoy his leisure time as they provide power for his small recreation and transportation vehicles.

As the engine operates burning gasoline, *heat* and *power* are generated. Some of the generated power is lost due to heat and *friction;* most of the power is converted from a reciprocating movement of the piston into a rotating force, or torque, at the crankshaft. The crankshaft is coupled to a tool or a machine to perform *work.* The engine develops power that is measured in terms of horsepower. The greater the output power of the engine with respect to the input power, the greater is the *efficiency* of the engine.

This chapter concerns itself with the mathematics and measurements relating to engine *work, potential energy, power, torque, horsepower,* and *volumetric efficiency.* A method of converting *torque values* from foot-pounds to inch-pounds and vice versa is also included so that conversions can be made if you have a torque wrench calibrated in one of the measurements. Finally, since you may not be familiar with the metric system, Section 10-9 describes metric to English and English to metric conversions. References and use of Appendixes B to F on metric to English and English to metric conversions are also explained.

This chapter does not attempt to teach arithmetic or basic algebra; if you need review in these subject areas, refer to applicable texts. It does, however, describe the equations and related unknowns when only some of the specifications are available or when certain measurements can be made physically on the engine. In addition to the equations, example problems are given to familiarize you with the method of solution for the unknown.

10-1. WORK

Work is the moving of an object by force over a distance. Since engines ease the efforts of man by doing work for him—by moving an object by force over a distance—we are concerned with the mathematical relationship of work. It is expressed as:

$$\text{work} = \text{distance in feet or inches} \times \text{force in pounds}$$

$$W = D \times F \text{ (foot-pounds or inch-pounds)} \qquad (10\text{-}1)$$

where: W = work in foot-pounds or inch-pounds
D = distance in feet or inches
F = force in pounds

Example 1. How much work is done when an engine-driven hoist lifts 240 pounds to a height of 35 feet?

$$W = D \times F$$
$$W = 35 \times 240$$
$$W = 8400 \text{ foot-pounds}$$

10-2. POTENTIAL ENERGY

Potential energy is energy having the potential to do work. A body is said to have potential energy if, by virtue of its position or state, it is able to do work. Potential energy is calculated by:

$$\text{potential energy} = \text{distance in feet times force in pounds}$$

$$P.E. = D \times F \text{ (foot-pounds or inch-pounds)} \qquad (10\text{-}2)$$

where: $P.E.$ = potential energy in foot-pounds or inch-pounds
D = distance in feet or inches
F = force in pounds

Example 2. If a weight of 40 pounds is raised to a height of 7 feet, what is its potential energy?

$$P.E. = D \times F$$
$$P.E. = 7 \times 40$$
$$P.E. = 280 \text{ foot-pounds}$$

10-3. POWER

Power is the amount of work done per unit of time, or the rate at which work is done. In mechanics, power is normally expressed in foot-pounds per second or foot-pounds per minute.

$$\text{power} = \text{work in foot-pounds per unit of time}$$
$$P = \frac{W}{t} \text{ (foot-pounds per second or foot-pounds per minute)} \quad (10\text{-}3)$$

where: P = power in foot-pounds per second or foot-pounds per minute
W = work in foot-pounds
t = time in minutes or seconds

Example 3. Find the power of an engine capable of lifting 250 pounds to a height of 20 feet in 10 seconds.

$$P = \frac{W}{t}$$
$$P = \frac{D \times F}{t}$$
$$P = \frac{20 \times 250}{10}$$
$$P = 500 \text{ foot-pounds/second}$$

10-4. TORQUE

Torque is a turning *effort;* there may or may not be motion. For example, when a nut is tightened on a thread, a wrench is used to apply a force in a turning or rotating direction. When the nut is relatively loose, a light force moves the nut through a large turning distance, but as the nut gets tighter, the distance moved decreases. When a *torque wrench* is used to tighten a nut, a calibrated dial indicates the amount of torque applied to the nut. Similarly torque is the twisting force of the engine's crankshaft to drive wheels, pulleys, etc. The maximum torque of an engine is developed at a point somewhat below maximum engine speed. Torque is measured in foot-pounds or in inch-pounds and is defined as the product of the force and the perpendicular distance from the axis of rotation to the line of action of the force. This distance is known as the lever arm.

torque = distance in feet or inches times force in pounds

$$T = D \times F \text{ (foot-pounds or inch-pounds)} \qquad (10\text{-}4)$$

where: T = torque in foot-pounds or inch-pounds
D = distance in feet or inches
F = force in pounds

Example 4. What is the torque applied to a nut through a 12 inch wrench with a 22-pound force?

$$T = D \times F$$
$$T = 12 \times 22$$
$$T = 264 \text{ inch-pounds}$$

The torque developed by an engine can also be determined at various revolutions per minute. It is necessary to know the braking horsepower (Section 10-5).

torque = 33,000 times braking horsepower divided by two pi times the number of revolutions per minute

$$T = \frac{33{,}000 \times \text{bhp}}{2\pi N} \qquad (10\text{-}5)$$

where: T = torque in pound-feet
bhp = braking horsepower
N = number of revolutions per minute

Example 5. What is the torque developed by an engine with a brake horsepower of 12 at 2000 rev/min?

$$T = \frac{33{,}000 \times \text{bhp}}{2\pi N}$$
$$T = \frac{33{,}000 \times 12}{2 \times 3.14 \times 2000}$$
$$T = 31.51 \text{ pound-feet}$$

10-5. HORSEPOWER

Horsepower is the standard unit by which power is measured; it is equal to 33,000 pounds lifted one foot in one minute or to 550 pounds lifted one foot in one second. Horsepower is used to measure the ability of an engine to perform work; thus, a one horsepower engine can lift 33,000 pounds a height of one foot in one minute; it can also raise 3300 pounds one foot in one-tenth of a minute. An engine rated at 4 horsepower can theoretically do the work of four horses; however that presumes 100 percent efficiency of the engine which is not possible because of the friction of moving parts.

horsepower = distance moved in feet times force in pounds divided by 33,000 (33,000 is used when time is in minutes; 550 for seconds) times the time to move the distance in minutes or seconds.

$$HP = \frac{D \times F}{33{,}000 \times T\text{ (minutes)}} = \frac{D \times F}{550 \times T\text{ (seconds)}} \quad (10\text{-}6)$$

where: HP = horsepower
D = distance in feet
F = force in pounds
T = time in minutes or seconds

Example 6. What is the horsepower of an engine that is capable of lifting 200 pounds to a height of 55 feet in 10 seconds?

$$HP = \frac{D \times F}{550 \times T}$$

$$HP = \frac{55 \times 200}{550 \times 10}$$

$$HP = 2$$

Some states in the United States tax engines for licensing on the *rated,* or SAE, horsepower. This is a formula that was developed long ago when less efficient engines were made, but an easy formula was needed that was felt usable on all engines. The taxable horsepower is a lower number than the actual developed horsepower.

$$\text{SAE HP} = \frac{D^2 \times N}{2.5} \quad (10\text{-}7)$$

where: SAE HP = taxable horsepower
D = diameter of the cylinder bore in inches
N = number of cylinders

Example 7. What is the SAE taxable horsepower of an eight cylinder engine having a bore of 3.5 inches?

$$\text{SAE HP} = \frac{D^2 \times N}{2.5}$$

$$\text{SAE HP} = \frac{(3.5)(3.5)(8)}{2.5}$$

$$\text{SAE HP} = 39.2$$

There are three other terms relating to power known as *brake horsepower* (bhp), *frictional horsepower* (fhp), and *indicated horsepower* (ihp). *Brake* horsepower is the horsepower used by engine manufacturers to specify the output of their engines; a braking device is used to hold the engine speed down as the horsepower is measured. The braking horsepower is the horsepower that the engine can produce at a given speed with the throttle wide open; it is measured with a dynamometer. A dynamo in the dynamometer is driven by the engine; an electrical current output from the dynamo is proportional to the speed of the engine.

The braking horsepower can be calculated by using the equation.

$$\text{bhp} = \frac{2\,\pi\,\text{FLN}}{33{,}000} \qquad (10\text{-}8)$$

where: bhp = braking horsepower
F = force as measured on the scale
L = length in feet of the lever arm connecting the brake to the scale
N = number of rev/min of engine speed

Example 8. What is the braking horsepower of an engine tested on a braking device if the engine is running at 2000 rpm and the scale reading is 25 pounds with a lever arm length of 3 feet?

$$\text{bhp} = \frac{2\,\pi\,\text{FLN}}{33{,}000}$$

$$\text{bhp} = \frac{2 \times 3.14 \times 25 \times 3 \times 2000}{33{,}000}$$

$$\text{bhp} = 28.56$$

The power developed in the engine to overcome friction is the *frictional horsepower.* It is measured by driving the engine (no fuel in the engine) with an electric motor; the horsepower required to drive the engine is the frictional horsepower. Most all of the friction in the engine is from the piston and rings moving within the cylinder.

The *indicated horsepower* is the sum of the brake horsepower and the frictional horsepower. The indicated horsepower is the power produced by the burning fuel within the engine; it is measured with an oscilloscope.

The three horsepower ratings described, braking (bhp), indicated (ihp), and frictional (fhp) are related:

$$\text{bhp} = \text{ihp} - \text{fhp} \qquad (10\text{-}9)$$

The horsepower that the engine delivers (braking hp) is equal to the power produced in the engine (indicated hp) minus the power lost because of friction (frictional hp).

10-6. VOLUMETRIC EFFICIENCY

Volumetric efficiency is the ability of the combustion chamber to receive a full charge of air-fuel mixture during the intake stroke. The higher the ability—the efficiency—the higher is the output horsepower. As the engine speed increases the volumetric efficiency decreases because of the inability of the engine to receive a full air-fuel charge.

Volumetric efficiency is calculated by the equation:

$$\text{volumetric efficiency} = \frac{\text{amount of air-fuel that could enter}}{\text{amount of air-fuel that does enter}} \qquad (10\text{-}10)$$

Efficiency is calculated by the equation:

efficiency = output power divided by input power times 100. The answer is expressed as a percent.

$$E = \frac{\text{output}}{\text{input}} \times 100 \qquad (10\text{-}11)$$

Finally, the *mechanical efficiency* of the engine is calculated by:

$$\text{mechanical efficiency} = \frac{\text{braking horsepower}}{\text{indicated horsepower}}$$

$$ME = \frac{\text{bhp}}{\text{ihp}} \qquad (10\text{-}12)$$

10-7. SUMMARY

Torque, engine speed, and volumetric efficiency are interrelated. Consider the torque at mid-speed range. Now increase the speed to such a point that the engine is operating so fast that it cannot get as large an amount of air-fuel mixture into the combustion chamber. The combustion is therefore less powerful and this causes a decrease in torque to the crankshaft. Thus when engine speed increases beyond a certain point, the torque decreases. The braking horsepower also decreases at high engine speeds because of the increased frictional horsepower.

10-8. TORQUE VALUE CONVERSIONS

Many of the fasteners (bolts, nuts, etc.) used in small gasoline engines must be torqued (tightened) to a certain value (usually a range is given). Sometimes the engine specification states torque values in inch-pounds, sometimes in foot-pounds, and sometimes in both. If your torque wrench has a scale for only one of the sets of values and the manufacturers' specifications are in the other values, you must convert from the one value to the other.
Convert as follows:

To Convert	*Into*	*Operation*
foot-pounds	inch-pounds	multiply by 12
inch-pounds	foot-pounds	divide by 12 (or multiply by 0.0833)

Example 9. The cylinder head screws on a particular 6 hp outboard engine are to be torqued to between 5 and 7 foot-pounds; a torque wrench calibrated in inch-pounds is the only wrench available. What is the range of torque values in inch-pounds for this engine?

$$5 \times 12 = 60$$

$$7 \times 12 = 84$$

Answer: 60 to 84 inch-pounds

Example 10. The connecting rod screws of a certain 25 horsepower engine are to be tightened to 180 to 186 inch-pounds. What is the torque value in foot-pounds?

$$\frac{180}{12} = 15$$

$$\frac{186}{12} = 15\frac{1}{2}$$

Answer: 15 to 15½ foot-pounds

10-9. METRIC TO ENGLISH AND ENGLISH TO METRIC CONVERSIONS

The English measurement system is a system of measures and weights in common usage in the United States and several other English speaking countries. The English system consists of units such as inches, feet, yards, ounces, pounds, Fahrenheit, etc. This system is known to most of us and we're content to use it. However, this English measurement system is gradually being replaced by the metric system.

The metric system is a system of weights and measures based on the decimal system; that is, the metric system is based on the number 10. It was introduced and adopted by law in France and subsequently has been adopted by all countries except the United States and a few other countries as the common system of weights and measures. It is adopted by all countries as the system used in scientific work. There is great pressure now for the world to become standardized with one measurement system—this one system is known as the International System of Units (SI). It is important that one world wide system becomes a reality because of the ever increasing trade of machines, tools, and manufactured products between countries. It is likewise important that you become familiar with the International System of Units so that you can make conversions readily between the metric and English systems.

The International System of Units is a modernized version of the metric system. It is generally superior for most scientific work and is the common language for scientific and technical data. Nearly all scientific experiments in the United States as well as abroad are performed using metric units.

The chief advantage of the metric measurement system over the English system is that all units are divisible into ten parts. This enables fractional distances, areas, volumes, capacities, and weights (such as meters, liters, and grams) to be expressed as decimals. Decimals are easier to manipulate in addition, subtraction, multiplication, and division than are fractions.

In your work with linear dimensions (as part dimensions) and liquid measures (as fuels and oils), you may encounter the metric system sooner than you'd like to believe. Hence you should become familiar with the information in the following paragraphs and the conversion factors in Appendixes B through F.

The metric linear measurement system is based on the *meter*. The meter is further divided into tenths (1/10) called *decimeters,* hundredths (1/100) called *centimeters,* and thousandths (1/1000) called *millimeters.* The most useful conversion factors from the English system to the metric system are:

1 meter = 39.37 inches = 1.094 yards
1 yard = 0.9144 meters
1 foot = 0.3048 meters
1 inch = 2.54 centimeters

Thus a football field (100 yards) is 91.44 meters long, a 6 foot man is 1.83 meters tall, and a well proportioned young lady is 87-58-78 centimeters.

Many manufacturers realize today that the metric system is fast upon us and they are taking steps to convert. Tool manufacturers are making tools graduated in English, graduated in metric, and graduated in English and metric. You are advised to buy measurement tools that have some edges that are graduated in English and other edges that are graduated in metric. These tools will then be useful no matter which system you are working with on a particular project.

To use the metric system—based on the number 10—and to be able to readily convert from English to metric and metric to English, it is important that you understand *scientific notation*—a method of expressing the *powers* of 10. Scientific notation is a way of expressing a small or a large number such that the number is expressed as a number equal to or greater than one (1) but less than ten (10) and is multiplied by the number of times that a number (in this case, 10) is multiplied times itself. The power is often referred to as an *exponent*. Thus 3.1417 is expressed in scientific notation because it is expressed as a number equal to or greater than one but less than 10. (It is also multiplied by the appropriate power of ten, which in this case is zero—thus 10^0, where 10^0 always equals one. A basic law in mathematics states that any number raised to the zero power is one. Thus $10^0 = 1$; $2^0 = 1$; $5^0 = 1$, etc.)

For example, the speed of light of approximately 186,300 miles per second is expressed in scientific notation as 1.863×10^5. The scientific notation is read as "one point eight six three times ten to the fifth power." The expression actually means $1.863 \times 10 \times 10 \times 10 \times 10 \times 10$. This is 1.863 times ten times itself (10) five times. The 5 (in 10^5) is the power of 10; the 5 is known as the *exponent*.

Some other numbers that are familiar to you are expressed in powers of ten—scientific notation—as follows:

5,280 ft/mile $= 5.280 \times 10^3$ ft/mile
60 sec/min $= 6.0 \times 10$ sec/min (note: 10 and 10^1 equal each other)

Conversion units are the units that define the type of measure or weight of the number with which we are working. For example, in the expression 5.2 minutes, the conversion unit is minutes.

We know that to convert minutes to seconds, there must be a conversion factor and associated conversion unit. We know that 1 minute equals 60 seconds. Since both 1 minute and 60 seconds equal the same amount of time, the formula can be stated as follows:

$$1 \text{ minute} = 60 \text{ seconds}$$

By dividing both sides of the equation by 60 seconds, we have

$$\frac{1 \text{ minute}}{60 \text{ seconds}} = \frac{60 \text{ seconds}}{60 \text{ seconds}}$$

$$\frac{1 \text{ minute}}{60 \text{ seconds}} = 1$$

Likewise, we could divide by 1 minute:

$$1 \text{ minute} = 60 \text{ seconds}$$

$$\frac{1 \text{ minute}}{1 \text{ minute}} = \frac{60 \text{ seconds}}{1 \text{ minute}}$$

$$1 = \frac{60 \text{ seconds}}{1 \text{ minute}}$$

We have shown, by example, that when two quantities with different conversion units are equal, they can be divided by one of the conversion units and set equal to one.

By this method of setting conversion units equal to one, conversions from one set of conversion units to another can be easily made. For example, convert 5.2 minutes to seconds. The conversion unit of 1 minute equals 60 seconds can be rewritten as:

$$1 \text{ minute} = 60 \text{ seconds}$$

$$1 = \frac{60 \text{ seconds}}{1 \text{ minute}} = \frac{1 \text{ minute}}{60 \text{ seconds}}$$

In this problem, it is desirable to get rid of minutes and convert to seconds. Hence:

$$5.2 \text{ minutes} \times \frac{60 \text{ seconds}}{1 \text{ minute}}$$

For conversions, always set up for multiplication as shown above. Arrange the conversion factor units so that they will divide out, just as numbers do:

$$5.2 \text{ minutes} \times \frac{60 \text{ seconds}}{1 \text{ minute}}$$

$$= \frac{5.2 \times 60 \text{ seconds}}{1}$$

$$= 112.0 \text{ seconds}$$

For another example, convert 1.72 yards into inches. By the use of conversion tables, the following is known:

$$1 \text{ yard} = 3 \text{ feet}$$
$$1 \text{ foot} = 12 \text{ inches}$$

By setting the above conversions equal to one by dividing by one conversion factor unit or the other, the problem becomes:

$$\frac{1.72 \text{ yards}}{1} \times \frac{3 \text{ feet}}{\text{yard}} \times \frac{12 \text{ inches}}{1 \text{ foot}}$$

Thus the numerical answer is 61.92 and the unit is inches. All other conversion units have divided out.

You should study the above technique and apply it to other conversions with which you are now familiar. Its use will be invaluable in making conversions from one system of measures and weights to another.

Appendix B provides measurement conversions from metric to English and from English to metric. For example, if you want to compare the bore of two cylinders A and B that measure 2-5/16 inches and 60 millimeters, respectively, one of the dimensions must be converted to the other system. Let us convert from English to metric first and compare the dimensions in metric; then we'll convert from metric to English and compare the dimensions in English.

$$A = 2\text{-}5/16 \text{ in.} \qquad\qquad B = 60 \text{ mm}$$

$$A = 2.3125 \text{ in. (use Appendix C)}$$

$$A = 2.3125 \cancel{\text{in.}} \times \frac{25.400 \text{ mm}}{\cancel{\text{in.}}} \text{ (use Appendix B)}$$

$$A = 58.74 \text{ mm} \qquad\qquad B = 60 \text{ mm}$$

$$B = 60 \text{ mm} \qquad\qquad A = 2\text{-}5/16 \text{ in.}$$

$$B = 60 \cancel{\text{mm}} \times \frac{0.039 \text{ in.}}{\cancel{\text{mm}}}$$

$$B = 2.34 \text{ in.}$$

$$B = 2\text{-}11/32 \text{ in.} \qquad\qquad A = 2\text{-}5/16 \text{ in.}$$

As another example, how many pints of oil are required in a crankcase if the crankcase capacity is specified as 0.25 liters?

$$0.25 \cancel{\text{liters}} \times \frac{2.113 \text{ pints}}{\cancel{\text{liter}}}$$

$$= 0.53 \text{ pints}$$

Appendix C provides conversions from fractions of an inch to millimeters and to decimals of an inch. For example, 3/16 inch equals 0.1875 inch equals 4.762 millimeters.

If you need to convert millimeters to inches, use Appendix D. For example, 15 millimeters equals 0.59055 inches. Using Appendix C, 0.59055 inches is about 19/32 inch. Using Appendix D again, 54.32 millimeters is converted to inches as follows:

$$54.00 \text{ mm} = 2.12598 \text{ in.}$$

$$0.32 \text{ mm} = \underline{0.01260 \text{ in.}}$$

$$2.13858 \text{ in. (by addition)}$$

If necessary, this can be converted to approximate fractions using Appendix C as follows:

$$
\begin{aligned}
2.00000 &= 2 \\
0.13858 &= \underline{\;\;9/64\;\;} \\
& \;\;\;\;\text{2-9/64 in. (by addition)}
\end{aligned}
$$

Appendix E provides conversions within the English system of weights and measures and Appendix F provides conversions within the metric system of weights and measures.

11

Rotating Engines and Beyond

The rotary engine has been perfected to the point that it is being produced by foreign and American automobile engine manufacturers for some models of the smaller automobiles. A few foreign and American manufacturers have also built small rotary engines, but on a limited production or experimental basis at this time because of some design problems that have not been overcome.

The rotary engine can be used for the same applications as the reciprocating engine. Present rotary engines range from a 14 ounce ½ horsepower engine for model airplanes to a 400 horsepower engine for a sports car that travels to 190 miles per hour. Between these are engines of 6 to 300 horsepower to drive boats, motorcycles, snowmobiles, automobiles, garden tractors, industrial pumps, lawn mowers, and compressors. Some of these engines are only prototypes; others are in limited production.

This chapter describes the rotary engine—its uses, operation, features, performance, advantages, problems, and its history. Other engines beyond the rotary engine in the foreseeable future are introduced and include the gas turbine, steam, Stirling, stratified charge modification, diesel, and electric engine.

11-1. OPERATION OF THE ROTARY ENGINE

The rotary engine, often called the Wankel engine after its inventor Dr. Felix Heinrich Wankel, has two major moving assemblies: a three sided rotor and a main shaft. Like its counterpart, the piston in a reciprocating piston engine, the rotor performs the actions of intake, compression, power, and exhaust.

The rotor (Fig. 11-1) has three faces; its shape resembles a rounded equilateral triangle. At each apex between two faces, there is an apex seal that presses against the mirror-like finish of the inside of the housing and seals the pressures of compression and combustion between the rotor face and the housing (Fig. 11-2). The apex seals are bars about ¼ inch thick and are as long as the rotor is thick. The bar is spring loaded to take up any slack caused by heat expansion, imperfect machining, and wear. The seals are comparable to the piston compression rings of a reciprocal engine. Each face contains a combustion recess (Fig. 11-1) which is similar to the combustion chamber of the reciprocating engine. The sides of the rotor contain side seals and oil seals. The internal surface of the rotor mates with the eccentric of the crankshaft (Fig. 11-3); the moving internal timing gear meshes with the stationary gear (Fig. 11-2) that is fixed to the housing.

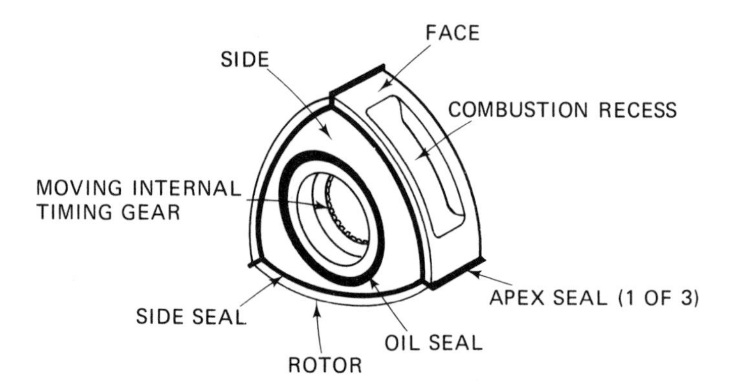

FIG. 11-1. The rotor is one of the two major moving assemblies of the rotary engine; the main shaft (crankshaft) is the other moving assembly.

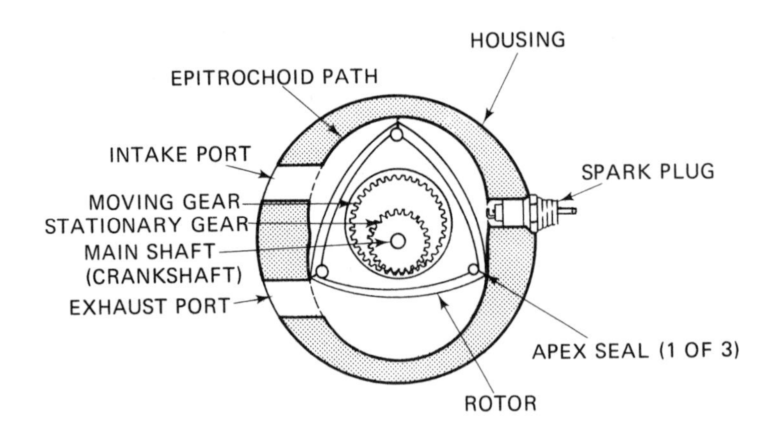

FIG. 11-2. The apex seals of the rotor follow an epitrochoidal path around the inside of the housing.

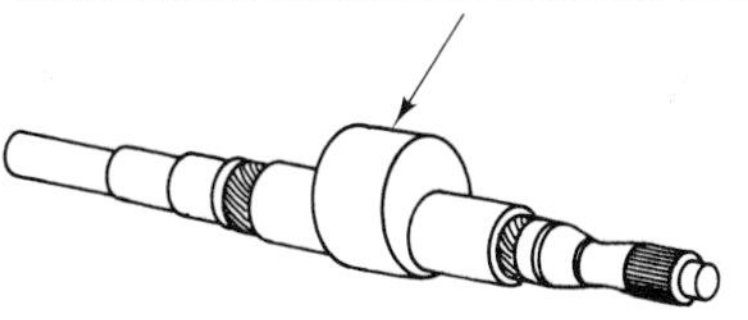

FIG. 11-3. The crankshaft has an eccentrically mounted circular journal that is turned by the rotor.

Much the same as the cylinder of the two-cycle reciprocating engine, the rotary engine housing (Fig. 11-2) contains an intake port, an exhaust port, and a spark plug. The internal shape of the housing resembles a fat figure eight; the shape is called an *epitrochoid*. The outside of the housing on smaller engines has fins for air cooling. On larger engines, water is circulated in coolant cavities between the inner and outer walls of the housing.

Centered within the rotor, behind the moving gear, is a large circular journal that is mounted eccentrically on a straight shaft that passes through the middle of the stationary gear in the housing. This shaft is the crankshaft. The power of the rotating rotor is applied as direct torque to the crankshaft. For each rotation of the rotor, the crankshaft rotates three times. Thus, if the crankshaft speed is 6000 rev/min, the rotor speed is only 2000 rev/min.

The actions of intake, compression, power, and exhaust are shown in Figure 11-4 for one face of the rotor. The rotor apex seal tips follow an epitrochoidal path around the inside of the housing making differently sized and shaped areas (volumes) between the rotor and the housing as the rotor moves. When the leading apex seal of a rotor face passes the intake port, a mixture of air, oil and gasoline enters the area between the rotor face and the housing. The mixture continues to enter the area (Fig. 11-4A) until the trailing apex seal passes the intake port (Fig. 11-4B). The air-fuel mixture is now trapped in the area between the apex seals, the face, and the housing and the mixture can expand or be compressed as this area changes. As the rotor continues its clockwise rotation, the volume of air-fuel mixture is compressed to make it more volatile for combustion. At the correct precise instant, the ignition spark is generated and jumps between the spark plug electrodes igniting the mixture for the power sequence (Fig. 11-4C). The pressure of combustion felt on the combustion recess in the rotor face drives the rotor around. When the leading apex seal passes the exhaust port (Fig. 11-4D), the exhaust gases escape. The trailing apex seal of the face sweeps the gases out as the rotor continues.

The cycle of operation has only been described for *one* face of the rotor; there are three faces. Hence some of the actions of one cycle are taking place at the same time as actions of other cycles. For example in Figure 11-4C, the action at face 1 is power while the action at face 2 is exhaust and at face 3 is intake. It is easily seen that for one complete rotation of the rotor, three power impulses—one for each face—are provided. Thus for a crankshaft speed of 6000 rev/min, the rotor speed is 2000 rev/min and there are 6000 power impulses per minute. A flywheel is attached to the crankshaft, as in piston engines, to keep the crankshaft rotating at a smooth constant speed.

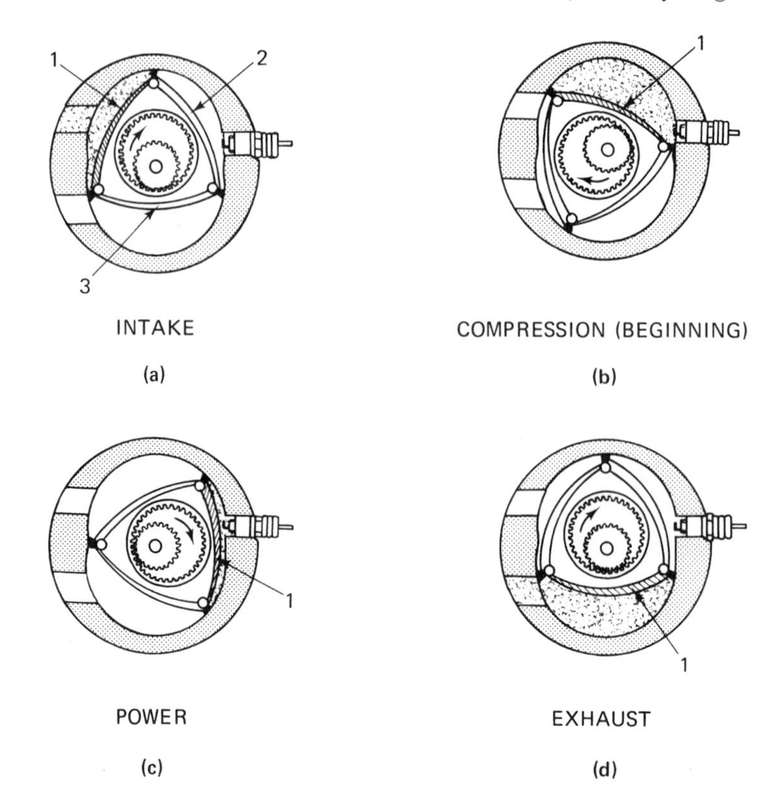

Fig. 11-4. The rotary engine operating cycle has the same four actions as the reciprocating engine.

A little oil is mixed with the gasoline to lubricate the apex seals. The oil also cools the rotor. Oil-to-gasoline ratios vary by manufacturer from 1:25 to 1:50. Oil changes are eliminated because the rotary engine is free of blow-by gases that contaminate the oil in reciprocating engines. Connecting rods and valves are not needed in the rotary engine.

11-2. ROTARY ENGINE ADVANTAGES

For an equivalent horsepower reciprocal engine, the rotary engine is more compact, of lighter weight, and is mechanically simpler. The size of the rotary engine is about one-half the size of a reciprocal engine, it weighs about one-third to one-half as much, and it has about 40 percent less parts. The rotary engine does not vibrate like a reciprocal engine; it can always be perfectly balanced. The rotary engine is less expensive to make (about one-half to two-thirds the price of a reciprocal engine of equivalent horsepower) yet at the present time the cost of the rotary engine is about the same as for the reciprocal engine because of the limited production.

Prognosticators estimate that by 1980, about 75 to 95 percent of the engines in the United States will be rotary engines. The rotary should potentially be more reliable and have lower maintenance costs because it has less parts (less *moving* parts in particular). When slightly more perfected and when industry can change its manufacturing, more and more rotary engines will be built.

11-3. ROTARY ENGINE PROBLEMS

The rotary engine has four problems that design engineers are working on to improve: leakage at the apex seals, fuel economy, exhaust emissions, and manufacturing operations. The most critical leakage site is at the three apex seals of the rotor as the seals travel along the epitrochoid housing. New seal materials and housing surfaces are continually being developed. Some automobile engine manufacturers are currently claiming seals that will last for 100,000 miles.

Fuel consumption of the prototype rotary engine is higher than the consumption of the piston reciprocating engine. Reasons for higher fuel consumption are that combustion tends to be slower and occurs later in the cycle and hence is less efficient than it is in the piston engine, some of the fuel leaks past the apex seals, and combustion in the trailing portion of the chamber is slow and irregular. Some of the latest rotary engine designs are using two spark plugs to overcome this latter problem. Rotary automobile engines are getting about 25 percent less miles per gallon than reciprocating engines of equivalent horsepower. Oil consumption in automobile rotary engines is rated at about one quart per 1500 to 2000 miles. With the elimination of blow-by gases and with the addition of a quart of oil every 1500 to 2000 miles, automobile engine manufacturers are estimating that by 1980, oil changes will be required only every 50,000 miles.

Exhaust emission problems of the rotary engine are similar to the problems of piston engines. The rotary engine puts out more unburned hydrocarbons and greater carbon monoxide, but less nitrogen emissions than the piston engine. One or two thermal reactors—catalysts—are used to alleviate emission problems.

New and highly specialized manufacturing operations are required to mass produce the rotary engine with its epitrochoidal shaped rotor housing requiring a super finished surface. The apex seals are also sophisticated parts and are made from exotic materials.

11-4. A BRIEF HISTORY OF ROTARY ENGINES

A number of people experimented and tested variations of the rotary combustion engine as early as 1799. It was a German inventor, Dr. Felix Heinrich Wankel, however, who developed the working rotary engine that today threatens to revolutionize the engine industry. Wankel worked for years on the rotary engine. Finally in 1951 he contacted the NSU Engine Manufacturing Company in Germany; through joint efforts, the production of the rotating combustion engine finally materialized. In 1958 the Curtiss-Wright Corporation of the United States became the licensee of the NSU/Wankel Engine in North America. Since then a number of United States and foreign engine manufacturers have entered into legal agreements with Curtiss-Wright and NSU/Wankel for permission to develop and manufacture rotary engines.

11-5. BEYOND THE ROTARY ENGINE

Designers are continually striving for a better engine; they are not completely satisfied with the reciprocating engine nor the rotary engine, nor convinced that either engine can meet pollution emission requirements. Some of the ideas of the future include a gas turbine, a steam engine, the Stirling engine, a stratified charge modification, a diesel engine, and electric engines.

The gas turbine is a revolutionary engine in the aircraft field and has replaced virtually all piston engines over 500 horsepower. It is on the brink of breaking into the heavy duty automotive market of large trucks and buses. The gas turbine engine is a simple engine; it is essentially a torch that blows against a set of fan blades and causes them to turn. The gas turbine is fueled with any liquid combustible, usually kerosene for aircraft.

The steam engine is an external-combustion engine. It is of interest because of its clean burning. The most popular steam engine of today is the Rankine cycle engine; it involves a furnace boiler with plumbing which leads to either a piston engine or a turbine wheel. After most of the power of the steam is used, some remaining heat can be extracted via a regenerator and then added to water that is on its way to the boiler. The spent steam is directed to a condenser where it is further cooled until it is a liquid again. The liquid is then pumped to the boiler for reheating.

The Stirling engine is another external-combustion engine that is continuously fired and involves positive displacement (piston), compression, and expansion. The Stirling engine utilizes a completely sealed working fluid and operates at low speeds. This engine has low emissions, low noise levels, low fuel consumption, and low maintenance requirements. This engine has not been designed for mass production.

A recent development known as *compound vortex controlled combustion* is a principle of *modified stratified charge* that can be applied to engines now marketed for automobiles. This method provides a rich mixture near the spark plug for positive combustion and ignites a very lean mixture elsewhere in the combustion chamber for efficient and clean burning of the fuel charge. The modified stratified charge has low emissions, good mileage per gallon of fuel, and can use a range of fuels.

The diesel engine has been used in trucks, buses, and heavy machinery, but has not gained acceptance with automobiles because of high costs, smelly exhaust, poor acceleration, and difficulty in starting in cold weather. The diesel is however economical and durable, and requires little maintenance. It has no ignition system.

As was mentioned in Chapter 1, electric powered engines are impractical today except for small powered engines for short time periods.

Appendixes

A

Hand Tools

If you are planning to become a small gasoline engine technician, you will need the proper tools to perform maintenance correctly and efficiently. Chances are that since you have an interest in this mechanical field you probably already have an assortment of hand tools. What you need now are the additional hand tools to complete your set. This appendix describes the hand tools and special tools used by a small gasoline engine technician.

In purchasing hand tools, consider their use. Will they be used daily? Weekly? Only occasionally? Is accuracy required? Will they be subjected to abuse? Do you plan to use them for their intended purpose? Will the tool become part of your permanent workshop?

If your plans are to use the tools regularly, buy quality tools from a reputable manufacturer. Watch your newspaper—there are often sales that will lower your total costs—but be aware that inexpensive tools are often low in quality. Choose inexpensive tools only for abnormal or infrequent use.

Buy the best tools that you can afford. Buy the tools that you need most first and buy the others as your budget permits. Don't reject the idea of buying used tools—many a good buy has been made from someone who has lost interest in a hobby or job and wants to sell his tools. Watch for sales in the classified section of your newspaper.

It is advisable to buy certain tools in sets because the price of the set is often considerably less than the prices of the individual pieces. For example, screwdrivers often come in sets of four or five sizes that cover the normal range of screw sizes. Since you'll eventually need all of the sizes, why not (especially when you see a brand name set on sale) initially buy a set?

Once you have purchased a tool, adopt the following rule that is used by professionals: *use quality tools only for their intended purposes.* This will ensure that screwdriver tips, plier teeth, wrench openings, and so on, are in proper condition at all times and are ready for you to use in performing their intended function in the most efficient and accurate manner.

The following common hand tools are needed to repair small gasoline engines:

awl
drill, electric, with set of twist drills
epoxy
files, metal, assorted rough and fine cut
gauge, feeler (blade type)
gauge (spark plug, wire type)
hacksaw
hammer, ball peen
hammer, soft-faced
pliers, needle nose, 6 in.
pliers, snap ring (optional)
punches, pin (set)
screwdrivers, conventional (set)
screwdrivers, Phillips head (set)
sealant, vulcanizing silicone rubber
torch, propane
tweezers
vise, bench, 4 inch jaw
wrench, adjustable open end
wrench, combination (set)—English sizes and metric sizes
wrench, locking plier (vise grips)
wrench, socket, ratchet drive, (optional)
wrench, spark plug
wrench, torque (inch pound calibrations)

NOTE

For a complete guide to the selection, description, use, and care of hand tools, refer to *Everyone's Book of Hand and Small Power Tools,* by George R. Drake, Reston Publishing Company Inc., Reston, Virginia 22090.

Completely equipped small engine repair shops will have most, if not all, of the following special tools. Figure A-1 illustrates these special tools; sometimes two different models are illustrated, but bear in mind that different engine manufacturers have different special tools. Some manufacturers' tools can be used interchangeably on other manufacturers' engines; others cannot.

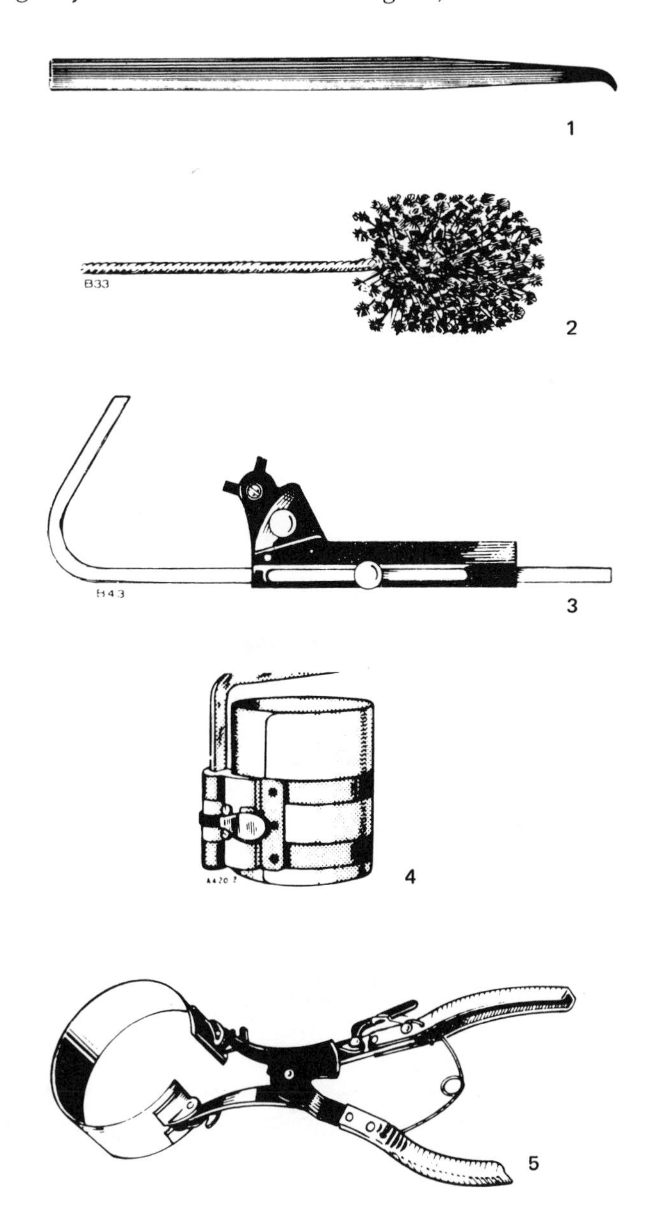

FIG. A-1. Well-equipped repair shops have most of these special tools.

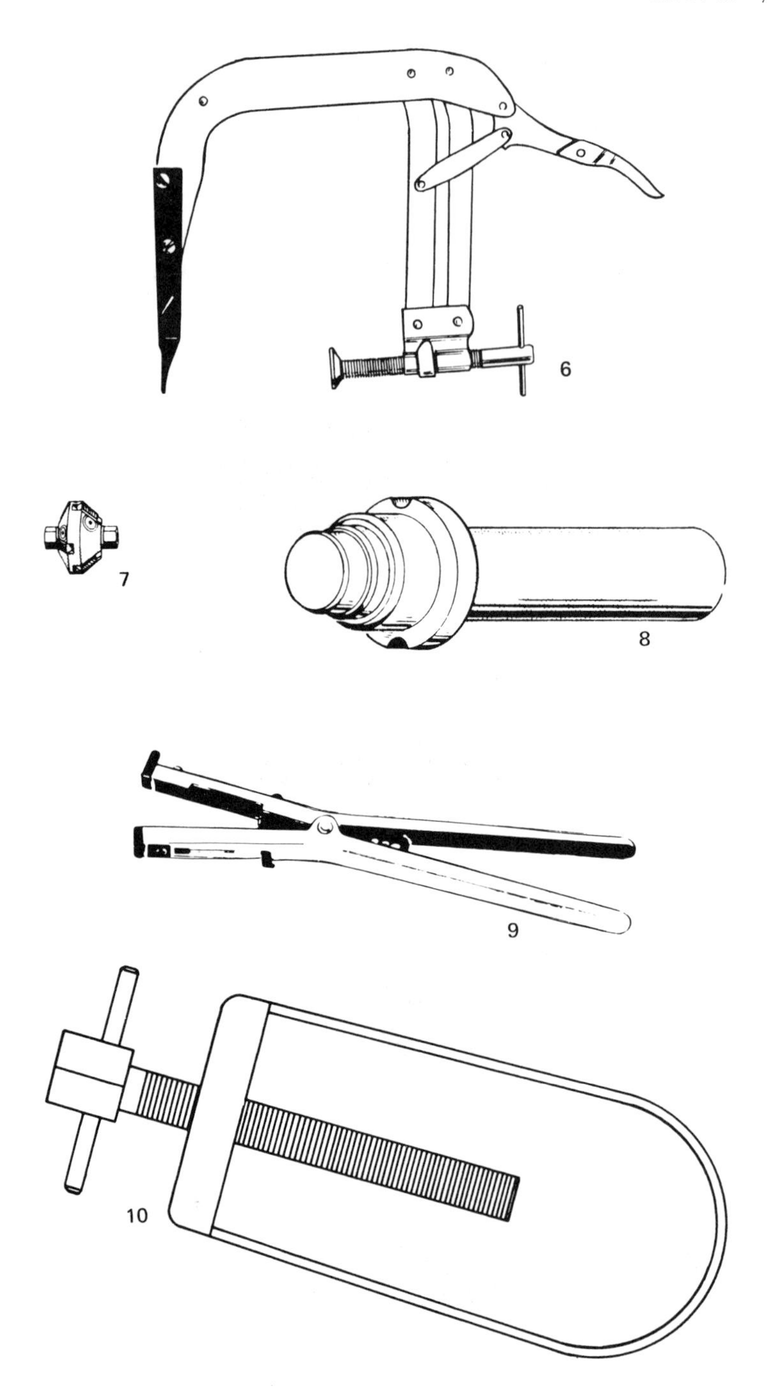

Fig. A-1. Repair shop tools (Continued).

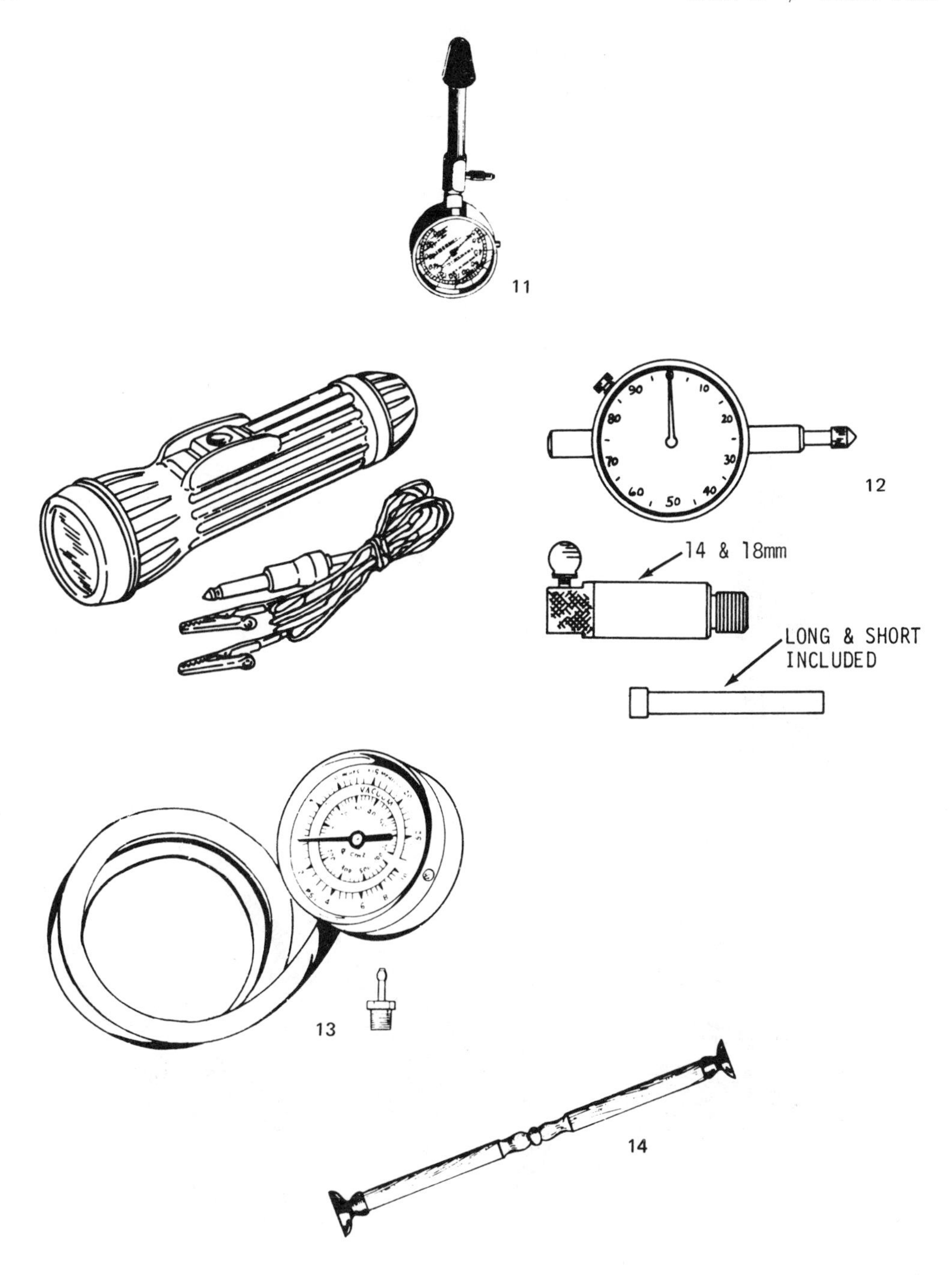

Fig. A-1. Repair shop tools (Continued).

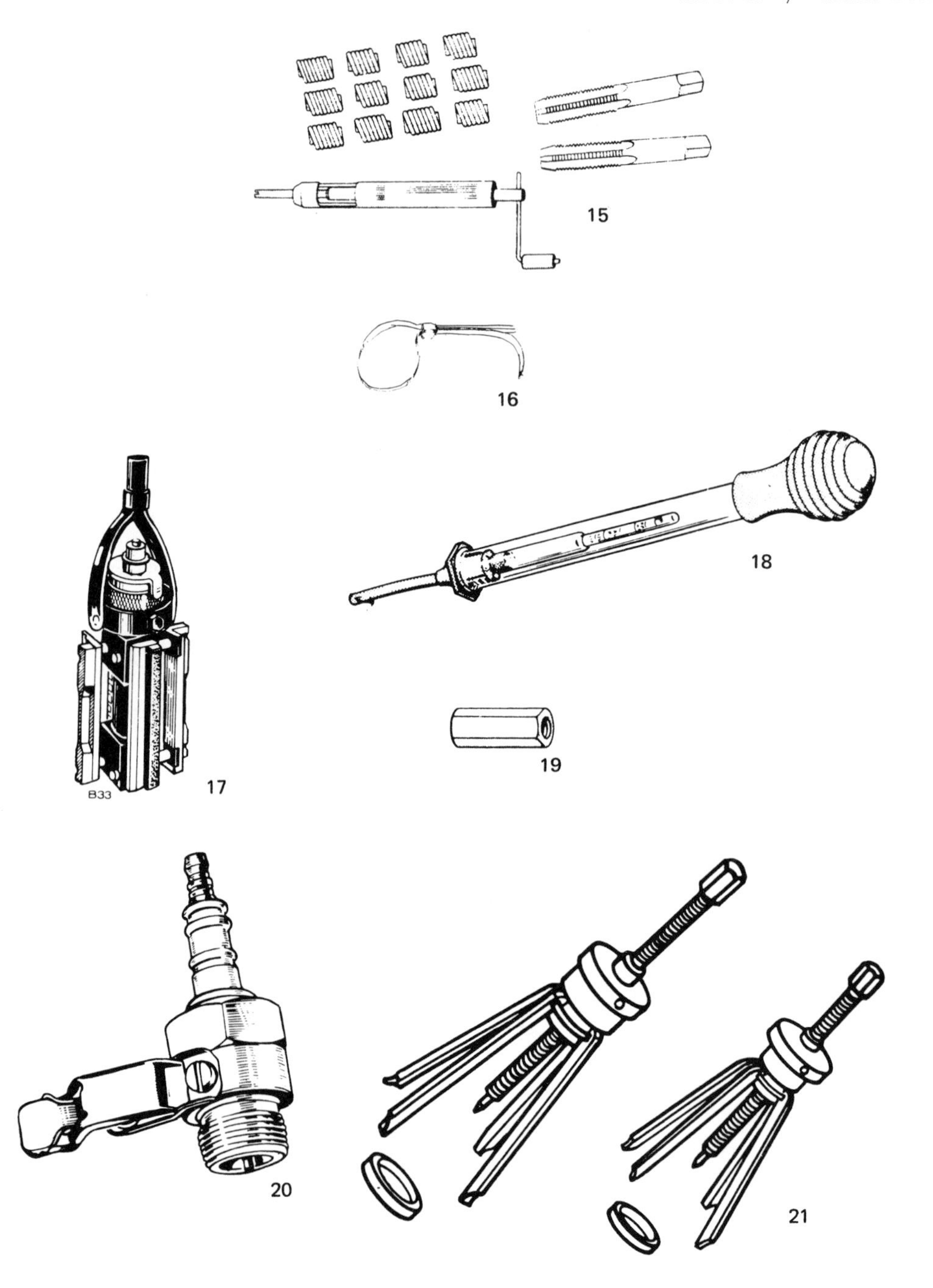

Fig. A-1. Repair shop tools (Continued).

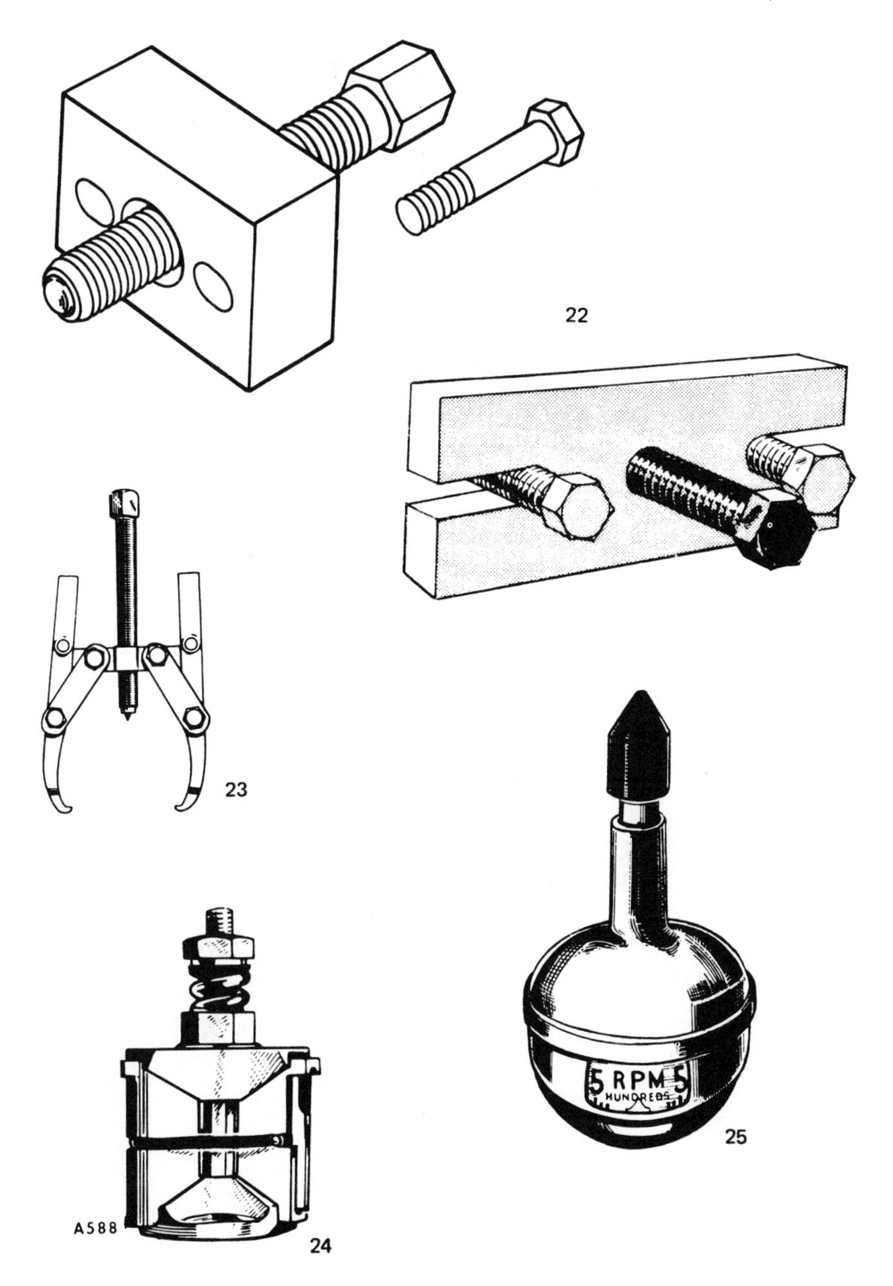

Fig. A-1. Repair shop tools (Continued).

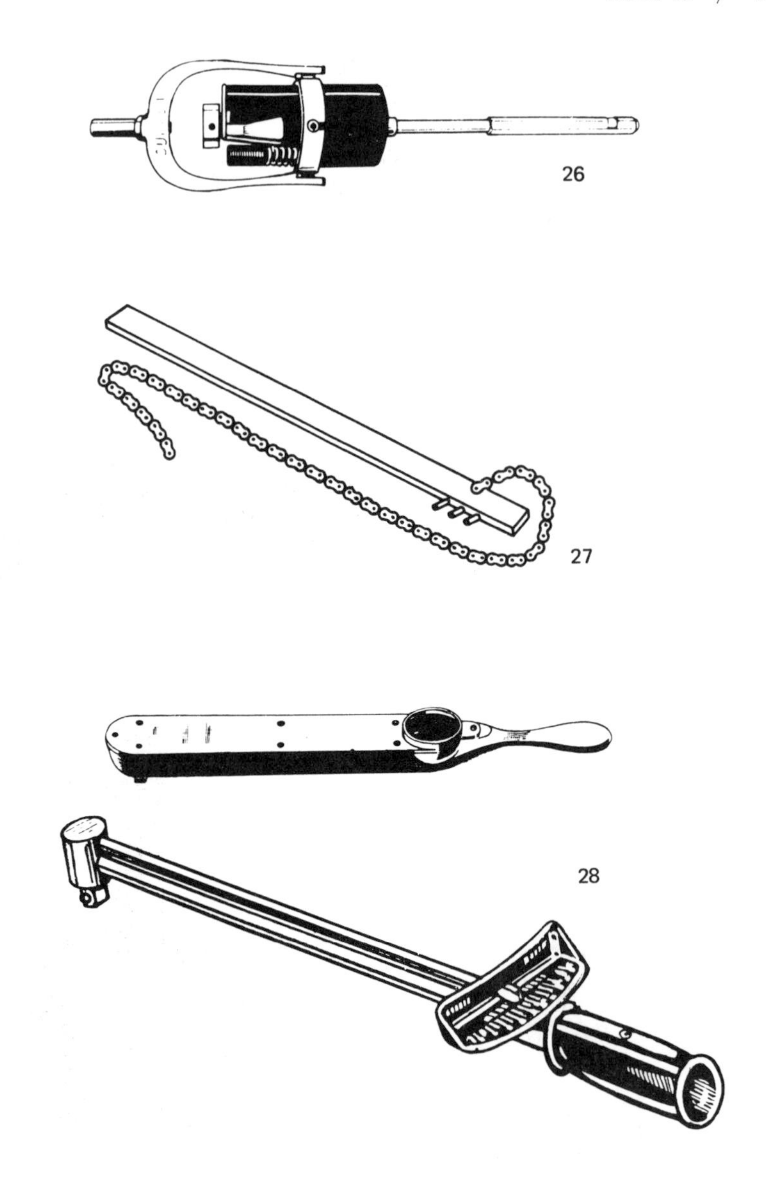

Fig. A-1. Repair shop tools (Continued).

Tool	*Use*	*Figure Reference*
Bar, flywheel puller	Simplifies removal of flywheel from crankshaft.	1
Brush, cylinder wall microfinishing	Deglazes cylinder walls. Produces ideal finish for rings to prevent blow-by and to lower oil consumption.	2
Cleaner, piston groove	Removes carbon from piston ring grooves.	3
Compressor, piston ring	Compresses all rings evenly for easy installation of piston into cylinder.	4
Compressor, piston ring, pliers type	Compresses all rings evenly for easy installation of piston into cylinder.	5
Compressor, valve spring	Used to remove and install valve springs.	6
Cutter, valve seat	Cuts valve seats.	7
Driver, main and cam bearing, combination	Used for installing crankshaft and camshaft bearings.	8
Expander, piston ring	Expands piston rings for removal and installation of rings on piston.	9
Extractor, piston pin	Removes piston pin from piston and connecting rod.	10
Gauge, compression	Measures engine compression.	11
Gauge, ignition timing	Used to set correct ignition timing.	12
Gauge, vacuum and pressure	Measures vacuum and pressures.	13
Grinding tool, valve, lapping	Used to lap valve and valve seat.	14
Heli-coil, thread repair kit	Repairs stripped bolt/screw threads.	15
Holder, flywheel	Used to hold flywheel when removing flywheel nut. See wrench, chain.	16
Hone, cylinder	Reconditions cylinder bore.	17
Hydrometer, battery	Checks battery specific gravity.	18
Nuts, impact	Used to remove flywheel from crankshaft taper.	19
Plug, spark, test	Simulates normal compression firing.	20
Puller, bearing	Used to remove bearings from engine.	21
Puller, flywheel	Removes flywheel from crankshaft taper.	22
Puller, gear	Removes gear from crankshaft.	23
Reamer, ridge	Removes ridge from top of cylinder wall.	24
Tachometer	Requires contact with moving part. Determines speed in rev/min. (vibrating tachometers also available).	25
Valve guide, honing set	Hones valve guides.	26
Wrench, chain	Prevents flywheel from turning while flywheel nut is loosened. See holder, flywheel.	27
Wrench, torque	Measures torque applied to bolts/screws.	28

The following tools are very special and are only needed in well equipped repair shops. The list included here is more for reference than for necessity.

Bearing clearance guide—"Plasti-gage" simplifies fitting bearing clearances of split bearings.

Bearing driver—Used to remove and install bearings.

Cam bearing driver—Simplifies installation of precision sleeve cam bearings.

Connecting rod aligning set—Checks rods with or without pistons.

Crankshaft run out gauge—Used for checking engines for bent crankshafts.

Main bearing driver—Simplifies installation of precision sleeve main bearings.

Oil seal guide and driver—Simplifies seal installation and prevents damage to seal surfaces.

Reamer—Used to ream bearings in blocks.

Rolling tool—Used to peen or roll metal around outside diameter of valve seat insert after installation.

Timing advance mechanical cover driver—Used for proper installation of spark advance cover.

Valve guide driver—Assures correct installation of valve guides.

Valve guide reamer—Used to enlarge valve guide holes so oversize stemmed valve can be used.

Valve insert driver—Used to drive valve seat inserts in place.

Valve lock replacer—Simplifies installation of split valve keepers.

Valve seat driver—Simplifies installation of valve seat into cylinder block.

Valve seat staker—Assures tight fit of seat.

A flashlight continuity jig (timing light) can be purchased inexpensively or can be easily made by rewiring a flashlight as shown in Figure A-2. Solder all connections and use alligator clips on the ends of the leads.

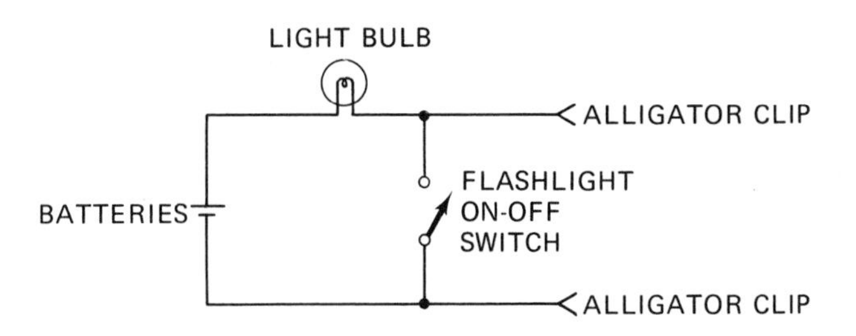

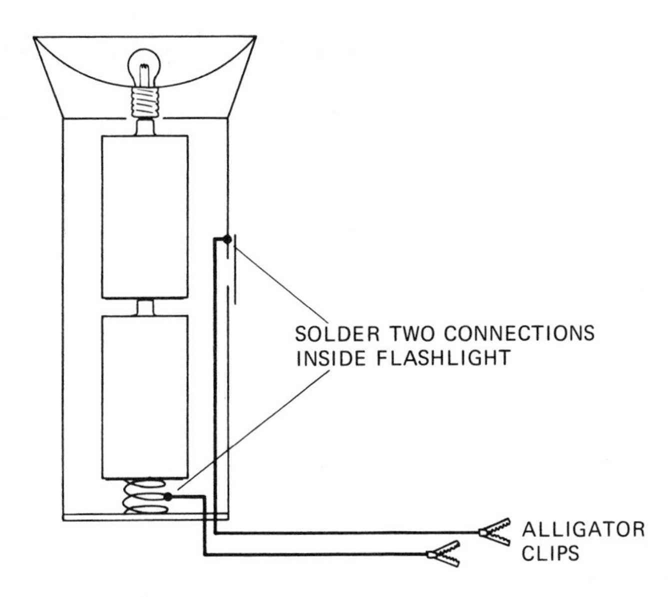

FIG. A-2. A flashlight continuity jig (timing light) can be easily made.

The test spark plug shown in Figure A-1 Item 20 can easily be made. Drill and tap a hole into the spark plug; bolt an alligator clip onto the plug. Cut the side electrode off of the plug. In use connect the alligator clip to the bare metal of the engine (ground). Connect the high-tension spark plug wire to the test plug. If a spark can jump from the center electrode to the side during an ignition test (Section 9-3) the ignition system is operating correctly.

B

Metric to English and English to Metric Conversion Factors

To Convert	*Into*	*Multiply by*
	Metric to English	
millimeters	inches	0.039
millimeters	feet	3.281×10^{-3}
centimeters	inches	0.394
centimeters	feet	3.281×10^{-2}
meters	inches	39.370
meters	feet	3.281
cubic centimeters	cubic inches	0.016
cubic meters	cubic feet	35.316
square millimeters	square inches	1.550×10^{-3}
square centimeters	square inches	0.155
square meters	square feet	10.764
liters	cubic inches	61.023
liters	pints	2.113
liters	quarts	1.057
liters	U.S. gallons	0.264
grams	ounces, Avoirdupois	0.035
kilograms	pounds	2.205
kilograms per square centimeter	pounds per square inch	14.223
kilograms per cubic meter	pounds per cubic foot	0.062
calories, gram (mean)	British Thermal Units (mean)	3.968×10^{-3}

To Convert	*Into*	*Multiply by*
	English to Metric	
inches	millimeters	25.400
inches	centimeters	2.540
inches	meters	0.025
feet	millimeters	304.800
feet	centimeters	30.480
feet	meters	0.305
cubic inches	cubic centimeters	16.389
cubic inches	cubic meters	1.639×10^{-5}
cubic feet	cubic meters	0.028
square inches	square millimeters	645.2
cubic inches	liters	0.016
pints	liters	0.473
quarts	liters	0.946
U.S. gallons	liters	3.785
ounces, Avoirdupois	grams	28.349
pounds	kilograms	0.454
pounds per square inch	kilograms per square centimeter	0.070
pounds per cubic foot	kilograms per cubic meter	16.019
British Thermal Units	gram–calories	252.0

C

Inch-Millimeter Equivalents of Decimal and Common Fractions

Inch	½'s	¼'s	8ths	16ths	32nds	64ths	Millimeters	Decimals of an Inch[a]
						1	0.397	0.015 625
					1	2	0.794	0.031 25
						3	1.191	0.046 875
				1	2	4	1.588	0.062 5
						5	1.984	0.078 125
					3	6	2.381	0.093 75
						7	2.778	0.109 375
			1	2	4	8	3.175[a]	0.125 0
						9	3.572	0.140 625
					5	10	3.969	0.156 25
						11	4.366	0.171 875
				3	6	12	4.762	0.187 5
						13	5.159	0.203 125
					7	14	5.556	0.218 75
						15	5.953	0.234 375
		1	2	4	8	16	6.350[a]	0.250 0
						17	6.747	0.265 625
					9	18	7.144	0.281 25
						19	7.541	0.296 875
				5	10	20	7.938	0.312 5
						21	8.334	0.328 125
					11	22	8.731	0.343 75
						23	9.128	0.359 375
			3	6	12	24	9.525[a]	0.375 0

Inch	*½'s*	*¼'s*	*8ths*	*16ths*	*32nds*	*64ths*	*Millimeters*	*Decimals of an Inch*[a]
						25	9.922	0.390 625
					13	26	10.319	0.406 25
						27	10.716	0.421 875
				7	14	28	11.112	0.437 5
						29	11.509	0.453 125
					15	30	11.906	0.468 75
						31	12.303	0.484 375
	1	2	4	8	16	32	12.700[a]	0.500 0
						33	13.097	0.515 625
					17	34	13.494	0.531 25
						35	13.891	0.546 875
				9	18	36	14.288	0.562 5
						37	14.684	0.578 125
					19	38	15.081	0.593 75
						39	15.478	0.609 375
			5	10	20	40	15.875[a]	0.625 0
						41	16.272	0.640 625
					21	42	16.669	0.656 25
						43	17.066	0.671 875
				11	22	44	17.462	0.687 5
						45	17.859	0.703 125
					23	46	18.256	0.718 75
						47	18.653	0.734 375
		3	6	12	24	48	19.050[a]	0.750 0
						49	19.447	0.765 625
					25	50	19.844	0.781 25
						51	20.241	0.796 875
				13	26	52	20.638	0.812 5
						53	21.034	0.828 125
					27	54	21.431	0.843 75
						55	21.828	0.859 375
			7	14	28	56	22.225[a]	0.875 0
						57	22.622	0.890 625
					29	58	23.019	0.906 25
						59	23.416	0.921 875
				15	30	60	23.812	0.937 5
						61	24.209	0.953 125
					31	62	24.606	0.968 75
						63	25.003	0.984 375
1	2	4	8	16	32	64	25.400[a]	1.000 0

[a] Exact.

D

Decimal Equivalents of Millimeters (0.01 to 100 mm)

mm.	*Inches*	*mm.*	*Inches*	*mm.*	*Inches*	*mm.*	*Inches*	*mm.*	*Inches*
0.01	0.00039	0.41	0.01614	0.81	0.03189	21	0.82677	61	2.40157
0.02	0.00079	0.42	0.01654	0.82	0.03228	22	0.86614	62	2.44094
0.03	0.00118	0.43	0.01693	0.83	0.03268	23	0.90551	63	2.48031
0.04	0.00157	0.44	0.01732	0.84	0.03307	24	0.94488	64	2.51968
0.05	0.00197	0.45	0.01772	0.85	0.03346	25	0.98425	65	2.55905
0.06	0.00236	0.46	0.01811	0.86	0.03386	26	1.02362	66	2.59842
0.07	0.00276	0.47	0.01850	0.87	0.03425	27	1.06299	67	2.63779
0.08	0.00315	0.48	0.01890	0.88	0.03465	28	1.10236	68	2.67716
0.09	0.00354	0.49	0.01929	0.89	0.03504	29	1.14173	69	2.71653
0.10	0.00394	0.50	0.01969	0.90	0.03543	30	1.18110	70	2.75590
0.11	0.00433	0.51	0.02008	0.91	0.03583	31	1.22047	71	2.79527
0.12	0.00472	0.52	0.02047	0.92	0.03622	32	1.25984	72	2.83464
0.13	0.00512	0.53	0.02087	0.93	0.03661	33	1.29921	73	2.87401
0.14	0.00551	0.54	0.02126	0.94	0.03701	34	1.33858	74	2.91338
0.15	0.00591	0.55	0.02165	0.95	0.03740	35	1.37795	75	2.95275
0.16	0.00630	0.56	0.02205	0.96	0.03780	36	1.41732	76	2.99212
0.17	0.00669	0.57	0.02244	0.97	0.03819	37	1.45669	77	3.03149
0.18	0.00709	0.58	0.02283	0.98	0.03858	38	1.49606	78	3.07086
0.19	0.00748	0.59	0.02323	0.99	0.03898	39	1.53543	79	3.11023
0.20	0.00787	0.60	0.02362	1.00	0.03937	40	1.57480	80	3.14960
0.21	0.00827	0.61	0.02402	1	0.03937	41	1.61417	81	3.18897
0.22	0.00866	0.62	0.02441	2	0.07874	42	1.65354	82	3.22834
0.23	0.00906	0.63	0.02480	3	0.11811	43	1.69291	83	3.26771
0.24	0.00945	0.64	0.02520	4	0.15748	44	1.73228	84	3.30708
0.25	0.00984	0.65	0.02559	5	0.19685	45	1.77165	85	3.34645
0.26	0.01024	0.66	0.02598	6	0.23622	46	1.81102	86	3.38582
0.27	0.01063	0.67	0.02638	7	0.27559	47	1.85039	87	3.42519
0.28	0.01102	0.68	0.02677	8	0.31496	48	1.88976	88	3.46456
0.29	0.01142	0.69	0.02717	9	0.35433	49	1.92913	89	3.50393
0.30	0.01181	0.70	0.02756	10	0.39370	50	1.96850	90	3.54330
0.31	0.01220	0.71	0.02795	11	0.43307	51	2.00787	91	3.58267
0.32	0.01260	0.72	0.02835	12	0.47244	52	2.04724	92	3.62204
0.33	0.01299	0.73	0.02874	13	0.51181	53	2.08661	93	3.66141
0.34	0.01339	0.74	0.02913	14	0.55118	54	2.12598	94	3.70078
0.35	0.01378	0.75	0.02953	15	0.59055	55	2.16535	95	3.74015
0.36	0.01417	0.76	0.02992	16	0.62992	56	2.20472	96	3.77952
0.37	0.01457	0.77	0.03032	17	0.66929	57	2.24409	97	3.81889
0.38	0.01496	0.78	0.03071	18	0.70866	58	2.28346	98	3.85826
0.39	0.01535	0.79	0.03110	19	0.74803	59	2.32283	99	3.89763
0.40	0.01575	0.80	0.03150	20	0.78740	60	2.36220	100	3.93700

E

English System of Weights and Measures

Linear Measure (Length)

1000 mils = 1 inch (in.)
12 inches = 1 foot (ft.)
3 feet = 1 yard (yd.)
5280 feet = 1 mile

Square Measure (Area)

144 square inches (sq.in.) = 1 square foot (sq.ft.)
9 square feet = 1 square yard (sq.yd.)

Cubic Measure (Volume)

1728 cubic inches (cu.in.) = 1 cubic foot (cu.ft.)
27 cubic feet = 1 cubic yard (cu.yd.)
231 cubic inches = 1 U.S. gallon (gal.)
277.27 cubic inches = 1 British imperial gallon (i.gal.)

Liquid Measure (Capacity)

4 fluid ounces (fl.oz.) = 1 gill (gi.)
2 pints = 1 quart (qt.)
4 quarts = 1 gallon

Dry Measure (Capacity)

2 pints = 1 quart
8 quarts = 1 peck (pk.)

Weight (Avoirdupois)

27.3438 grains = 1 dram (dr.)
16 drams = 1 ounce (oz.)
16 ounces = 1 pound (lb.)
100 pounds = 1 hundredweight (cwt.)
112 pounds = 1 long hundredweight (l.cwt.)
2000 pounds = 1 short ton (S.T.)
2240 pounds = 1 long ton (L.T.)

Weight (Troy)

24 grains = 1 pennyweight (dwt.)
20 pennyweights = 1 ounce (oz.t.)
12 ounces = 1 pound (lb.t.)

Angular or Circular Measure

60 seconds = 1 minute
60 minutes = 1 degree
57.2958 degrees = 1 radian
90 degrees = 1 quadrant or right angle
360 degrees = 1 circle or circumference

F

Metric System of Weights and Measures

Linear Measure (Length)

1/10 meter = 1 decimeter (dm)
1/10 decimeter = 1 centimeter (cm)
1/10 centimeter = 1 millimeter (mm)
1/1000 millimeter = 1 micron (μ)
1/1000 micron = 1 millimicron (mμ)
10 meters = 1 dekameter (dkm)
10 dekameters = 1 hectometer (hm)
10 hectometers = 1 kilometer (km)
10 kilometers = 1 myriameter

Square Measure (Area)

1 are = 1 square dekameter (dkm^2)
1 centare = 1 square meter (m^2)
1 hectare = 1 square hectometer (hm^2)

Cubic Measure (Volume)

1 stere = 1 cubic meter (m^3)
1 decistere = 1 cubic decimeter (dm^3)
1 centistere = 1 cubic centimeter (cm^3)
1 dekastere = 1 cubic dekameter (dkm^3)

Capacity

1/10 liter = 1 deciliter (dl)
1/10 deciliter = 1 centiliter (cl)
1/10 centiliter = 1 milliliter (ml)
10 liters = 1 dekaliter (dkl)
100 liters = 1 hectoliter (hl)
1000 liters = 1 kiloliter (kl)
1 kiloliter = 1 stere (s)

Weight

1/10 gram = 1 decigram (dg)
1/10 decigram = 1 centigram (cg)
1/10 centigram = 1 milligram (mg)
10 grams = 1 dekagram (dkg)
100 grams = 1 hectogram (hg)
1000 grams = 1 kilogram (kg)
10,000 grams = 1 myriagram
100,000 grams = 1 quintal (q)
1,000,000 grams = 1 metric ton (t)

G

Battery Test Procedures

INTRODUCTION—To determine the ability of a battery to function properly requires testing. The accuracy of the testing changes with temperature, specific gravity, age of the battery, etc. An accurate test has more than one step.

Step 1: Visual inspection

Step 2: Specific gravity check (hydrometer)

Step 3: 421 test*

Step 4: Load test*

Testing Cautions: Wear safety glasses. Do not break live circuits at battery terminals. When testing, be certain to remove gases at battery cover caused by charging.

*"421" and Load Test equipment is available from many automotive test equipment suppliers.
Your Delco wholesaler is able to advise you as to price and availability of this equipment.

STEP 1

VISUAL INSPECTION
CHECK FOR OBVIOUS DAMAGE SUCH AS A CRACKED OR BROKEN CASE THAT SHOWS LOSS OF ELECTROLYTE

- OBVIOUS DAMAGE
  - REPLACE BATTERY
- NO OBVIOUS DAMAGE
  - CHECK ELECTROLYTE LEVEL
    - BELOW TOP OF PLATES IN ONE OR MORE CELLS
      - ADD WATER TO JUST ABOVE SEPARATORS BUT NOT TO SPLIT RING.

        CHARGE FOR 15 MIN. @ 15-25 AMPS WITH NON-FLAME ARRESTOR VENT CAPS REMOVED BUT LEAVE FLAME ARRESTOR CAPS IN PLACE
    - ELECTROLYTE LEVEL ABOVE TOP OF PLATES IN ALL CELLS—PROCEED TO STEP 2

STEP 2

SPECIFIC GRAVITY (HYDROMETER) TEST

- 50 POINTS OR MORE VARIATION BETWEEN HIGHEST AND LOWEST CELL
  - REPLACE BATTERY
- LESS THAN 50 POINTS VARIATION BETWEEN HIGHEST AND LOWEST CELL
  - SPECIFIC GRAVITY 1.100 @ 80° OR MORE IN ALL CELLS
    - IF "421" TESTER IS AVAILABLE GO TO STEP 3. IF NOT AVAILABLE GO TO STEP 4. (6 & 8 VOLT PROCEED TO STEP 4)
  - SPECIFIC GRAVITY LESS THAN 1.100 IN ONE OR MORE CELLS
    - PROCEED TO STEP 4

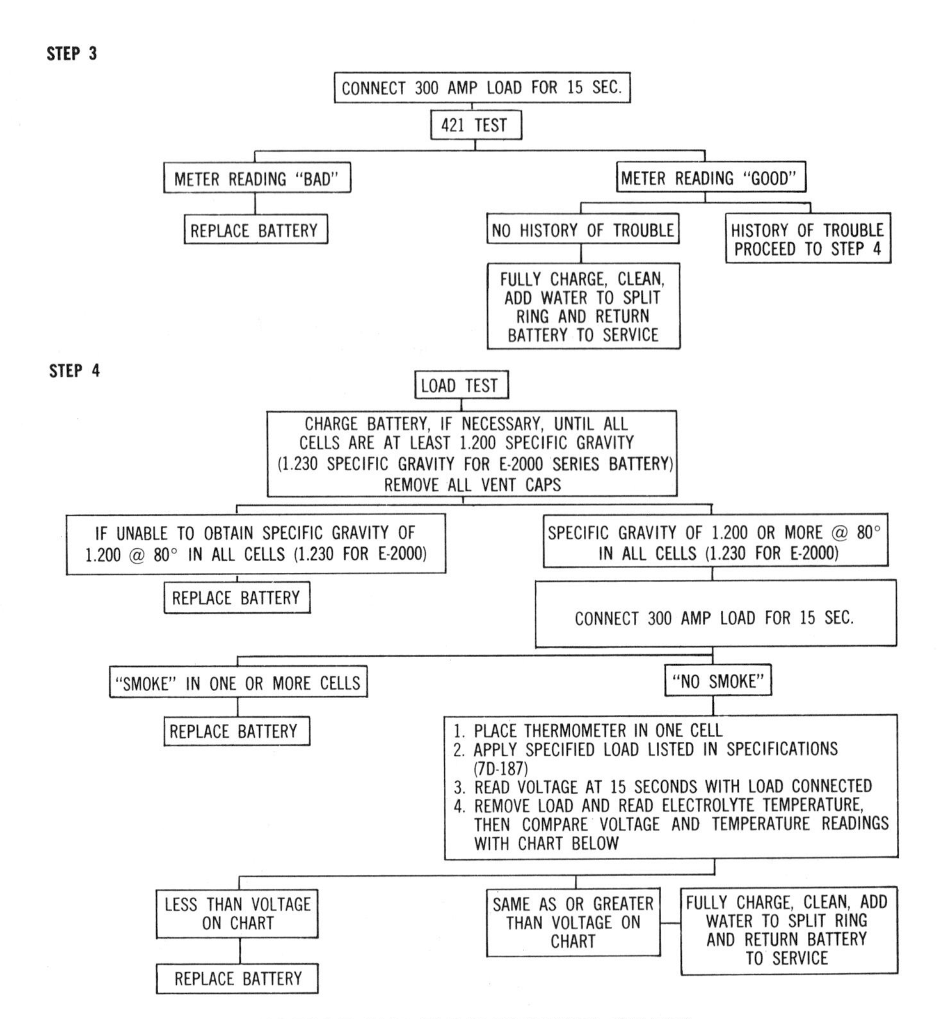

VOLTAGE AND TEMPERATURE CHART

Voltage must not drop below minimum listed at given temperature when battery is subjected to the proper load for 15 seconds and is 1.200 specific gravity @ 80° or more.

MINIMUM VOLTAGES									
Voltage	Electrolyte Temperature Down To:								
	80°	70°	60°	50°	40°	30°	20°	10°	0°
6-Volt	4.8	4.8	4.7	4.7	4.6	4.5	4.4	4.3	4.2
8-Volt	6.4	6.4	6.3	6.2	6.1	6.0	5.9	5.7	5.6
12-Volt	9.6	9.6	9.5	9.4	9.3	9.1	8.9	8.7	8.5

Illustrated Parts of Typical Small Gasoline Engines

The illustrations in this appendix show the parts and electrical circuits of typical small gasoline engines. See if you can identify the parts in the illustrations by correct name. If you have difficulty or want to check your answer, refer to the keyed parts lists. The following illustrations are included:

Illustration	*Subject*	*Page*	*Assembled View in Text—Figure No.*
H-1	Two-stroke cycle, retractable rope starter, magneto-ignition, lawn mower engine.	327	1-3A
H-2	Two-stroke cycle, electrical starter, self propelled, magneto-ignition, lawn mower engine.	329	1-3B
H-3	Four-stroke cycle, two cylinders, horizontally opposed pistons. Used for industrial applications.	331	1-16
H-4	Two-stroke cycle, single cylinder, multipurpose saw.	332	1-17
H-5	Two-stroke cycle, single cylinder engine.	337	—

Illustration	*Subject*	*Page*	*Assembled View in Text—Figure No.*
H-6	Location diagram of an outboard engine, two-cycle, two cylinders, 9.9 horsepower at 5000 rev/min.	339	—
H-7	Location diagram, four-cycle engine.	340	—
H-8	Electrical diagram of a trackster.		—
H-9	Electrical wiring diagram of a manual start two-cycle, two cylinder, outboard engine.	342	H-6
H-10	Electrical wiring diagram of an electric start, two-cycle, two cylinder, outboard engine.	342	H-6

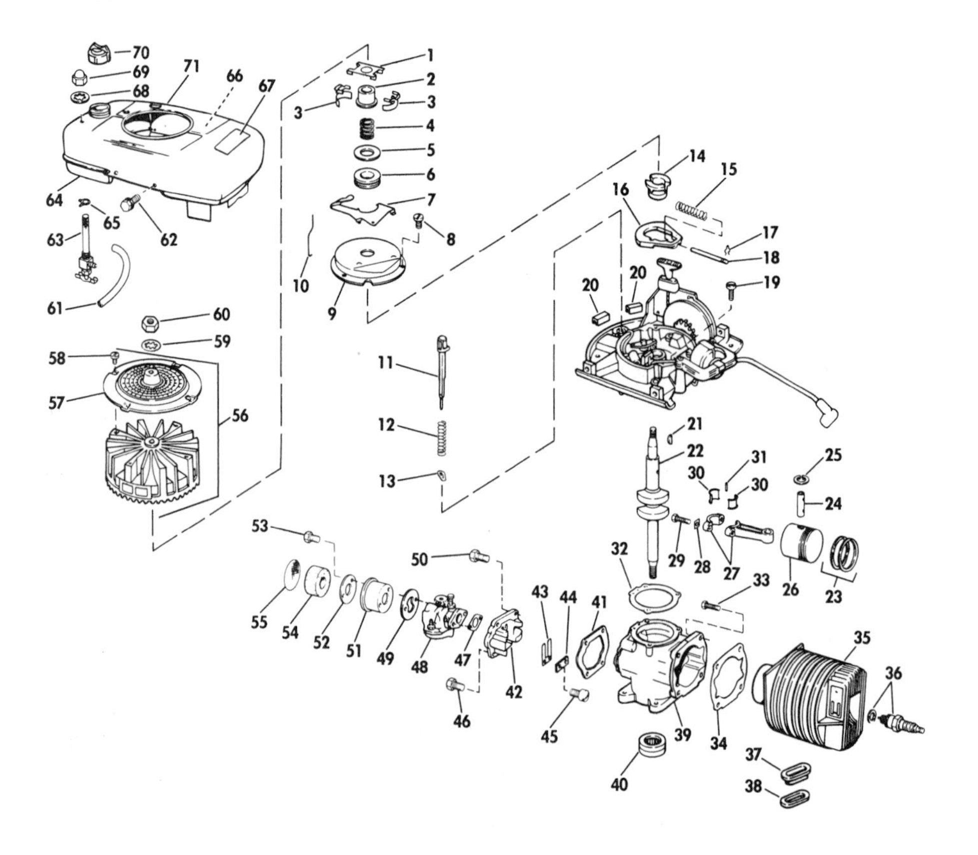

Fig. H-1. Illustrated parts breakdown of a two-stroke cycle lawn mower engine. The engine, magneto, retractable rope starter, and carburetor are illustrated.

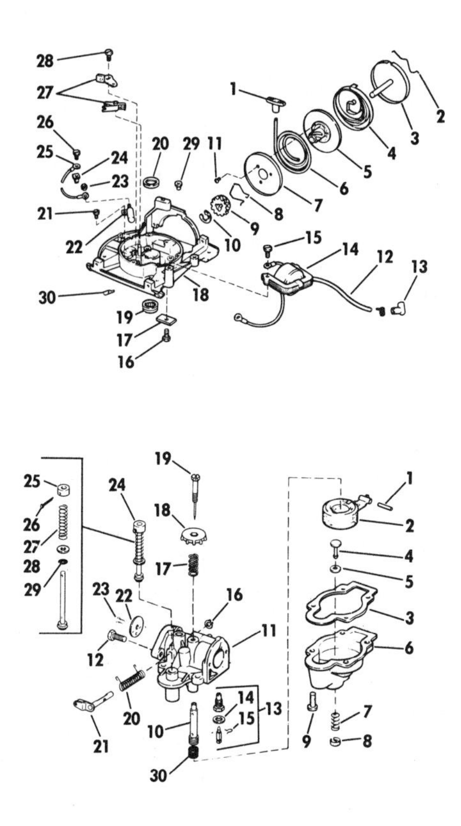

Fig. H-1. Illustrated parts breakdown of a two-stroke cycle lawn mower engine (Continued).

Ref. No.	Description	No. Req'd.
+	Governor Assembly	1
1	. Yoke - Governor	1
2	. Collar - Governor	1
3	. Weight - Governor	2
4	. Spring - Governor	1
5	. Thrust Washer	1
6	. Collar - Thrust	1
7	. Lever - Governor	1
8	Screw - Attaching Dust Cover	3
9	Dust Cover	1
10	Rod - Governor	1
11	Bar - Shut-Off and Primer	1
12	Spring - Shut-Off Switch	1
13	Blade - Switch	1
14	Cam - Spark Advance	1
15	Spring - Spark Advance	1
16	Flyweight - Spark Advance	1
17	Retainer - Spring	1
18	Pin - Spark Advance	1
19	Screw - Arm. Plate to Crankcase	3
20	Pad - Rubber, Arm. Plate	2
21	Key - Crankshaft	1
22	Crankshaft	1
+	Piston & Connecting Rod Assy.	1
23	. Ring Set (2 Rings)	1
24	. Wrist Pin	1
25	. Retaining Ring - Wrist Pin	2
26	. Piston	1
27	. Connecting Rod Assy.	1
28	. . Lockplate	2
29	. . Screw	2
30	. . Liners - Connecting Rod	1
31	. . Needle Set	1
32	Gasket - Crkc. to Arm. Pl. .005 in.	A/R
32	Gasket - Crkc. to Arm. Pl. .010 in.	A/R
33	Screw - Machine, Cyl. to Crkc.	4
34	Gasket - Cyl. to Crankcase	1
35	Cylinder and Sleeve Assy.	1
36	Spark Plug and Gasket	1
37	Sleeve - Exhaust Gasket	1
38	Gasket - Exhaust	1
39	Crankcase and Seal Assy.	1
40	. Oil Seal	1
41	Gasket - Reed Plate	1
42	Reed Plate Assembly	1
43	. Reed	1
44	. Plate - Back-Up	1
45	. Screw - Machine	2
46	Scr. - Reed Pl. to Crkc.	3
47	Gasket - Carburetor	1
48	Carburetor Assembly	1
49	Gasket	1
50	Scr. - Reed Pl. to Crkc.	1
51	Cup - Air Filter	1
52	Plate - Back-Up	1
53	Screw - Filter to Carburetor	2
54	Element - Air Filter	1
55	Screen - Air Filter	1
56	Flywheel and Screen Assy.	1
57	. Screen and Cup	1
58	. Screw	4
59	Lockwasher - Flywheel	1
60	Nut - Flywheel	1
61	Gas Line	1
62	Screw - Shroud to Armature Plate	4
+	Shroud and Gas Tank Assy.	1
63	. . Shut-Off Valve	1
64	. Gas Tank Assy.	1
65	. . Clamp	1
66	. Decal - LAWN-BOY	1
67	. Decal - Instruction	1

Ref. No.	Description	No. Req'd.
68	. Lockwasher	2
69	. Nut	2
70	. Cap	1
71	. Shroud and Shield Assy.	1
	CARBURETOR	
+	Carburetor Assy.	1
1	. Pin - Float Arm Hinge	1
2	. Float and Arm Assy.	1
3	. Gasket - Float Chamber	1
+	. Float Chamber & Dump Valve Assy.	1
4	. . Valve Stem	1
5	. . Valve Seat	1
6	. . Chamber - Float	1
7	. . Valve Spring	1
8	. . Valve Keeper	1
9	. Screw - Float Chamber	4
10	. Nozzle	1
11	. Carburetor Body Assy.	1
12	. . Screw	2
13	. . Float Valve & Seat	1
14	. . . Washer - Float Valve Seat	1
15	. . . Clip	1
16	. Grip Ring	1
17	. Spring - Needle Adj.	2
18	. Knob - Needle Adj.	1
19	. Needle Assy.	1
20	. Spring - Throttle	1
21	. Throttle Shaft & Lever Assy.	1
22	. Disc - Throttle	1
23	. Screw - Throttle Shaft Disc	1
24	. Primer Assy.	1
25	. . Knob Primer	1
26	. . Cotter Pin	1
27	. . Spring Compression	1
28	. . Washer - Plain	1
29	. . "O" Ring	1
30	. Filter	1
	MAGNETO AND STARTER	
1	Handle - Starter	1
2	Spring - Starter Rope Retainer	1
+	Starter Assembly	1
3	. Spring - Cup - Pin Assy.	1
4	. Spring - Starter	1
5	. Pulley - Starter	1
6	. Rope - Starter	1
7	. Plate - Starter Pulley	1
8	. Spring - Starter Pinion	1
9	. Pinion - Starter	1
10	. Retainer - Push-On	1
11	. Screw - Threaded Cutting Starter	3
12	High Tension Lead Assy.	1
13	. Cover & Terminal Assy.	1
14	Coil & Lamination Assy.	1
15	Scr. - Lamination Mtg.	2
16	Scr. - Starter Attach.	1
17	Clamp - Starter Pin	1
18	Arm. Pl., Brg. & Oil Seal Assy.	1
19	. Brg.	1
20	. Oil Seal	1
21	Scr. - Cond. Mtg.	1
22	Cond. & Nut Assy.	1
23	. Nut	1
24	Grommet - Nylon	1
25	Terminal & Lead Assy.	1
26	Scr. - Shut-Off Sw.	1
27	Brkr. Base & Arm Assy.	1
28	Scr. - Breaker Base	1
29	Grommet	1
30	Clip	1

FIG. H-1. Illustrated parts breakdown of a two-stroke cycle lawn mower engine (Continued).

Ref. No.	Description	No. Req'd.
1	Nut	1
2	Lockwasher	1
3	Flywheel and Screen Assy.	1
4	. Screw	4
5	. Screen	1
+	Governor Assy.	1
6	. Yoke	1
7	. Collar	1
8	. Weight	2
9	. Spring	1
10	. Thrust Washer	1
11	. Collar	1
12	. Lever	1
13	Rod	1
14	Screw - Arm. Plt. to C'case	3
15	Bracket	1
16	Screw	3
17	Spring	1
18	Tab	1
19	Key	1
20	Piston & Connecting Rod Assy.	1
21	. Connecting Rod & Brg.	1
22	. . Screw	2
23	. . Lockplate	2
24	. . Liners	1
25	. . Needle Brg. Set	1
26	. Retaining Ring	2
27	. Wrist Pin	1
28	. Piston	1
29	. Ring Set	1
30	Spark Plug	1
31	Cylinder	1
32	Gasket	1
33	Screw	4
34	Gasket	1
35	Crankshaft	1
36	Gasket	1
37	Crankcase, Brg. & Seal	1
38	. Seal, Driveshaft	1
39	Driveshaft	1
40	Washer	4
41	Screw	4
42	Seal, Crankshaft	1
43	Cover	1
44	Gasket	1
45	Washer, Thrust	1
46	Gear	1
47	Reed Plate	1
48	. Screw - Reed to Plate	2
49	. Plate	1
50	. Reed	1
51	Screw	3
52	Gasket	1
53	Carburetor	1

Ref. No.	Part No.	Description	No. Req'd.
54	607577	Gasket	1
55	607578	Case - Air Filter	1
56	602775	Screw	2
57	607580	Element	1
58	607579	Washer	1
59	602192	Screw	1
60	605076	Washer	2
61	606007	Screw	2
62	679609	Air Baffle Assembly	1
63	607745	. Screw	2
64	608027	. Fastener	1
65	607844	. Cap	1
66	607311	. Base	1
67	607310	. Bulb	1
68	679691	. Bracket	1
69	607607	Screw	3
70	607575	Tubing	1
71	605120	Grommet	7
72	679607	Knob & Rod	1
73	607845	Grommet	1
74	604857	Screw	2
75	679610	Bracket	1
+	681151	Shroud, Tank, Valve and Decal Assy.	1
76	681064	. Fuel Shut Off Assy.	1
77	608657	. Decal - Left Side	1
78	608658	. Decal - Right Side	1
79	608659	. Decal - Lawn-Boy	1
80	608662	. Decal - Caution	1
81	607581	Gas Line	1
82	605119	Screw	7
83	679456	Fuel Cap	1
84	608663	Plug	1
+	679674	Electric Starter Assy.	1
85	679675	. Motor and Bracket Assy.	1
86	608034	. Pinion	1
87	681098	. Gear Assy.	1
88	608622	. . Retainer	1
89	607683	. Gasket, Gear Hsg. Cover	1
90	607684	. Gear Housing Cover	1
91	607685	. Screw, Gear Hsg. Cover	3
92	607686	. Starter Pinion	1
93	607687	. Spring	1
94	607816	. Stop Washer, Pinion	3

Ref. No.	Part No.	Description	No. Req'd.
95	607692	Screw	3
96	607693	Flat Washer	3
97	608584	Rope Guard	1
		CARBURETOR	
+	679710	Carburetor	1
1	602850	. Gasket	1
2	679623	. Float Chamber & Dump Valve Assy.	1
+	681077	. . Dump Valve Assy.	1
3	607570	. . . Valve Stem	1
4	607571	. . . Valve Seat	1
5	607572	. . . Valve Keeper	1
6	607573	. . . Valve Spring	1
7	306552	. Screw	4
8	678113	. Float & Arm Assy.	1
9	678882	. Float Valve & Seat Assy.	1
10	603145	. . Clip	1
11	301996	. . Washer	1
12	604295	. Disc	1
13	604296	. Spring	1
14	678119	. Shaft	1
15	607574	. Screw	1
16	607126	. Spring	1
17	607575	. Tubing	1
18	605522	. Screw	1
19	679888	. Carburetor Body Assy.	1
20	301996	. . Washer	1
21	607568	. . Plug	1
22	300096	. Float Pin	1
23	604286	. Grip Ring	1
		MAGNETO AND STARTER	
1	607526	Screw	2
2	679927	C.D. Ignition Pack	1
3	678117	. High Tension Lead	1
4	580339	. . Cover and Terminal Assy.	1
5	679890	Arm. Plate Brg. & Seal	1
6	605020	. Seal	1
7	679277	. Bearing	1
8	604261	Clamp	1
9	604857	Screw	1
10	606863	Handle	1
+	679614	Starter Assy.	1
11	604259	. Retainer	1
12	602821	. Screw	3
13	604244	. Plate	1
14	607874	. Rope	1
15	604646	. Pulley	1
16	604257	. Spring	1
17	679615	. Cup & Pin Assy.	1
18	604260	. Spring	1
19	607547	. Pinion	1

+ Not Shown

FIG. H-2. Illustrated parts breakdown of a two-stroke cycle self-propelled lawn motor engine. The engine, magneto, electric starter, and carburetor are illustrated.

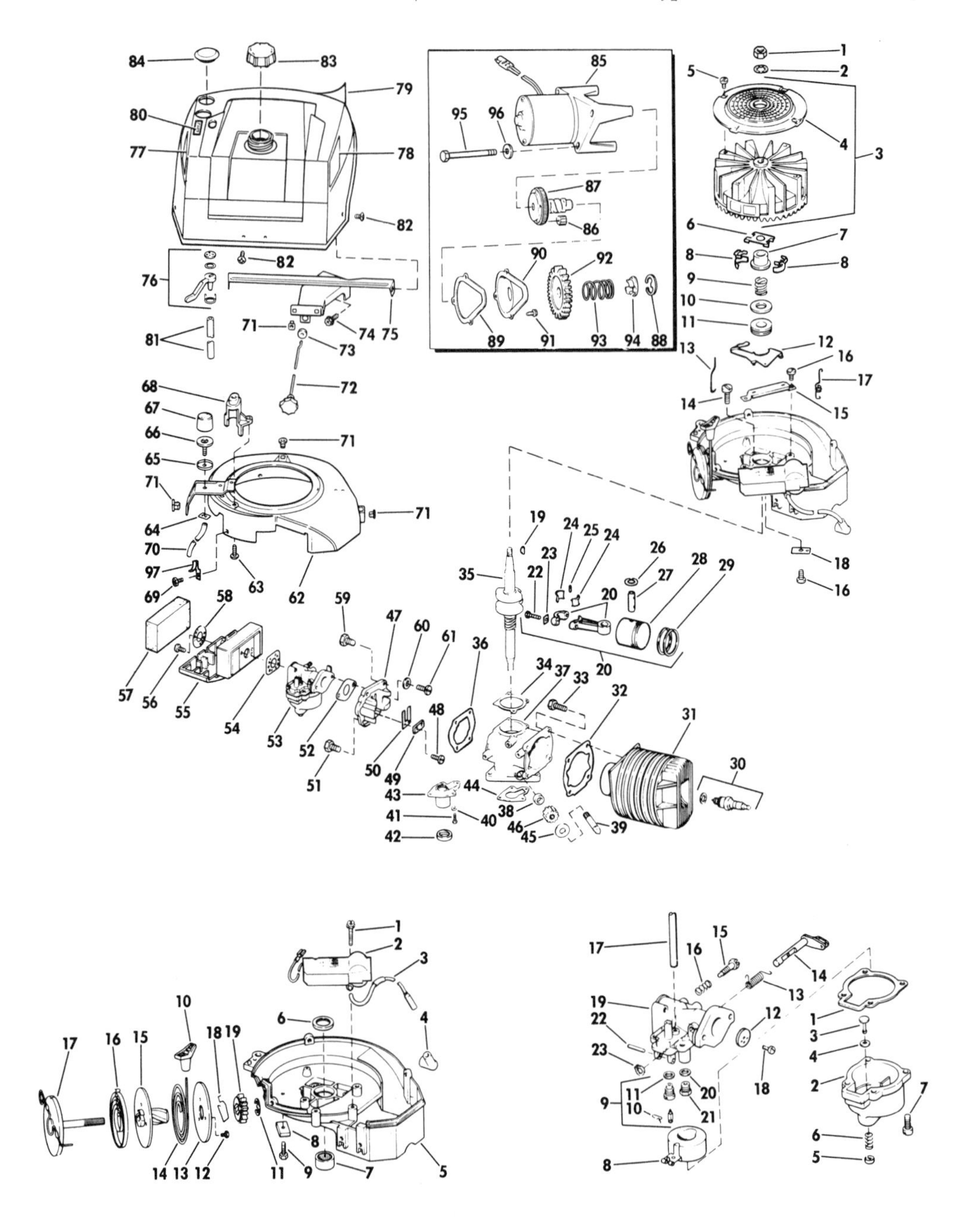

FIG. H-2. Illustrated parts breakdown of a two-stroke cycle self-propelled lawn motor engine (Continued).

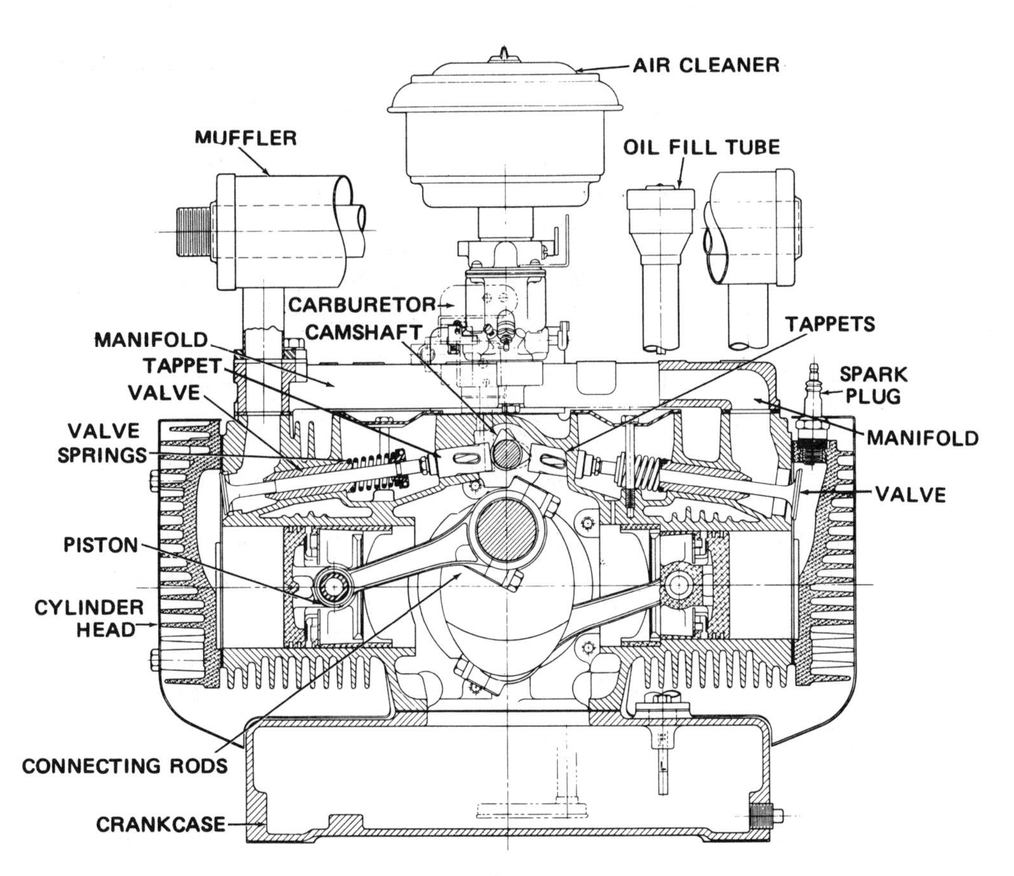

FIG. H-3. Location diagram for parts of a four-cycle, two-cylinder engine. The cylinders have horizontally opposed piston, 3¼ inch bores, and 3 inch strokes. The engine develops approximately 13 bhp at 2700 rev/min. The compression ratio is 5.5 to 1.

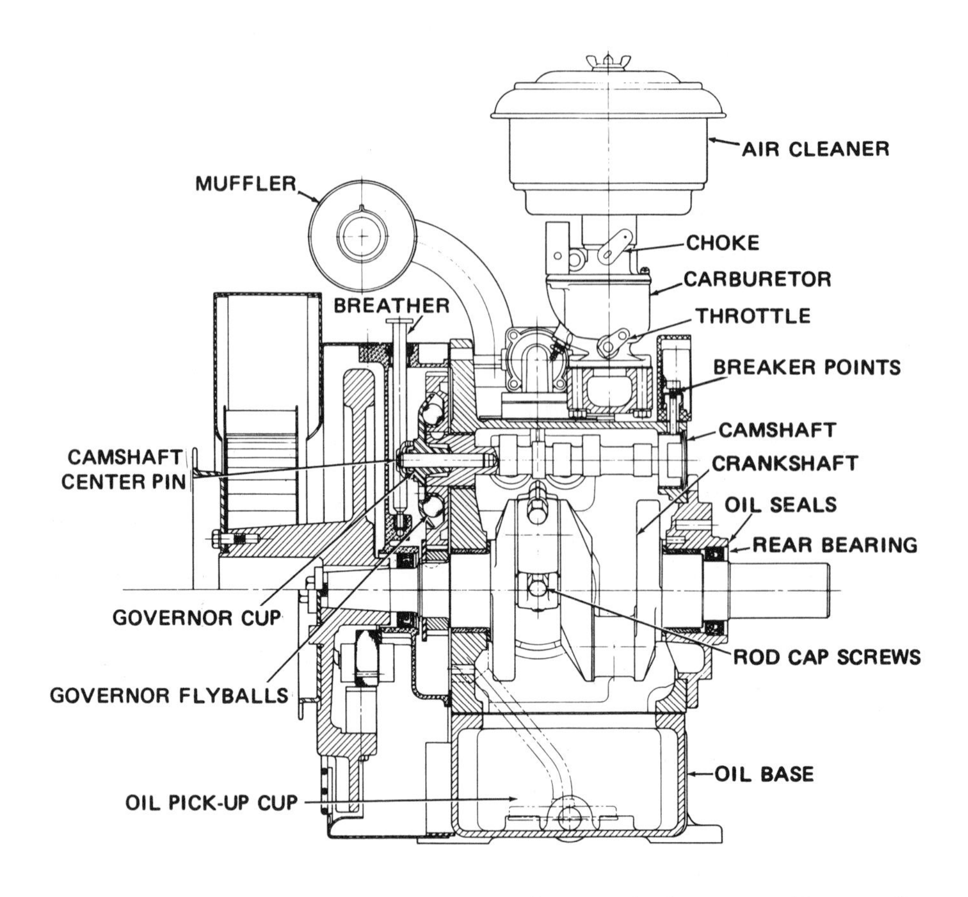

Fig. H-3. Location diagram for parts of a four-cycle, two-cylinder engine (Continued).

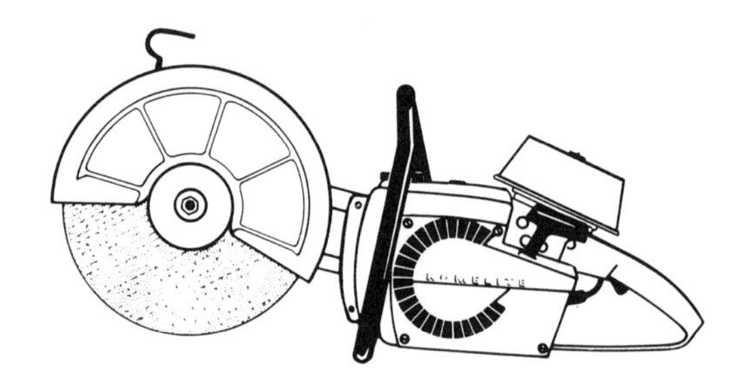

Fig. H-4. Illustrated parts breakdown of a two-stroke cycle multipurpose saw.

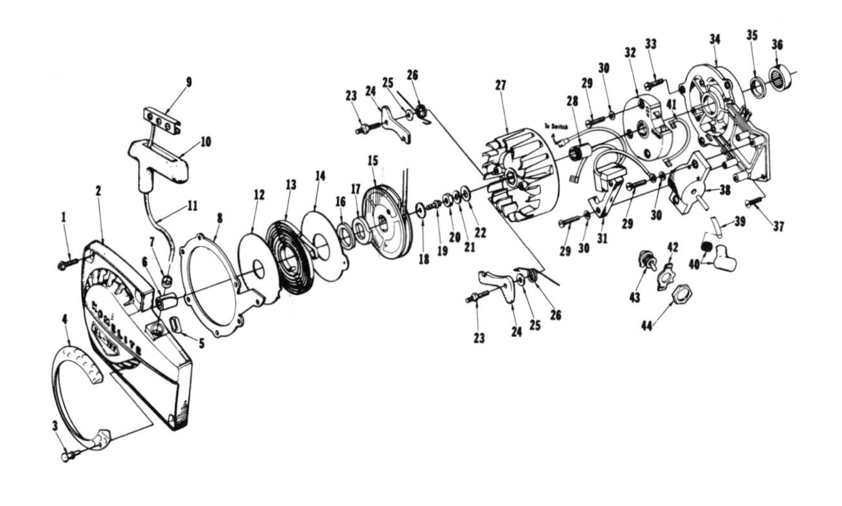

No.	Description	Qty.
1	SCREW-hex, spinlock, 12-24 x 5/8	4
2	STARTER HOUSING	1
	Includes:	
3	STUD-drive	11
4	SCREEN-starter housing	1
5	BUSHING-rewind spring	1
6	BUSHING-starter post	1
7	BUSHING-starter rope	1
8	RING-air flow	1
9	INSERT-rope retaining	1
10	GRIP-starter rope	1
11	ROPE-starter (48" long) #5	1
12	SHIELD-spring (inner)	1
13	SPRING-rewind	1
14	SHIELD-spring (outer)	1
15	PULLEY & CUP ASSEMBLY	1
	Includes:	
16	BUSHING-spring lock	1
17	LOCK-rewind spring	1
18	WASHER-retaining	1
19	SCREW-hex, 10-32 x 1/2	1
20	NUT-hex, jam, 3/8-24	1
21	WASHER-lock, 3/8	1
22	WASHER-flat	1
23	STUD-shoulder	2
24	PAWL-starter	2
25	WASHER-lock, #12	2
26	SPRING-starter pawl	2
27	ROTOR-magneto	1
28	CAP-dust	1
29	SCREW-pan, 8-32 x 3/4	7
30	WASHER-lock, #8	7
31	COIL-core and generator	1
32	MODULE-ignition	1
33	SCREW-hex, 1/4-20 x 5/8	3
*34	PLATE-back	1
	Includes:	
35	SEAL	1
36	BEARING-roller	1
37	SCREW-rd. hd. 1/4-20 x 5/8	1
38	COIL-transformer	1
39	LEAD-high tension	1
40	TERMINAL-spark plug	1
41	LEAD-transformer	1
42	PLATE-"ON-OFF"	1
43	SWITCH	1
	Includes:	
44	NUT	1

*Denotes new parts

FIG. H-4. Parts of a two-stroke cycle multipurpose saw (Continued).

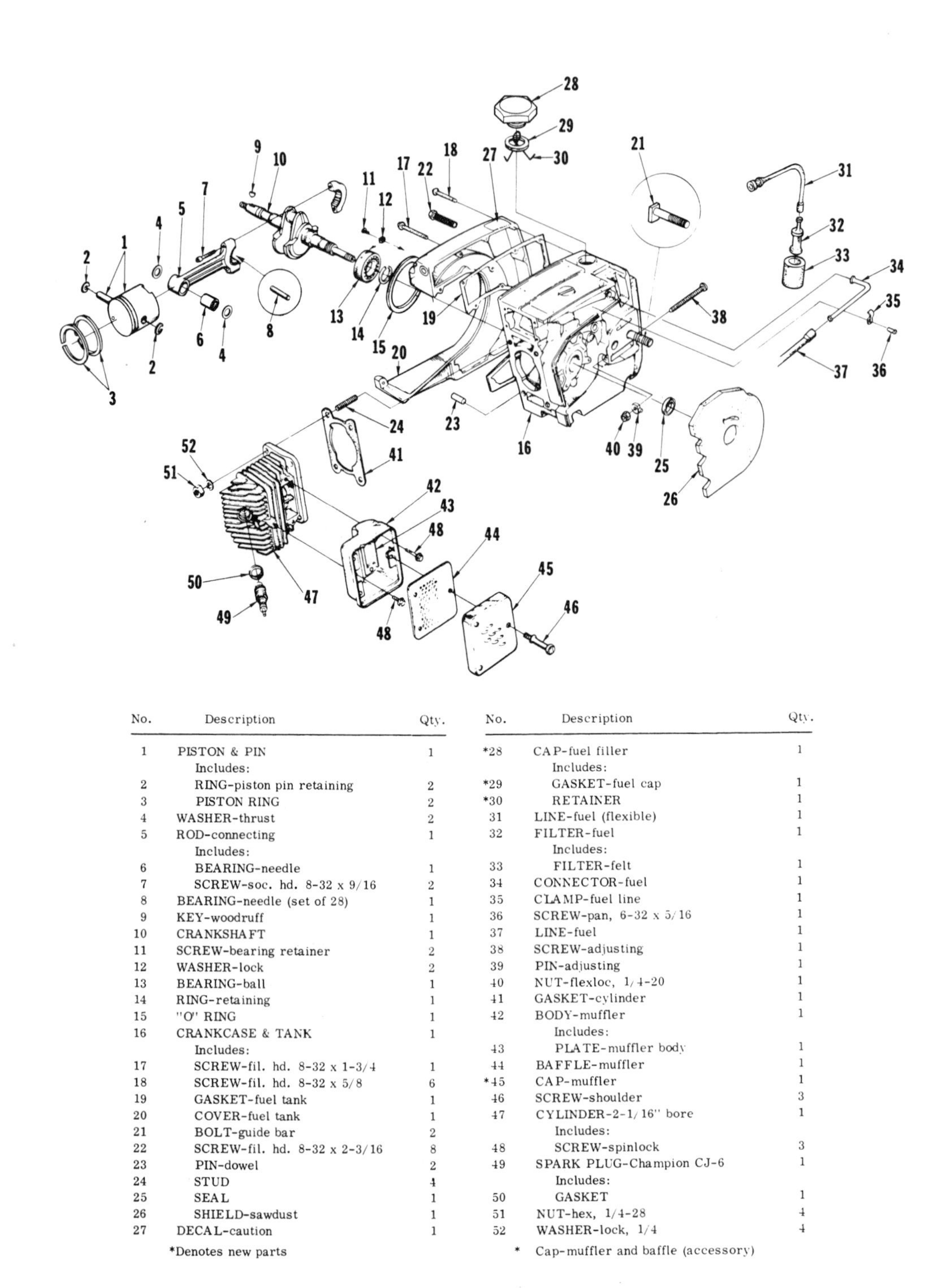

No.	Description	Qty.
1	PISTON & PIN	1
	Includes:	
2	RING-piston pin retaining	2
3	PISTON RING	2
4	WASHER-thrust	2
5	ROD-connecting	1
	Includes:	
6	BEARING-needle	1
7	SCREW-soc. hd. 8-32 x 9/16	2
8	BEARING-needle (set of 28)	1
9	KEY-woodruff	1
10	CRANKSHAFT	1
11	SCREW-bearing retainer	2
12	WASHER-lock	2
13	BEARING-ball	1
14	RING-retaining	1
15	"O" RING	1
16	CRANKCASE & TANK	1
	Includes:	
17	SCREW-fil. hd. 8-32 x 1-3/4	1
18	SCREW-fil. hd. 8-32 x 5/8	6
19	GASKET-fuel tank	1
20	COVER-fuel tank	1
21	BOLT-guide bar	2
22	SCREW-fil. hd. 8-32 x 2-3/16	8
23	PIN-dowel	2
24	STUD	4
25	SEAL	1
26	SHIELD-sawdust	1
27	DECAL-caution	1

*Denotes new parts

No.	Description	Qty.
*28	CAP-fuel filler	1
	Includes:	
*29	GASKET-fuel cap	1
*30	RETAINER	1
31	LINE-fuel (flexible)	1
32	FILTER-fuel	1
	Includes:	
33	FILTER-felt	1
34	CONNECTOR-fuel	1
35	CLAMP-fuel line	1
36	SCREW-pan, 6-32 x 5/16	1
37	LINE-fuel	1
38	SCREW-adjusting	1
39	PIN-adjusting	1
40	NUT-flexloc, 1/4-20	1
41	GASKET-cylinder	1
42	BODY-muffler	1
	Includes:	
43	PLATE-muffler body	1
44	BAFFLE-muffler	1
*45	CAP-muffler	1
46	SCREW-shoulder	3
47	CYLINDER-2-1/16" bore	1
	Includes:	
48	SCREW-spinlock	3
49	SPARK PLUG-Champion CJ-6	1
	Includes:	
50	GASKET	1
51	NUT-hex, 1/4-28	4
52	WASHER-lock, 1/4	4

* Cap-muffler and baffle (accessory)

Fig. H-4. Parts of a two-stroke cycle multipurpose saw (Continued).

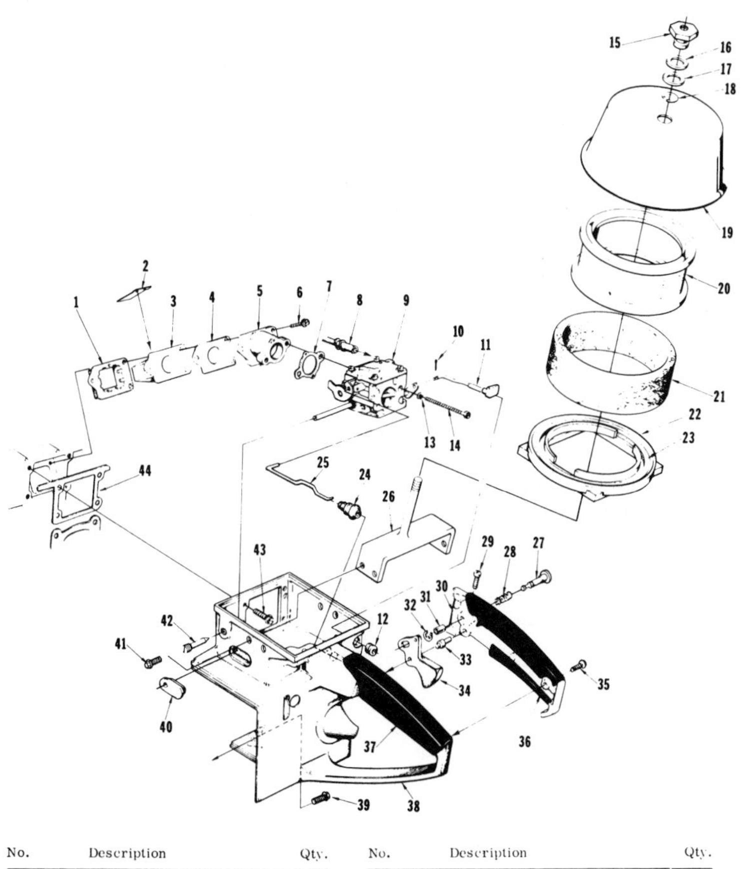

No.	Description	Qty.
1	RETAINER-reed valve	1
2	REED-valve	4
3	SEAT-reed valve	1
4	GASKET-reed valve seat	1
5	MANIFOLD-intake	1
6	SCREW-pan hd. 8-32 x 1-1/8	3
7	GASKET-carburetor	1
8	LINE-fuel	1
* 9	CARBURETOR	1
10	PIN-cotter	1
11	ROD-choke	1
	Includes:	
12	GROMMET	1
13	WASHER-lock #10	2
14	SCREW-hex hd. 10-32 x 1-7/8	2
*15	NUT-cover mounting	1
16	WASHER-nylon	1
*17	WASHER-rubber	1
18	RING-retaining	1
*19	COVER-air filter	1
	Includes:	
*	GASKET-air filter (not shown)	1
20	ELEMENT-air filter	1
*21	ELEMENT-air filter (foam)	1

*Denotes new parts

No.	Description	Qty.
*22	RING-air filter mounting	1
	Includes:	
*23	GASKET-air filter	1
24	BOOT-throttle rod	1
25	ROD-throttle	1
*26	BRACKET-cover mounting	1
27	PIN-throttle latch	1
28	SPRING-throttle latch	1
29	SCREW-fil. hd. 8-32 x 5/8	1
30	COVER-throttle handle	1
	Includes:	
31	BUSHING-nylon	1
32	RING-retainer	1
33	PIN-trigger latch	1
34	TRIGGER-throttle	1
35	SCREW-pan hd. spinlock, 8-32 x 1/2	2
36	GRIP-throttle handle cover	1
37	GRIP-throttle handle	1
*38	HANDLE-throttle	1
39	SCREW-hex	1
*40	GROMMET-"LO"	1
41	SCREW-pan hd. 8-32 x 1/4	4
*42	SCREW-idle adjustment	1
43	SCREW-hex, spinlock, 12-24 x 5/8	4
44	GASKET-throttle handle	1

FIG. H-4. Parts of a two-stroke cycle multipurpose saw (Continued).

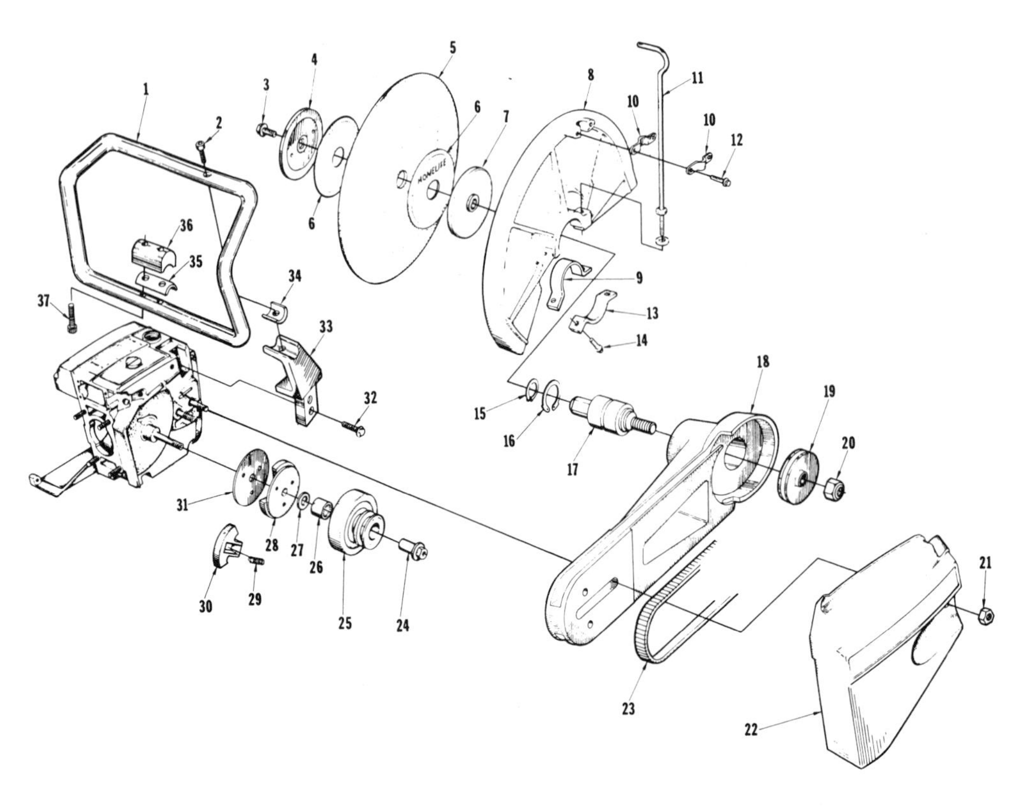

No.	Description	Qty.
* 1	HANDLE BAR	1
* 2	SCREW-soc. hd. 1/4-2 x 1	1
3	BOLT-blade	1
* 4	WASHER-outer	1
5	WHEEL-masonary (10 wheels in package)	1
	WHEEL-metal	1
	WHEEL-stainless	1
	WHEEL-transite	1
	WHEEL-aluminum	1
	WHEEL-copper	1
	WHEEL-ductile iron	1
	BLADE-saw	1
6	SPACER-blotter (included with wheel pkg.)	2
* 7	WASHER-inner	1
* 8	GUARD-blade	1
	Includes:	
* 9	SLEEVE-blade guard	1
	KIT-hook-bolt & clamp	1
	Includes:	
10	CLAMP	2
11	BOLT-hook	1
12	SCREW-hex, spinlock, 1/4-20 x 1/2	2
13	CLAMP	1
14	SCREW-hex, spinlock, 1/4-20 x 13/16	1
*15	RING-retaining	1
*16	RING-retaining	1
*17	BEARING-fan and pump shaft	1
*18	ARM-blade	1
*19	PULLEY-driven	1
*20	NUT-elastic stop, 7/16-14	1
21	NUT-hex, 3/8-16	2
22	COVER-drivecase	1
23	BELT-drive	1
*24	RACE-inner	1
*25	PULLEY-bearing and drum assembly	1
	Includes:	
*26	BEARING	1
27	WASHER-thrust (inner)	1
*	CLUTCH	1
	Includes:	
*28	PLATE-clutch	1
*29	SPRING-clutch	3
*30	SHOE-clutch	3
31	COVER-clutch	1
32	SCREW-hex, spinlock, 12-24 x 3/4	2
*33	BRACKET-handle bar	1
*34	SPACER-shock mount	1
*35	SPACER-shock mount	1
*36	SPACER-handle bar	1
*37	SCREW-soc. cap, 1/4-20 x 1-7/8	2

*Denotes new parts †optional

FIG. H-4. Parts of a two-stroke cycle multipurpose saw (Continued).

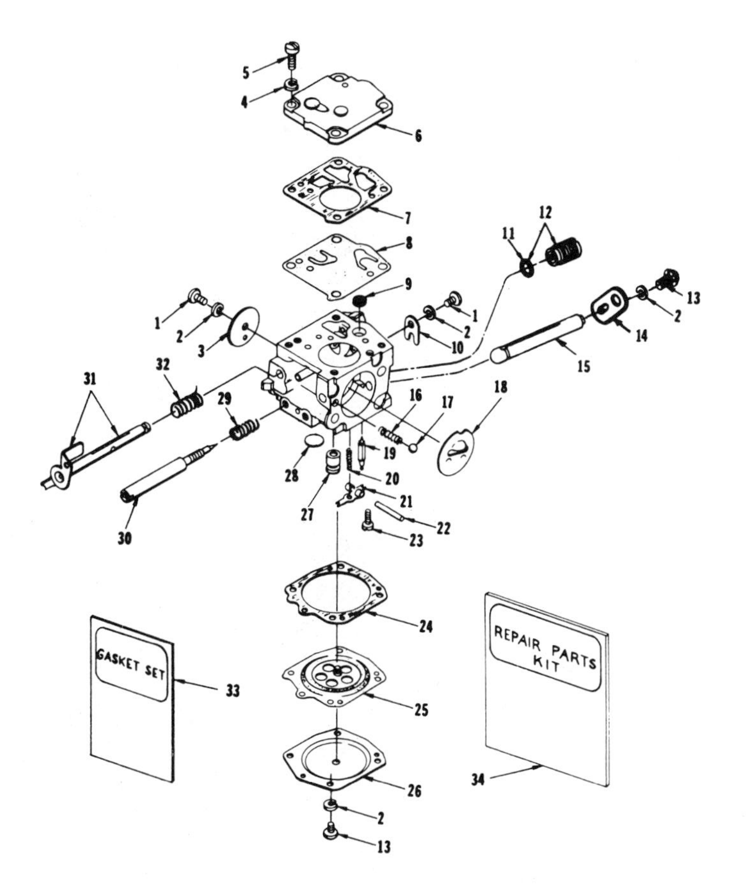

No.	Description	Qty.
1	SCREW-rd. 4-40 x 3/16	2
2	WASHER-lock #4	7
3	SHUTTER-throttle	1
4	WASHER-lock #6	4
5	SCREW-fil. 6-32 x 3/8	4
6	COVER-fuel pump diaphragm	1
7	GASKET-pump diaphragm	1
8	DIAPHRAGM	1
9	SCREEN-fuel inlet	1
10	CLIP-throttle shaft retaining	1
11	GASKET-governor valve	1
*12	GOVERNOR VALVE	1
13	SCREW-rd. 4-40 x 1/4	5
14	LEVER-choke	1
15	SHAFT-choke	1
16	SPRING-choke friction	1
17	BALL-choke friction	2
18	SHUTTER-choke	1

No.	Description	Qty.
19	NEEDLE-inlet	1
20	SPRING-control lever	1
21	LEVER-inlet control	1
22	PIN-control lever fulcrum	1
23	SCREW-oval 4-40 x 1/4	1
24	GASKET-main diaphragm	1
25	DIAPHRAGM-main control	1
26	COVER-main diaphragm	1
*27	VALVE-check	1
28	PLUG-extension (large)	1
29	SPRING-adjustment needle	1
30	SCREW-idle adjustment	1
31	SHAFT-throttle w/lever	1
32	SPRING-throttle shaft return	1
33	GASKET SET	1
34	REPAIR KIT	1
	DIAPHRAGM & GASKET SET	1

* Denotes new parts

† Included in Repair Kit

Fig. H-4. Parts of a two-stroke cycle multipurpose saw (Continued).

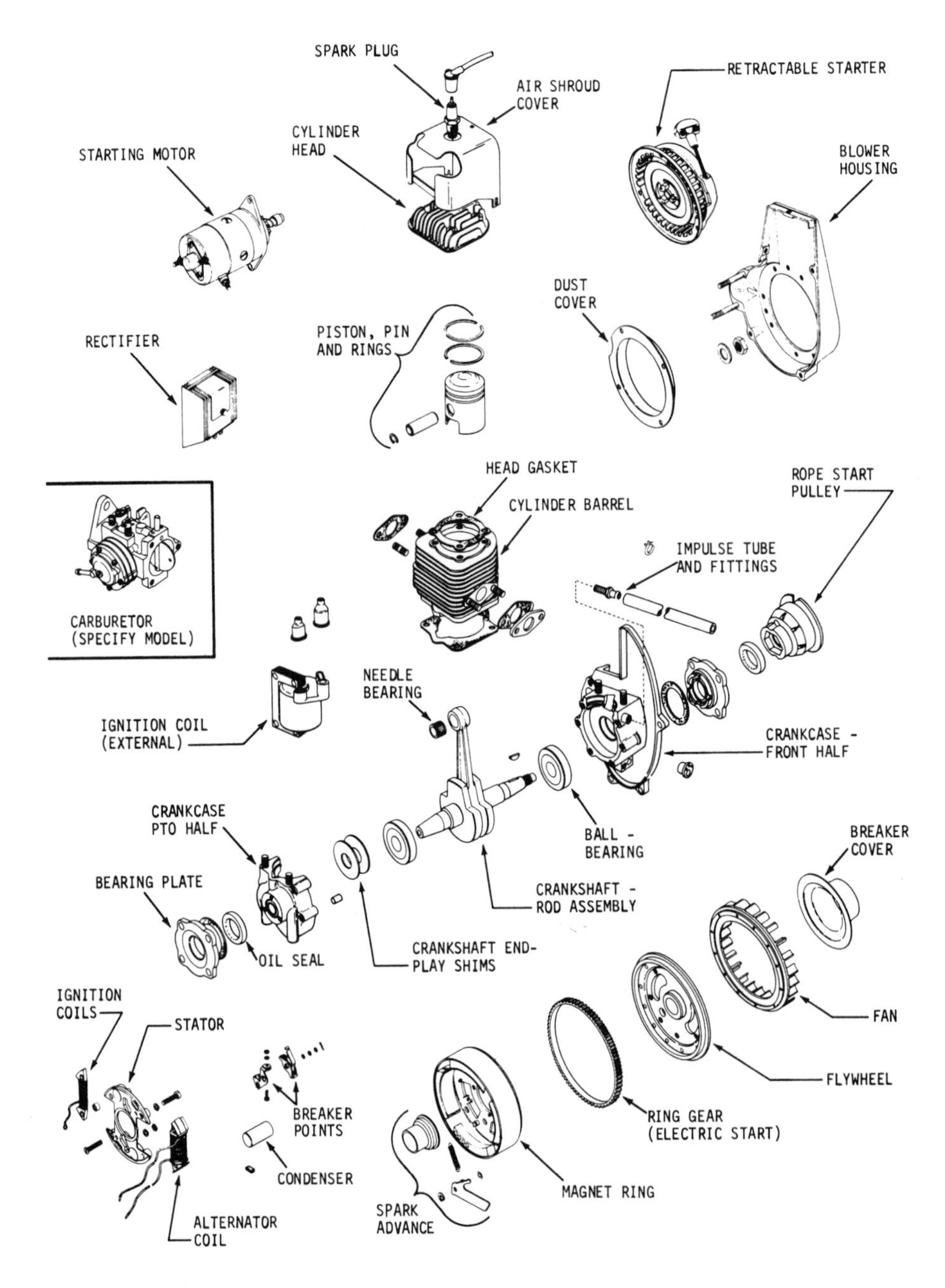

FIG. H-5. Illustrated parts breakdown of a two-stroke cycle single-cylinder engine.

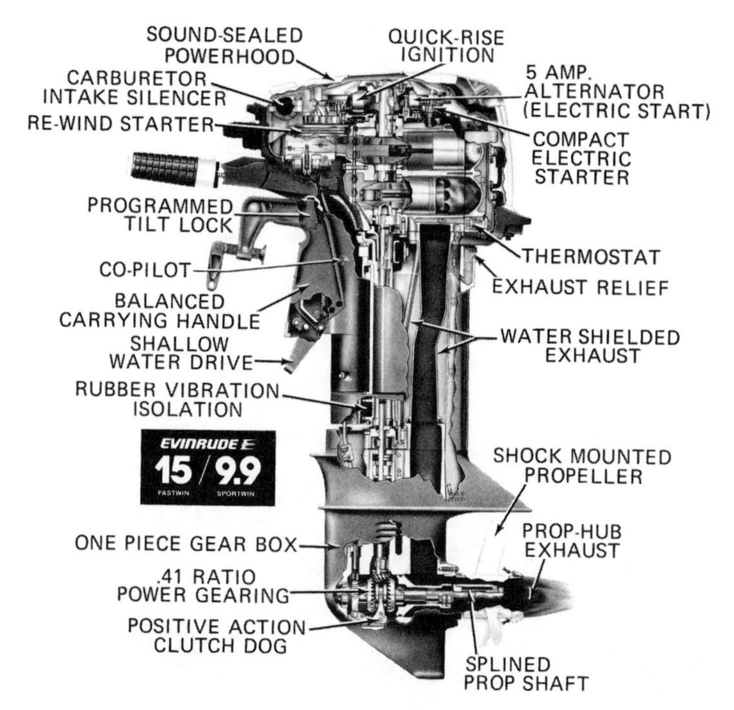

FIG. H-6. Location diagram, outboard engine.

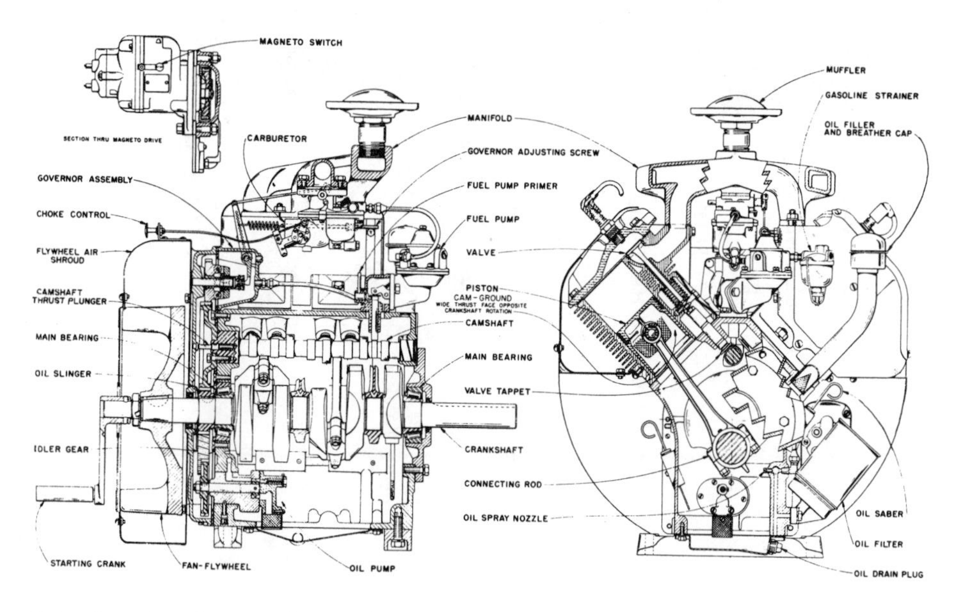

FIG. H-7. Location diagram, four-cycle engine.

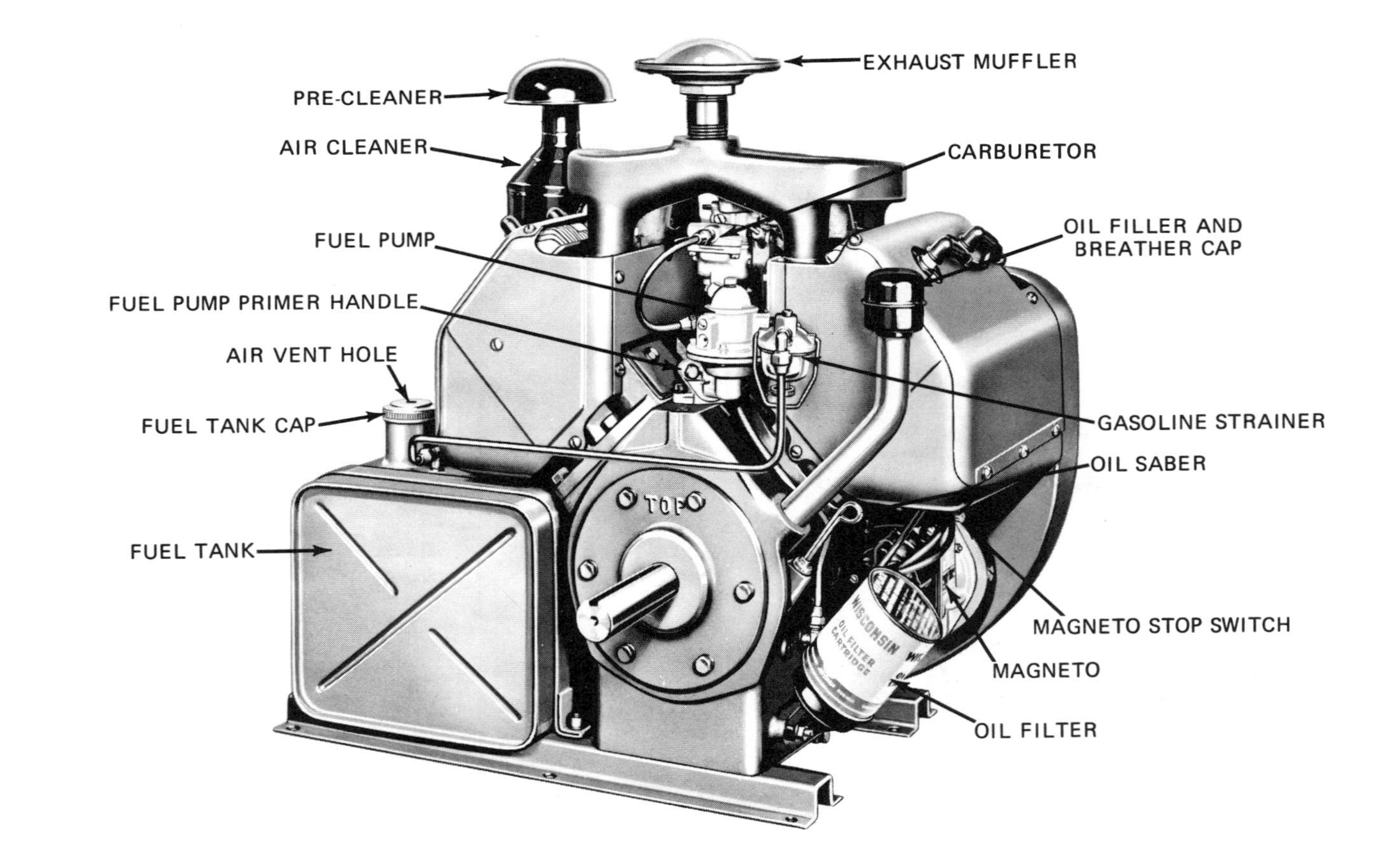

FIG. H-7. Location diagram, four-cycle engine (Continued).

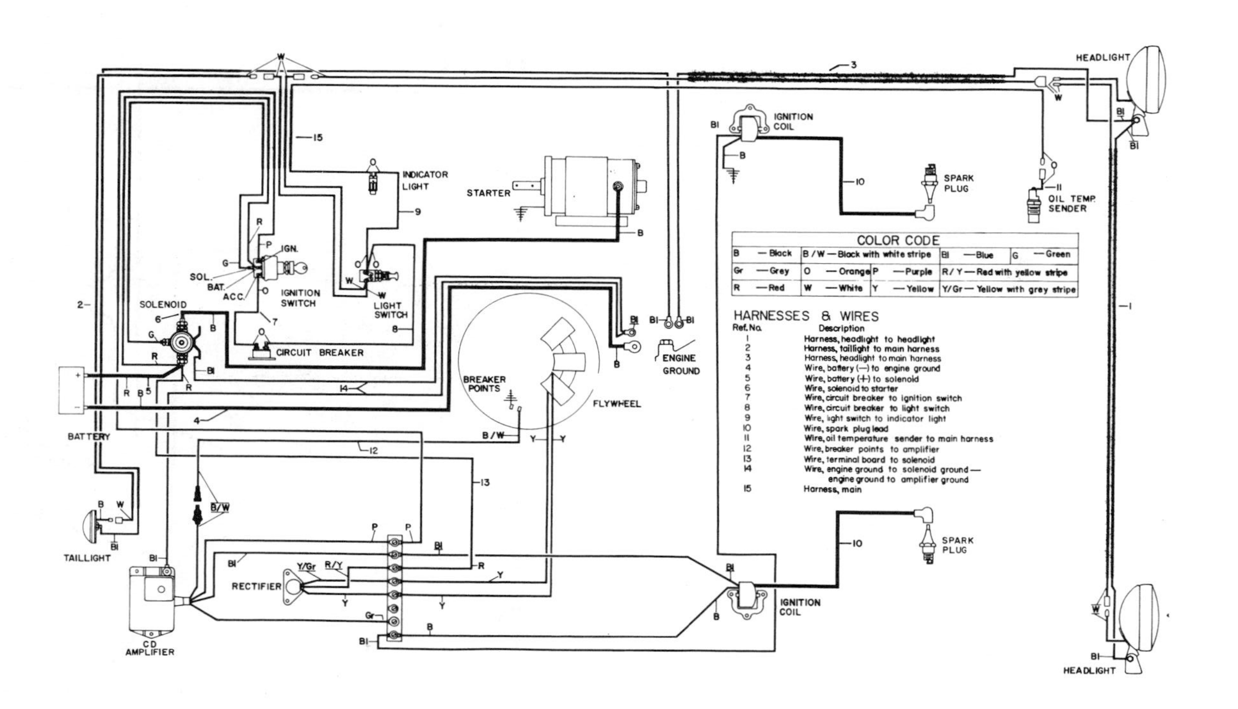

FIG. H-8. Electrical diagram of a trackster.

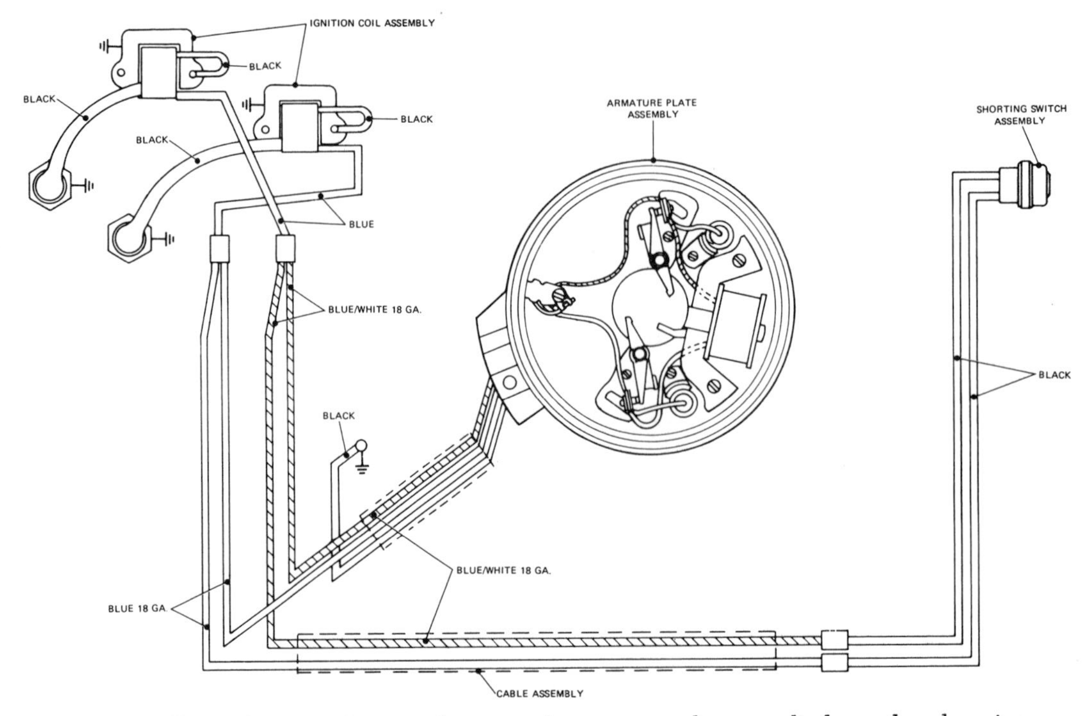

Fig. H-9. Electrical wiring diagram of a manual-start, two-cycle, two-cylinder outboard engine.

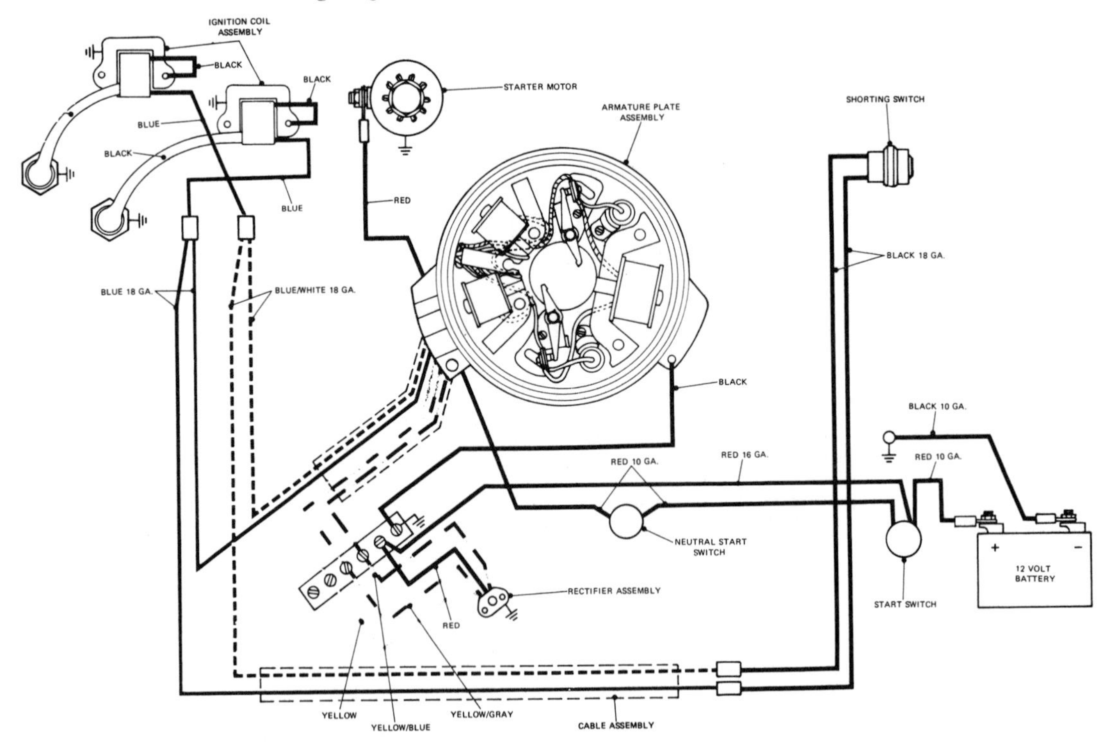

Fig. H-10. Electrical wiring diagram of an electric-start, two-cycle, two-cylinder outboard engine.

I

Clearances, Torque Data, Specifications, and Lubrication Charts of Sample Engines

This appendix provides *sample* charts of service part clearance, bolt/nut/screw torque values, engine specifications, and lubrication charts of a few typical small gasoline engines to illustrate the type of data provided by manufacturers. Remember, *use the engine manufacturer's data on the engine you are working on; the information contained in the tables herein is for reference only.*

The following tables are contained in this appendix:

Table	*Title*	*Manufacturer*
I-1	Clearances	Clinton
I-2	Torque Data	Clinton
I-3	Clearances, Lawn Mower Engine	OMC, Lawn-Boy
I-4	Torque Data, Lawn Mower Engine	OMC, Lawn-Boy
I-5	Specifications, Lawn Mower Engine	OMC, Lawn-Boy
I-6	Clearances, 15hp Outboard Engine	Evinrude
I-7	Torque Data, Outboard Engines	Evinrude
I-8	Specifications, 15hp Outboard Engine	Evinrude
I-9	Lubrication Chart, 15hp Outboard Engine	Evinrude

NOTE

To avoid distortion of a part when tightening two or more screws, first tighten all screws together to one-third of the specified torque, then to two-thirds of the specified torque, then torque down completely. Recheck the torque on cylinder head screws and spark plugs after the motor has been run, has reached operating temperature, and has cooled comfortably to the touch.

Torque standard screws as follows:

Size	*Inch-Pounds*	*Foot-Pounds*
No. 6	7–10	
No. 8	15–22	
No. 10	25–35	2–3
No. 12	35–40	3–4
¼″	60–80	5–7
5-16″	120–140	10–12
⅜″	220–240	18–20

Table I-1
CLEARANCES

Valve or Tappet Guide Bore	Min. Max.	0.2495 0.2510
Valve Stem to Guide Clearance	Min. Max.	0.0015 0.0045
Valve Clearance, Intake & Exhaust	Min. Max.	0.009 0.011
Camshaft to Axle Clearance	Min. Max.	0.001 0.003
Camshaft Axle Clearance P.T.O.	Min. Max.	0.001 0.003
Camshaft Axle Clearance Flywheel End	Min. Max.	0.001 0.003
Point Setting	Min. Max.	0.018 0.021
Spark Plug Gap	Min. Max.	0.025 0.028
Comp. at Cranking Speed, P.S.I.	Min.	65
Carburetor Float Setting (Clinton)	Min. Max.	5/32 11/64
Carburetor Float Setting (Carter)	Min. Max.	11/64 13/64
Magneto Air Gap	Min. Max.	0.007 0.017
Magneto Edge Gap (Phelon)	Min. Max.	5/32 9/32
Magneto Edge Gap (Clinton)	Min. Max.	7/64 1/4
Oil Recommended A.P.I. Rating		SB SD
Fuel Recommended		Reg. Gas
Cylinder Bore Dia.	Min. Max.	2.3745 2.3755
Piston Skirt Dia.	Min. Max.	2.3690 2.3700
Piston Skirt to Cylinder Clearance	Min. Max.	0.0045 0.0065
Piston Ring to Groove Clearance	Min. Max.	0.002 0.005
Ring End Gap in Cylinder	Min. Max.	0.007 0.017
Connecting Rod Bore Crankshaft End	Min. Max.	0.8140 0.8145
Connecting Rod to Crankshaft Clearance	Min. Max.	0.0015 0.0030
Connecting Rod to Wrist Pin Clearance	Min. Max.	0.0004 0.0011
Crankshaft Rod Pin Diameter	Min. Max.	0.8119 0.8125
Crankpin Out-of-Round	Max.	0.001
Crankshaft Main Diameter P.T.O. End	Min. Max.	0.8733 0.8740
Crankshaft Main Diameter Flywheel End	Min. Max.	0.8120 0.8127
Crankshaft to Main Bearing Clearance	Min. Max.	0.0018 0.0035
Crankshaft End Play	Min. Max.	0.008 0.018
Block or Bearing Plate Main Bearing Bore P.T.O. End	Min. Max.	0.8758 0.8768
Bearing Plate or Block Bearing Bore (Flywheel) End	Min. Max.	0.8145 0.8155

NOTES

1. All clearances in inches.
2. P.T.O. means power take off.
3. Comp. is compression.
4. API is American Petroleum Institute.

Table I-2
TORQUE DATA

		All 4 Cycle Aluminum Vertical & Horizontal Shaft	Chainsaws	Outboards
Connecting Rod Aluminum	Min. Max.	100 125	80 90	80 90
Connecting Rod Forged Steel	Min. Max.		90 100	90 100
Bearing Plate P.T.O. End	Min. Max.	75 85		
Bearing Plate Flywheel End	Min. Max.		80 90	75 95
Back Plate to Block	Min. Max.			
Head Bolts	Min. Max.	125 150		
Base Bolts	Min. Max.	75 85		
End Cover or Gear Box	Min. Max.			
Speed Reducer Mounting	Min. Max.	110 150		
P.T.O. Housing or Mounting Flange	Min. Max.	75 85		
Carb. Reed Plate or Manifold to Blk.	Min. Max.	60 65	50 60	50 60
Carb. to Reed Plate or Manifold	Min. Max.	35 50	50 60	50 60
Blower Housing	Min. Max.	60 70	80 90	75 85
Muffler to Block	Min. Max.	110 120	60 70	
Flywheel	Min. Max.	375 400	250 300	250 300
Flywheel Touch & Stop for Brake	Min. Max.	650 700		
Spark Plug	Min. Max.	275 300	230 270	230 270
Stator Plate	Min. Max.	50 60	50 60	45 65

NOTES

1. All torque values in inch = pounds.
2. P.T.O. is power take off.
3. Carb. is carburetor.

Table I-3
CLEARANCES, LAWN MOWER ENGINE

CRANKSHAFT	
Top Journal	0.6692 - 0.6689
Crank Pin	0.6865 - 0.6860
Bottom Journal	0.6692 - 0.6689
CONNECTING ROD	
Wrist Pin Hole	0.3659 - 0.3654
Crank Pin Hole	0.6880 - 0.6875
WRIST PIN	
Diameter	0.3650 - 0.3642
PISTON	
Diameter	1.7470 - 1.7465
Wrist Pin Hole	0.3655 - 0.3650
PISTON RINGS	
Diameter	1.740 ± 0.000
End Gap (In Cyl.)	0.020 ± 0.005
Thickness	0.0935 - 0.0930
CYLINDER	
Inside Diameter	1.751 - 1.750
Sub-Base Bushing	0.683 - 0.680
SUB-BASE HOUSING	
Sub-Base Bushing	0.683 - 0.680
CRANKCASE	
Top Bearing	0.8770 - 0.8762
Bottom Bearing	0.8805 - 0.8780
ARMATURE PLATE	
Bearing	0.8770 - 0.8762

NOTES

1. Clearances are in inches.
2. Cyl. is cylinder.

Table I-4
TORQUE DATA, LAWN MOWER ENGINE

DESCRIPTION	TYPE	SIZE	ASSEMBLY TORQUE INCH POUNDS	RECHECK TORQUE INCH POUNDS
NUT, Flywheel	3	7/16-20	335/400	320
NUT, Flywheel	3	7/16-20	190/225	180
SCREW, Shroud to armature plate	5,7	1/4-20	60/75	
SPARK PLUG	8,7	14MM	150/180	144
SCREW, Flywheel ring	6,7	10-24	20/25	20
TANK TO SHROUD	5	1/4-20	63/75	60
SCREW, Armature plate to crankcase	5,7,8	1/4-20	63/75	60
SCREW, Dust cover	5,7	10-24	20/25	20
BOLT, Shoulder, variable speed lever	6,7	10-24	20/25	20
SCREW, Lamination mounting	5,7	10-24	20/25	20
SCREW, Starter attachment	5,7	1/4-20	58/63	50
SCREW, Condenser mounting	5,7	10-24	20/25	20
SCREW, Condenser mounting	5,7	10-24	20/25	20
SCREW, Breaker base	5,7	10-24	20/25	20
SCREW, Shut off switch	6	8-18	6/8	6
SCREW, Starter pully	6,7	8-32	16/19	15
NUT, Condenser	3	8-32	10/13	10
SCREW, Cylinder to crankcase	5,7,8	5/16-18	105/115	90
SCREW, Reed plate to carburetor	4,8	1/4-28	63/75	60
NUT, Reed plate to carburetor	1,8	1/4-28	----------	
SCREW, Reed plate to crankcase	5,7,8	1/4-20	63/75	60
SCREW, Filter cup to carburetor	5,7,8	8-32	16/19	15
SCREW, Throttle shaft disc	6	2-56	5/7	5
SCREW, Choke and throttle disc	6	#2	3/5	
NOZZLE	5,7	#72 BRASS	16/19	15
NOZZLE	5,7	#72 BRASS	16/19	15
SCREW, Float chamber	5,7,8	8-32	12/15	10
SCREW, Start cap to base	5,7	10-24	20/25	20
SCREW, Starter to shroud	4	1/4-20	63/75	60
NUT, Starter to shroud	1	1/4-20	----------	
SCREW, Connecting rod	5,7	12-24	58/70	55
SCREW, Tank strap	6	#10	15/30	
SCREW, Pully to cap	5	1/4-20	63/70	60
SCREW, Governor lever	6,7	6-32	10/13	10
SCREW, Filter	5	10-24	10/15	
SCREW, Air filter	5	8-32	16/19	15
SCREW, Filter cup	5	10-24	10/15	
SCREW, Chamber intake to carburetor	5	8-32	16/19	15
SCREW, Reed to reed plate		6-32	10/13	10
SCREW, Scraper bracket	4	1/4-20	63/75	60
NUT, Scraper bracket to shroud	1	1/4-20	----------	
SCREW, Shroud mounting	5	10-24	25/30	
NUT, Shroud mounting, tinnerman	TINN. "U" TYPE		----------	
NUT, Acorn	3	10-24	10/15	10
SCREW, Tank bracket	6	#10	15/30	
SCREW, Baffle Left hand	5	10-24	25/35	
SCREW, Baffle Right hand	5	1/4-20	55/80	
SEAT, Float valve	5	5/16-24	19/23	18
SCREW, cable clamp	5	10-24	20/25	20
SCREW, Shut off switch	6	8-18	6/8	6
VALVE, Fuel shut off, 1/8-27-NPT				
NUT, Flywheel, zinc flywheel	3	7/16-20	190/225	180

TYPE:
1. NUT, Nylok
2. NUT, Conelok
3. NUT, Standard
4. SCREW, for nylok nut
5. SCREW, Standard machine
6. SCREW, Thread cutting
7. THREAD, Die cast
8. JOINT, Gasketed

1. NPT is American Standard Taper Pipe Tap.
2. Tinn. is tinnerman.
3. MM is millimeter.

Table I-5
SPECIFICATIONS, LAWN MOWER ENGINE

Horsepower			2.5 (3 H.P.)	3-1/2
Bore	1-3/4 in.	1-15/16 in.	2-1/8 in.	2-3/8 in.
Stroke	1-1/2 in.	1-1/2 in.	1-1/2 in.	1-1/2 in.
Displacement	3.603 cu. in.	4.43 cu. in.	5.22 cu. in.	6.65 cu. in.
Piston Diameter	1.7470-1.7465	1.9360-1.9355	2.1205-2.1200	2.3720-2.3715
Breaker Point Setting	0.020	0.020	0.020	0.020
Coil Air Gap	0.010	0.010	0.010	0.016
Capacitor Discharge Air Gap				0.010 0.010
Spark Plug	Champion J-11-J Gap-0.025	Champion J-14-J Gap-0.025	Champion J-14-J Autolite A11X or equivalent. Gap-0.025	Champion CJ-14 or equivalent. D-400 Series-0.025 D-600 Series-0.035
Governed Speed	3300 R.P.M.	3200 R.P.M.	3200 R.P.M.	3200 R.P.M.
Fuel Mixture	1/2 pint S.A.E. 40 service SB	1/2 pint S.A.E. 40 service SB	1/2 pint S.A.E. 40 service SB	1/2 pint S.A.E. 40 service SB

NOTES

1. SAE is Society of Automotive Engineers.
2. Specifications in inches unless otherwise indicated.

Table I-6
CLEARANCES, 15 HP OUTBOARD ENGINE

POWER HEAD		LOWER UNIT	
Piston ring gap	.015 Max. - .005 Min.	Propeller shaft in front gear bushing	.0087 Max. - .0002 Min.
Piston ring groove clearance, lower	.0035 Max. - .0025 Min.		
Piston pin to piston - loose end	.0005 Max. - .0000 Min.		
Cylinder and piston	.0053 Max. - .0040 Min.		
Crankshaft end play	Controlled by lower journal bearing		

NOTE Clearances in inches.

Table I-7
TORQUE DATA, OUTBOARD ENGINES

DESCRIPTION	2 HP		4 HP		6 HP		9.9, 15 HP		25 HP		40 HP		50 HP		70 HP		85, 115, 135 HP	
	INCH-POUNDS	FOOT-POUNDS	INCH-POUNDS	FOOT-POUNDS	INCH-POUNDS	FOOT-POUNDS	INCH-POUNDS	FOOT-POUNDS	INCH-POUNDS	FOOT-POUNDS	INCH-POUNDS	FOOT-POUNDS	INCH-POUNDS	FOOT-POUNDS	INCH-POUNDS	FOOT-POUNDS	INCH-POUNDS	FOOT-POUNDS
Flywheel nut		22-25		30-40		40-45		45-50		40-45		100-105		100-105		100-105		100-105
Connecting rod screws	60-66	5 - 5-1/2	60-66	5 - 5-1/2	60-66	5 - 5-1/2	48-60	4-5	180-186	15 - 15-1/2	348-372	29-31	348-372	29-31	348-372	29-31	348-372	29-31
Cylinder head screws	60-80	5-7	60-80	5-7	60-80	5-7	145-170	12-14	96-120	8-10	168-192	14-16	168-192	14-16*	168-192	14-16*	168-192	14-16*
NOTE: *After motor test re-torque to 16-18 ft.-lbs. - 50 and 70 hp only; 18-20 ft.-lbs. - 85, 115, 135 hp only.																		
Spark plug		17-1/2 - 20-1/2		17-1/2 - 20-1/2		17-1/2 - 20-1/2		17-1/2 - 20-1/2		17-1/2 - 20-1/2		17-1/2 - 20-1/2		17-1/2 - 20-1/2		17-1/2 - 20-1/2		17-1/2 - 20-1/2
Crankcase to cylinder screws - upper			60-80	5-7	60-80	5-7	145-170	12-14	120-144	10-12	150-170	12-14	216-240	18-20	216-240	18-20	216-240	18-20
Crankcase to cylinder screws - center			60-80	5-7	60-80	5-7	145-170	12-14	120-144	10-12	162-168	13-1/2 - 14	216-240	18-20	216-240	18-20	216-240	18-20
Crankcase to cylinder screws - lower			60-80	5-7	60-80	5-7	145-170	12-14	120-144	10-12	150-170	12-14	216-240	18-20	216-240	18-20	216-240	18-20
Side mount nuts - upper and lower					150-170	12-14			150-170	12-14	150-170	12-14						
Pivot Shaft Nut																130-150		130-150
Lower journal bearing retainer plate screws																	96-120	8-10
Stator screws													60-80	5-7	48-60	4-5	120-144	10-12
Driveshaft pinion nut														40-45		40-45		60-65
Starter through bolts - American Bosch							30-40						90-105		90-105		90-105	
Starter through bolts - Prestolite													110		110-122		110-122	
Starter drive assembly locknut - American Bosch														20-25		20-25		20-25
Starter drive assembly locknut - Prestolite														25-30		25-30		25-30
Crankcase head screws - upper																	120-144	10-12
Crankcase head screws - lower													96-120	8-10	96-120	8-10	96-120	8-10

Table I-8
SPECIFICATIONS, 15 HP OUTBOARD ENGINE

Model Numbers 15404 - Standard length (15" transom)
15405 - 5" longer (20" transom)
15454 - Standard length (15" transom)
15455 - 5" longer (20" transom)

Horsepower (B.I.A. certified 15 hp at 6000 rpm
Full throttle operating range 5500 to 6500 rpm

Engine type 2 cyl., 2 cycle, alternate firing
Bore and stroke 2.188" bore x 1.760" stroke
Piston displacement 13.20 cubic inches

Piston ring sets (2 per set)
standard
.030" oversize

Diameter of ring	2.1875 in. (standard)
Width of ring	Upper - .0700 - .0695 in. Lower - .0615 - .0625 in.
Lbs. compression recommended when compressed	Upper - .25 - 2.0 lbs. Lower - 2.5 - 5.0 lbs.
Piston less rings standard .030" oversize	
Crankshaft size Top journal Center journal Bottom journal	 .8125 - .8120 .8125 - .8120 .8125 - .8120
Connecting rod crank pin	1.06350 - 1.06300 in.
Carburetion	Single barrel, float feed, fixed high speed adjustable low-speed, manual choke
High speed orifice plug	Identification Number 60 Check with #.060" dia. drill
Float level setting	Flush with rim of casting
Inlet needle seat	.065 - .062 Use #52 drill as gage
Cooling system	Centri-matic (combination positive displacement and centrifugal pump) Thermostatically controlled
Propeller gear ratio	12:29
Propeller supplied with motor	3 blade, 9-1/2" dia. x 10" pitch
Propeller options	3 blade, 9-3/4" dia. x 7" pitch 2 blade weedless 10" dia. x 10" pitch
Speed control	On steering handle Remote control available
Gear shift control	Forward, neutral and reverse
Weight (without fuel tank)	Model 15404 - 65 lbs. Model 15405 - 66 lbs. Model 15454 - 73 lbs. Model 15455 - 74 lbs. (Fuel tank weight 11 pounds net)
Fuel capacity	6 gallons
Electrical system (Electric start models only)	5 amp flywheel alternator
Starter	Manual - Self-winding Electric - 12 volt, and rope
Starter amperage draw while cranking	55 AMPS Max.
Ignition	Low tension magneto
Spark plug	Champion UL4J, 14mm
Spark plug gap	.030 inch
Spark plug torque	17-1/2 - 20-1/2 foot-pounds
Breaker point	Gap .020 inch
Condenser Capacity	.25 - 29 Mfa.

COIL TEST SPECIFICATIONS

With Stevens Tester Model No. M.A.-75 or M.A.-80 with M.A.-14 Adapter in Series with High Tension Lead

Switch	Index Adjustment
B	25

Merc-O-Tronic

Operating Amperage	Primary Resistance Min. Max.	Secondary Continuity Min. Max.
1.7	0.8 - 1.2	60 - 70

Graham Tester Model

Maximum Secondary	Maximum Primary	Coil Index	Minimum Coil Test	Gap Index
20,000 ohms	14.0 ohms	50	24	45

COIL OHMMETER TEST

Primary (Low Ohms)	Secondary (High Ohms)
1.35 ± .3	13,500 ± 1500

*Horsepower established at sea level. Allow 2% reduction per 1000' above sea level.

Table I-9
LUBRICATION CHART, 15 HP OUTBOARD ENGINE

LUBRICATION POINT	LUBRICANT	FREQUENCY (PERIOD OF OPERATION)	
		FRESH WATER	#SALT WATER
1. Tilt Reverse Lock Lever Shaft, Clamp Screw Threads, and Throttle Shaft Gears	OMC Sea-Lube* Anti-Corrosion Lube	60 days	30 days
2. Idle Speed Adjustment, Magneto Linkage, and Manual Starter Drag Spring	↓	60 days	30 days
3. Shift Lever Detent Cam, Carburetor Linkage and Choke	↓	60 days	30 days
4. Gearcase	OMC Sea-Lube* Gearcase Lube Capacity 13.9 ozs.	Check level after first 10 hours of operation and every 50 hours of operation thereafter. Add lubricant if necessary. Drain and refill every 100 hours of operation or once each season, whichever occurs first.	Same as Fresh Water Same as Fresh Water
5. Electric Starter Pinion Shaft Helix	Lubriplate 777	60 days	30 days
6. Swivel Bracket and Motor Cover Latch Lever Shaft	OMC Sea-Lube* Anti-Corrosion Lube	60 days	30 days

LUBRIPLATE 777
OMC SEA-LUBE* GEARCASE LUBE
OMC SEA-LUBE* ANTI-CORROSION LUBE
OIL CAN SAE 90
GREASE GUN OMC SEA-LUBE* ANTI-CORROSION LUBE

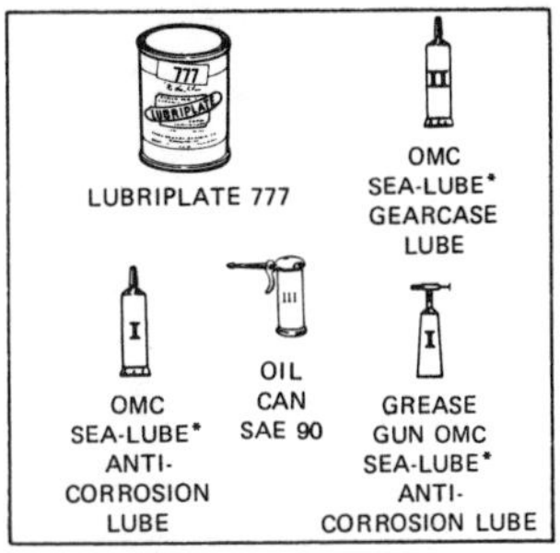

#Some areas may require more frequent lubrication.

Glossary

Additives—a substance added to oil in relatively small quantities to improve desirable properties and suppress undesirable properties.

Air Cleaner (filter)—a device to trap dust and other foreign material from the air before the air is drawn into the carburetor.

Air Horn—passageway for air.

Alternating Current—an electric current that reverses its direction at regularly recurring intervals (AC).

Alternator—a generator that produces alternating current electricity.

Ammeter—a device for measuring electric current in a circuit.

Ampere—the electric current produced by one volt applied across a resistance of one ohm.

Apex—the uppermost point.

Armature—a piece of soft iron or steel that connects the poles of a magnet or of adjacent magnets; a part which consists essentially of coils of wires around a metal core and in which electric current is induced in a generator or in which the input current interacts with a magnetic field to produce torque in a motor.

Atmospheric Pressure—the pressure due to the weight of the air above the earth. The pressure at sea level is 14.7 pounds per square inch.

Atom—the smallest particle of an element that can exist either alone or in combination.

Atomize—to reduce to a fine spray.

Bail—a supporting half loop of metal (as used to hold a glass bowl onto a fuel filter mount).

Battery—a number of primary or storage cells grouped together as a single source of direct current electricity. A cell is a single element of a battery.

Bearing—a support for a revolving shaft or axle which must be rigid and self-aligning and be designed to take up most of the wear and be replaceable.

Block (engine)—the cylinder casting of a gas engine. It includes the cylinder bores and provision for cooling (cooling fins for air-cooling or water jackets for water-cooling).

Blow-By Gases—gases of combustion that pass by the piston and piston rings into the crankcase.

Bottom Dead Center (BDC)—the extreme bottom position of the piston in the cylinder.

Breaker Points—a set of electrical contacts, one movable and one stationary, used to break the current flow in the primary of the ignition coil.

Bushing—a special type of bearing called a sleeve bearing. It may be solid or split.

Cam—a device mounted on a revolving shaft used for transposing rotary motion into an alternating reciprocating motion.

Camshaft—a shaft of a four-cycle engine to which cams are fastened (as to open the intake and exhaust valves).

Carbon—a nonmetallic element found in all organic substances. Carbon is the solid element of combustion.

Carbon Monoxide—a colorless odorless very toxic gas; CO. It is formed as a product of the incomplete combustion of carbon and is poisonous.

Carburetion—the mixing of air and liquid fuel in the proper proportion to form a combustible mixture for burning in an engine.

Carburetor—a device for converting liquid fuel into vapor and mixing it with air in such proportions as to form the most efficient combustible mixture.

Carburetor Bowl—the reservoir part of the carburetor which holds gasoline or fuel.

Carburetor Float—usually an airtight metal container which floats on the surface of the fuel in the bowl of the carburetor and controls the flow of gasoline from the main fuel line.

Catalyst—a substance that initiates a chemical reaction and enables it to proceed under milder conditions than otherwise possible.

Centrifugal Force—the force that tends to impel a thing or parts of a thing outward from a center of rotation.

Choke—blocks off the incoming air so that the mixture is rich to enable the starting of a cold engine.

Coil—a number of turns of wire in spiral form for developing an electromagnetic effect.

Commutator—a device for reversing the direction of electric current in any circuit.

Compound—composed of the chemical units of two or more elements.

Compression Ratio—the ratio of the space remaining in the cylinder at top dead center of the piston stroke to the space in the cylinder when the piston is at bottom dead center.

Compression Ring—a piston ring used to reduce compression and combustion pressure losses.

Compression Stroke—the stroke of an engine when the fuel is compressed in the combustion chamber or recess to make it more volatile.

Condenser—a capacitor; a device consisting of conducting plates or foils separated by thin layers of dielectric (as air or mica) with the plates on opposite sides of the dielectric layers oppositely charged by a source of voltage and the electrical energy of the charged system stored in the polarized dielectric.

Connecting Rod—a rod with bearings at both ends. Connects the piston to the crank of the crankshaft.

Continuity—uninterrupted connection; no electrical resistance between two points of an electric circuit.

Crank—a lever that rotates about the axis of the crankshaft. It connects with the connecting rod.

Crankcase—the housing of the crankshaft, main bearings, connecting rod, breather, and of the camshaft and cam gears in four-cycle engines.

Crankcase Breather—a device to rid the crankcase of blow-by gases.

Crankshaft—the main shaft of an engine to which the connecting rod is attached via the crank.

Crude Oil—petroleum in its natural state as it comes from the earth.

Cycle—a course or series of events or operations that recur regularly and usually lead back to the starting point; a complete series of intake, compression, power, and exhaust.

Cylinder—a circular body of uniform diameter, the extremities of which are equal parallel circles.

Cylinder Block—the main body of the engine which is bored to receive the piston. The cylinder block and crankcase are frequently cast as one piece.

Cylinder Bore—the internal diameter of an engine cylinder.

Cylinder Head—refers to the cylinder cover of an engine.

Detergent—a cleansing agent.

Detonation—rapid combustion in an internal-combustion engine that results in knocking.

Diaphragm—a sheet of metal or other material that is sufficiently flexible to permit vibration.

Diode—a rectifier that consists of a semiconducting crystal with two terminals; the resistance to electron flow in one direction is high, in the other direction is low.

Direct Current—an electric current flowing in one direction only and substantially constant in value (DC).

Eccentric—deviating from a circular path; located elsewhere than at the geometrical center.

Electrode—a conductor of electricity.

Electrolyte—a nonmetallic electric conductor in which current is carried by the movement of ions. The electrolyte in a storage battery is a solution of sulfuric acid and water.

Electromotive Force—something that moves or tends to move electricity; measured in volts.

Electron—a very small negatively charged particle of an atom. Electrons orbit around the nucleus of the atom.

Element—that form of matter that cannot be further broken down.

Emission—a putting into circulation; something sent forth by emitting.

Epitrochoid—the internal shape of the housing of a rotary engine. The shape resembles a fat figure eight.

Equilateral Triangle—a triangle having three equal sides and three equal enclosed angles.

Exhaust Stroke—the stroke that expels the burned gases of combustion from the combustion chamber or recess.

Exponent—a symbol written above and to the right of a mathematical expression to indicate the operation of raising to a power.

Farad—the unit of capacitance equal to the capacitance of a capacitor between whose plates there appears a potential of one volt when it is charged by one coulomb of electricity.

Float Valve—a valve that floats on top of the fuel in the fuel bowl and shuts off the supply of fuel as the level rises.

Flyweight—a light weight device that moves away from a center support because of a centrifugal force applied.

Flywheel—a heavy wheel used where reciprocal motion is converted into rotary motion to maintain uniformity of motion.

Foul—clog.

Friction—resistance to relative motion between two bodies in contact.

Fuel—a material used to produce heat or power by burning. Gasoline for four-cycle engines and a mixture of gasoline and oil for two-cycle and rotary engines.

Fuel Filter (strainer)—a device to remove dirt and other foreign matter from the fuel before the fuel enters the carburetor.

Fuel Pump—a device operated by a piston or diaphragm to create a vacuum which insures a supply of fuel to the carburetor.

Gasket—paper, metal, rubber, or other material used between two joints, as between the cylinder and cylinder head, to prevent leaking of a gas or liquid.

Generator—a machine that transforms mechanical energy into electrical energy.

Governor—a device to keep the engine operating at a constant speed with change in load and to prevent the engine from running at a speed above a predetermined speed.

Horsepower—the standard unit by which power is measured; it is equal to 33,000 pounds lifted one foot in one minute or to 550 pounds lifted one foot in one second.

Hydrocarbon—a compound that contains hydrogen and carbon. Often occurs in petroleum, natural gas, coal, and bitumens.

Hydrogen—a nonmetallic element that is the simplest and lightest of the elements. It is normally a colorless, odorless, highly flammable gas.

Ignition—the lighting of the fuel in a gas engine by a spark across the electrodes of a spark plug.

Induced Voltage—a voltage set up by a varying magnetic field linked with a wire, coil, or circuit.

Intake Stroke—the stroke of an engine when the fuel enters the combustion chamber or recess.

Journal—rotating machine part (as part of the crankshaft) supported in a bearing. Journals are cylinders that are true machined to resist deformation and to prevent excessive pressure in the bearings.

Lap—to work two surfaces together with or without abrasives until a very close fit is produced.

Lead—a heavy, soft, malleable, ductile, plastic, but inelastic bluish white metallic element found mostly in combination. It is used in batteries.

Lean Fuel Mixture—a lower than normal ratio of fuel to air.

Linkage—a system of links or bars which are joined together.

Liter—a metric system unit of capacity equal to the volume of one kilogram of water at 4° centigrade. One liter equals 1.057 liquid quarts.

Magnetism—a class of physical phenomena that includes the attraction for iron observed in lodestone and a magnet, is believed to be inseparably associated with moving electricity, is exhibited by both magnets and electric currents, and is characterized by fields of force.

Magneto Generator—an alternator that makes use of magnets in the flywheel to generate alternating current electricity.

Meter—the basic metric system unit of linear measure. One meter equals 39.37 inches.

Microfarad—one millionth of a farad. See farad.

Molecule—a combination of atoms. A molecule is the smallest unit of any compound.

Motor-Generator—combines the functions of both a starting motor and a generator into one unit. Converts electrical energy to mechanical energy to start an engine and then converts mechanical energy to electrical energy to recharge the battery.

Muffler—a mechanical device consisting of a hollow cylinder attached to the exhaust. The noise of the exhaust gases passing through it is partially deadened (muffled).

Needle Valve—a small valve in which the flow of fuel is regulated by the adjustment of a needle point which sets in a cone-shaped depression having a small hole at the bottom.

Neutron—an uncharged elementary particle that has a mass nearly equal to that of the proton and is present in all known atomic nuclei except the hydrogen nucleus.

Octane Rating—a measure of the antiknock qualities of gasoline.

Ohm—the unit of electrical resistance. It is equal to the amount of opposition offered by a conductor to the flow of one ampere of current when a pressure of one volt is applied across its terminals.

Ohmmeter—a device for measuring resistance.

Oil—a greasy liquid of animal, vegetable, or mineral origin; used as a lubricant.

Oil Control Ring—a type of piston ring designed to scrape oil from the cylinder wall. The oil drains to the crankcase.

Oxide—a binary compound of oxygen with an element.

Piston—the plunger which moves with a reciprocating motion within the cylinder of the engine.

Piston Pin—a hollow steel shaft, hardened and ground, which connects the upper end of the connecting rod to the piston. Also known as the "wrist" pin.

Piston Ring—a spring packing ring for a piston. See compression ring and oil control ring.

Pneumatic—moved or worked by air pressure.

Potential Energy—the energy that a piece of matter has because of its position or because of the arrangement of parts.

Power—the amount of work done per unit of time.

Power Stroke—the stroke of an engine that includes ignition and the resulting driving force of combustion against the piston or rotor.

Preignition—ignition in an internal-combustion engine before compression is completed.

Prime—to prepare for combustion by supplying with a small amount of fuel.

Proton—a particle in the nucleus of an atom having a positive electrical charge.

Reach (of a spark plug)—the linear distance from the shell gasket to the end of the threaded portion of the shell.

Reactor—an apparatus in which a chain reaction of fissionable material is initiated and controlled.

Reciprocating—a forward and backward or upward and downward movement (as the piston in the cylinder).

Rectifier—a device that changes alternating current into direct current. A diode is such a device.

Reed Valve (leaf valve)—a thin flexible valve located between the carburetor and the crankcase of two-cycle engines to admit the fuel from the carburetor into the crankcase.

Regulator—a device used to keep the DC output voltage and current of a generator at a constant level.

Resistor—a device that impedes the flow of electrons through an electrical circuit.

Rich Fuel Mixture—a greater ratio of fuel to air than normal.

Rotor—the rotating member in a rotary engine.

Scavenging—removing (cleaning) burned gases from the cylinder after a working stroke.

Shroud—a cover placed over a bladed, or vaned, flywheel to direct the flow of air over the cylinder and cylinder head for cooling.

Sludge—sediment in the bottom of the crankcase.

Solenoid—a coil of wire commonly in the form of a long cylinder that when carrying a current resembles a bar magnet so that a movable core is drawn into the coil when a current flows.

Solid-State—a system using semiconductors such as diodes, transistors, or silicon controlled rectifiers.

Solvent—a liquid used to dissolve a substance. In engine work mineral spirits, kerosene, diesel fuel, turpentine, or alcohol can be used to dissolve and remove oil and grease.

Spark Plug—a part that fits into the cylinder head of an internal-combustion engine and carries two electrodes separated by an air gap across which the current from the ignition system discharges to form the spark for combustion.

Specific Gravity—the density or weight of a substance; estimated relative to water.

Starter-Motor—a device to start an engine. Also called motor, starter, motor-starter, or motor-generator.

Stratified—arranged in layers.

Stroke—the movement of the piston from top dead center (TDC) to bottom dead center (BDC) or from BDC to TDC.

Stud—any of various infixed pieces (as a rod or pin) projecting from a machine and serving chiefly as a support or axis (as the stud in a spark plug).

Throttle—controls engine speed by controlling the flow of air-fuel mixture through the air horn.

Top Dead Center (TDC)—the extreme top position of the piston in the cylinder.

Torque—a force that produces or tends to produce rotation or torsion.

Transistor—an electronic device consisting of semiconductor materials with at least three electrodes; performs electronic switching and amplification functions.

Turbine—a rotary engine actuated by the reaction or impulse or both of a current of fluid (as water or steam) subject to pressure and usually made with a series of curved vanes on a central rotating spindle.

Vacuum—a space partially evacuated; a degree of minimum pressure below atmospheric pressure.

Valve—a device for regulating the flow of intake or exhaust gas to and from the engine cylinder.

Valve Port—opening into or from the combustion chamber which provides a channel for the passage of gases and a seal for the head of the valve.

Valve Seat—that part of the cylinder machined to receive the valve and to provide a seal against leakage of gases.

Valve Spring—a compression spring used to keep valves in a closed position.

Valve Stem—the shank of a valve. The stem slides in the valve guide.

Vapor—a substance in the gaseous state as distinguished from the liquid or solid state.

Vent Cap—the fuel tank cover. It has a tiny hole in it to vent the tank to atmospheric pressure.

Venturi—narrow passage that causes a low pressure at the area in the air passage just beyond the narrow passage.

Viscosity—the tendency of an oil (or other substance) to resist flowing.

Volatile—readily vaporizable at a relatively low temperature; tending to erupt into violence; explosive.

Volt—the electromotive force that will produce a current of one ampere through a resistance of one ohm.

Voltmeter—a device for measuring electrical voltage, or electromotive force.

Volumetric Efficiency—ability of the combustion chamber to receive a full charge of air-fuel mixture during the intake stroke.

Vortex—a mass of liquid with a whirling or circular motion that tends to form a cavity or vacuum in the center of a circle and to draw toward this cavity or vacuum bodies subject to its action.

Work—the moving of an object by force over a distance.

Wrist Pin—see piston pin.

Index

Access cover, valve, 26
AC charger, 169
AC generator, 170, 185
AC to DC converter, 169
Additives, 226
Air cleaner, 26, 87, 208
Air passage, 190
Air vane governor, 200
Alternator, 171
Ammeter, 127
Ampere, 112
Ampere-hours, 122
Analyzer, ignition, 140, 151, 152
Apex, 270
Armature, 116, 182
Atomic number, 110
Atoms, 109

Battery, 122
  charging, 106
  cleaning, 105
  electrolyte level, 104
  specific gravity, 106
  testing, 107, 295
Battery-ignition system, 136
Bearings, 42, 72
Belt tensions, 99
Block, cylinder, 24, 59
Block, short engine, 49
Blow-by gases, 26
Bolts, cylinder head, 25
Bore, cylinder, 24
Bottom dead center, 22
Brake horsepower, 261
Breaker points, 117, 148
Breather, crankcase, 26, 94
Bridge rectifier, 185
Bushings, 42

Cam gears, 26
Cams, 26
Camshaft, 21, 41, 62
Capacitive discharge ignition system, 132
Capacitor, 140
Cap, radiator, 45
Carbon monoxide, 15
Carburation system, 20
Carburetor, 28, 190
  adjustments, 210
  diaphragm type, 195
  float valve maintenance, 221
  maintenance, 213
  touch-and-start primer, 197
  vapor return line, 198

Cell, 122
Centrifugal force governor, 201
Chamber, combustion, 14, 25
Charger, AC, 169
Choke, 193
Classifications, of engines, 1
Cleaning the engine, 84
Clearances, typical, 315
Coil, 115, 152
Combustion, *See* external and internal combustion.
Combustion chamber, 14, 25
Commutator, 125
Components, electrical, 114
Compound, 109
Compound vortex controlled combustion, 274
Compression, 14, 18, 21, 242
Compression check, 103
Compression ratio, 32
Condenser, 114, 151
Conductor, 111
Connecting rod, 37, 50
Continuity, 127
Continuity flashlight jig, 284
Converter, AC to DC, 169
Cooling system, 14, 44, 74
Coulomb, 112
Crank, 38, 71
Crankcase, 26
Crankcase breather, 26, 94
Crankshaft, 38, 71
Current (electric), 111
Current regulator, 174
Cutout relay, 174
Cycle of operation, 3, 14, 17
  diesel, 3
  four-stroke cycle, 20
  Otto, 3
  rotary (Wankel), 271
  two-stroke cycle, 18
Cylinder block, 24, 59
Cylinder bore, 24
Cylinder head, 25, 50, 85
Cylinder head bolts, 25

DC generator, 171, 186
Decimal equivalents of millimeters, 290
Detonation, 227
Diaphragm carburetor, 195
Diaphragm pump, 195, 222
Diesel, 3
Diode, 125
Displacement, piston, 31
Dynamo, *See* generator, 169

Efficiency, volumetric, 262
Electrical components, 114
Electrical system, 176
Electricity, 111
Electrolyte, 104, 122
Element, 109
Energy, 1, 257
Engines, 17
  classifications, 1
  cycles, *See* cycle of operation.
  external combustion, 1, 3
  internal combustion, 1, 3, 17
  major parts, 12
  operating procedures, 78
  returning to service after storage, 80
  salt water operation, 80
  shutoff procedures, 79
  starting procedures, 76
  storage procedures, 80
  uses, 1, 4
Engine tests, basic, 240
English system of weights and measures, 291
English to metric conversions, 263, 287
Exhaust (action), 14, 18, 21
Exhaust ports, cleaning, 85
Exhaust valve, 21
External combustion, 1

Farad, 115
Filters, 90, 207
Flashlight, continuity jig, 284
Float bowl, 193

Float valve, 221
Flywheel, 40, 145
  magnets, 154
  puller, 146
  removal/replacement, 145
  vanes, 85
Foot-pounds to inch-pounds, 263
Four-stroke cycle, 20
Frictional horsepower, 261
Fuel filters (strainers), 90, 207
Fuel shutoff valve, 205
Fuel supply, tests, 242
Fuel system, 14, 189
  adjustments, 209
  air cleaners (filters), 208
  carburetors, 190, 210
  filters (strainers), 207
  gasoline, 225
  governors, 200, 213
  maintenance, 213
  tanks, 204, 224
  tests, 242
Fuel tanks, 204, 224
Fuel tank vent caps, 93

Gases, blow-by, 26
Gasoline, 15, 225
Gas turbine, 274
Gears, timing, 21
Generator, 169, 185
Glossary, 325
Governors, 200
  adjustments, 213
  maintenance, 224
  mechanical, or centrifugal force, 201
  pneumatic (air vane), 200
Guides, valve, 69

Hand tools, 275
Helicoils, 61
High tension lead (spark plug), 118
Horsepower, 259
Hydrocarbon, 225

Ignition, 20
Ignition analyzer, 140, 151, 152
Ignition coil, 152
Ignition system, 14, 128
  battery, 136
  breaker points, 148
  capacitive discharge, 132
  condenser, 114, 151
  flywheel magnets, 154
  flywheel (removal/replacement), 145
  ignition coil, 152
  magneto, 129
  maintenance, 139
  multicylinder engines, 139
  solid-state (capacitive discharge), 132
  spark plugs, 100, 119, 141, 241
  testing, 160, 240, 241
  timing advance mechanisms, 137, 155
Illustrated parts of typical small gasoline engines, 297
Inch-pounds to foot-pounds, 263
Indicated horsepower, 261
Induced voltage, 113
Inductance, 116
Intake, 14, 18, 21
Intake valve, 21
Internal combustion, 1, 3, 17

Journals, 42

Lands, piston, 29
Lapping, valves, 68
Lifter (tappets), valve, 69
Lubricating oils, 229, 231, 233
Lubricating system, 14, 229, 237
Lubrication
  four-cycle engine, 94, 235, 237
  nonengine parts, 99
  two-cycle engines, 97, 234
Lubrication chart, outboard engine, 323

Magnetism, 113
Magneto-generator, 172

Magneto-ignition system, 129
Magnets, flywheel, 154
Maintenance
  engine, 49
  engine tests, basic, 240
  fuel systems, 213
  ignition system, 139
  lubricating system, 237
  overhaul of engine, 255
  periodic, 82
  starting system, 177
  tune-ups, 82, 253
Metric system of weights and measures, 293
Metric to English conversions, 263, 286
Molecules, 109
Motors, *See* starter motor.
Motor-generator, 173
Motor-starter, *See* starter motor.
Muffler, 15, 42, 85
Multicylinder engines, 139

Neutrons, 110
Nonconductor, 111
Nucleus, 110

Octane rating, 226
Ohm, 112
Ohm's Law, 112
Oil, 14, 229, 231, 237
  changing, 94
  checking level, 94
Oil pump, 21
Oil seals, 26, 73
Oil slinger, 21
Operating procedures, 78
Operation, 12
Otto, 3
Overhauls, 253

Periodic maintenance, 82
Pin, piston (wrist), 35, 53
Piston, 29, 53
  displacement, 31
  lands, 29
  pin (wrist pin), 35, 53
  stroke, 31
Piston pin, 35, 53
Piston rings, 34, 53
Piston stop, 146
Plugs, *See* spark plugs.
Pneumatic governor, 200
Points, breaker, 117, 148
Poppet valve, 28
Ports, exhaust, cleaning, 85
Port, transfer, 18
Potential difference, 111
Potential energy, 257
Power, 258
Power (action), 14, 18, 21
Preignition, 44, 227
Primary, 115
Protons, 110
Puller, flywheel, 146
Pump, diaphragm, 195, 222
Pump, oil, 21

Radiator, 45
Ratio, compression, 32
Reciprocating motion, 12
Rectifier, 125
Rectifier, bridge, 185
Reed valve, 26, 28, 73
Regulator, 174, 186
Relay, cutout, 174
Resistor, 121
Return line, vapor, 198
Ring compressor, 58
Rings, piston, 34, 53
Rod, connecting, 37, 50
Rotary engine, 269
Rotary motion, 12, 18
Rotary valve, 28
Rotor, 270

SAE horsepower, 260
Safety, 15
Salt water operation, 80
Scavenging, 32
Seals, oil, 26, 73
Seats, valve, 65
Secondary, 115
Short block engine, 49
Shroud, cleaning, 85
Shutoff procedures, 79
Shutoff valve, 205
Slinger, oil, 21
Solenoid, 116
Solid-state ignition system, 132
Spark plugs
  checking, cleaning and regapping, 100
  condition, basic test, 241
  description, 119
  high tension lead, 118
  indicator of engine trouble, 141
  test tool, 285
Special tools, 277
Specifications, 315
Specific gravity, 106, 124
Springs, valve, 65
Starter motor, 165, 180
Starters, *See* starting system.
Starting procedures, 76
Starting system, 14, 161, 177
  AC generators, 184
  alternator, 171
  DC generators, 171, 186
  generators, 169, 184
  magneto-generator, 172
  maintenance, 177
  manual starters, 161, 177
    crank, 161, 177
    kick, 162, 177
    quick release spring, 164, 179
    rope, 162, 178
  motor-generator, 173
  regulators, 174, 186
  starter motor, 165, 180
Steam engine, 274
Stirling engine, 274
Stop, piston, 146
Storage procedures, 80
Stroke
  four-stroke cycle, 3, 20
  piston, 31
  two-stroke cycle, 3, 18
Stroke (of the piston), 31
Supporting systems, 14

Tank, fuel, 204, 224
Temperature indicator, 46
Tension, belt, 99
Tests, engine, 240
Test spark plug, 285
Thermostat, 45
Throttle, 192
Timing advance mechanisms, 137
Timing gears, 21
Timing, ignition, 155
Tools
  hand, 275
  special, 277
Top dead center, 22
Torque, 258
Torque data, 315
Torque value conversions, 263
Transfer port, 18
Transistor, 125
Troubleshooting, 244
  causes of engine failure, 252
  compression, basic test, 242
  engine fails to start or starts hard, 247
  engine missing under load or lack of power, 249
  engine noisy or knocks, 251
  engine surges or runs unevenly, 250
  engine vibrates excessively, 252
  fuel supply, basic test, 242
  ignition system, basic test, 240
  no ignition spark, 245
  spark plug condition, basic test, 241

Troubleshooting (*Contd.*)
- weak compression, 246
- weak ignition spark, 245

Tune-ups, 82, 253
Turbine, gas, 274
Two-cycle engine lubrication, 234
Two-stroke cycle, 18

Valves, 20, 35
- access cover, 26
- exhaust, 36
- float, 221
- guides, 69
- intake, 21, 36
- lapping, 68
- lifters, 69
- maintenance, 64
- poppet, 28
- reed, 26, 28, 73
- rotary, 28
- seats, 65
- shutoff, 205
- springs, 65
- timing, 22

Vanes, flywheel, 85
Vapor lock, 199
Vapor return line, 198
Vent cap, 93
Venturi, 190
Viscosity, 231
Voltage regulator, 174
Volts, 111, 112
Volumetric efficiency, 262

Wankel, Dr. Felix Heinrich, 269, 273
Water pump, 45
Work, 257
Wrist pin, *See* piston pin.